T0132681

NATURAL STATES

The Environmental Imagination in Maine, Oregon, and the Nation

Richard W. Judd and
Christopher S. Beach

Resources for the Future
Washington, DC

An RFF Press book
Published by Resources for the Future
1616 P Street NW
Washington, DC 20036–1400
USA
www.rffpress.org

Library of Congress Cataloging-in-Publication Data

Judd, Richard William.
  Natural states : the environmental imagination in Maine, Oregon, and the nation / by
Richard W. Judd and Christopher S. Beach.
          p.   cm.
        Includes bibliographical references and index.
        ISBN 1-891853-59-7 (library binding, dustjacket : alk. paper) --
        ISBN 1-891853-60-0 (pbk. : alk. paper)
          1. Environmentalism--Maine. 2. Environmentalism--Maine. 3.
        Environmentalism--United States. 4. Environmental policy--Maine.
        5. Environmental policy--Maine. 6. Environmental policy--United States.
        I. Beach, Christopher S. II. Title.
GE198.M2 J83 2003
333.7'2'09741—dc21

                                                          2003003696

f e d c b a

 Printed on recycled paper with 80% post-consumer content.
Printed with soy ink.

The paper in this book meets the guidelines for permanence and durability of the
Committee on Production Guidelines for Book Longevity of the Council on Library
Resources.

This book was typeset in *Giovanni* by Carol Levie. The cover was designed by Rosenbohm
Graphic Design. Interior design by Naylor Design Inc.

About Resources for the Future *and* RFF Press

Resources for the Future (RFF) improves environmental and natural resource policymaking worldwide through independent social science research of the highest caliber. Founded in 1952, RFF pioneered the application of economics as a tool to develop more effective policy about the use and conservation of natural resources. Its scholars continue to employ social science methods to analyze critical issues concerning pollution control, energy policy, land and water use, hazardous waste, climate change, biodiversity, and the environmental challenges of developing countries.

RFF Press supports the mission of RFF by publishing book-length works that present a broad range of approaches to the study of natural resources and the environment. Its authors and editors include RFF staff, researchers from the larger academic and policy communities, and journalists. Audiences for RFF publications include all of the participants in the policymaking process — scholars, the media, advocacy groups, NGOs, professionals in business and government, and the general public.

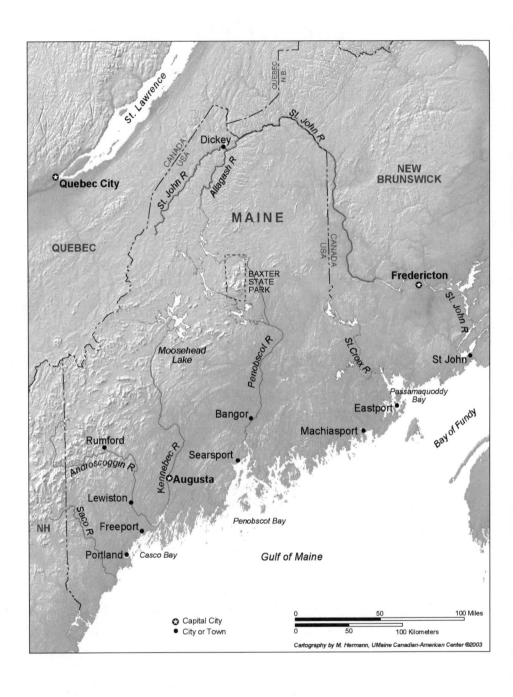

St. Lawrence

QUEBEC
N.B.

Dickey

St. John R

Quebec City

St. John R

Allagash R

NEW
BRUNSWICK

MAINE

QUEBEC

CANADA
USA

Fredericton

BAXTER
STATE
PARK

CANADA
USA

St. John R

Moosehead
Lake

Penobscot R

St. Croix R

St John

Passamaquoddy
Bay

Bangor

Eastport

Machiasport

Rumford

Androscoggin R

Kennebec R

Searsport

Bay of Fundy

Augusta

Lewiston

NH

Saco R

Freeport

Penobscot Bay

Portland

Casco Bay

Gulf of Maine

0   50   100 Miles

0   50   100 Kilometers

⊛ Capital City
● City or Town

Cartography by M. Hermann, UMaine Canadian-American Center ©2003

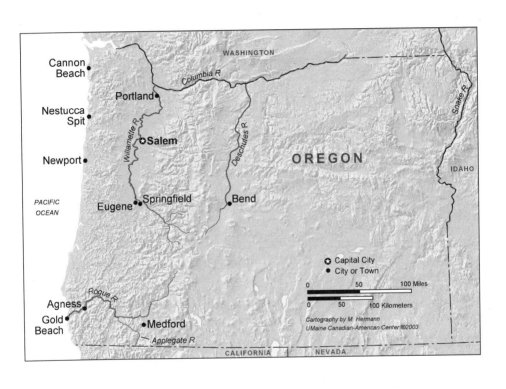

"We can always blow up the bridges to the mainland ... and return to the free life, the life without fences, without boundaries, the life that paid its own way ..."

—Robert P. Tristram Coffin, poet

CONTENTS

To Jed and Lily

PREFACE

An important element of this book came into focus one day in 1998 when after a week at the Oregon State Archives in Salem we took time off for a trip to eastern Oregon with family and friends. Along the way we stopped briefly at the elegant Black Butte Ranch resort near the Three Sisters Wilderness, on the eastern slope of the Cascades. In the lounge, as we sat self-consciously amid the region's *nouveau* elite, we gazed through a two-story window at three snow-tipped volcanic peaks rising out of a forested basin on the southern horizon. From our comfortable vantage, the distant slopes suggested a world completely free from human manipulation, an appealing vision of nature in its purest and most spectacular form.

In the foreground, stretching away from the lodge, was a tableau strikingly different from this remote wilderness scene: the closely cropped turf of a well-appointed golf course seemed preternaturally green in the mid-morning sun. To one side of the fairway, a few cattle grazing on lush mountain grass gave some credence to the resort's pretensions of being a "ranch."

Before us, we decided, were the artifacts of three distinct cultural moments. The cattle and their traditional pastoral setting suggested the romanticized vision of the West that in the 1950s served as a repository for the values Americans left behind when they crossed the threshold of the modern age. The pristine slopes in the distance, protected as part of the Three Sisters Wilderness, stood as a legacy of the 1960s belief in the virtues of contact with uninhabited nature. In the foreground the sculpted trappings of this stylish, ecologically correct resort offered up a generic but thoroughly picturesque recreational landscape crafted to the tastes of the West's new gentry — a cultural product of the 1970s. Each in turn derived from the American passion for nature, and each in its own way helped sustain a defense of those ideals we have come to call environmentalism.

We resolved, at that point, to explore these various constructs of nature and trace their role in the emergence of environmental politics. Today the majority of Americans think of themselves as environmentalists, according to one definition or another, and this percentage remains fairly constant across a varied political landscape. Gleaning a better understanding of what Americans mean when they call themselves environmentalists, and how they understand the nature they hope to protect, would help us explain the persistence of the environmental movement and its impact on state, national, and even global politics.

Historical studies often treat environmental politics and environmental ideals as separate topics. In this book, we trace their interaction. Politics drew inspiration from perceptions of the environment, and in turn politics encouraged new ways of thinking about the environment. This interaction involves what we have termed the environmental imagination: the core beliefs that animate the political defense of nature. Above all else, environmental imagination derives from the power of place, memory, and nature in people's lives.[1] We want to show how, over three decades, Americans have visualized, valued, and otherwise connected with nature and place, and how these connections gave rivers, coastlines, forests, and other distinguishing regional features transcendent meaning. This complex construct, combining domesticated and wild landscapes, emerged out of a nostalgic post–World War II literature of place and absorbed new values and meanings as Americans took up the task of saving threatened natural and cultural landscapes. Discovering the environmental imagination in the politics of this era tells us something about who we are as a society, and how we hope to find freedom, permanence, and authenticity by regaining a lost intimacy with nature and place.

Because the environmental imagination was such a multilayered concept, internal tensions were inevitable. It drew, for instance, from the pastoral tradition, which romanticizes rural America as a lightly humanized natural world in which the folk work in harmony with nature. But it also borrowed from the wilderness idea and its celebration of uninhabited nature. Pastoralism is a socially inclusive concept, a model for changing society as a whole, consonant with America's role as a redeemer nation. Wilderness appreciation, by contrast, involves an exclusive, personal, and usually temporary experience, a visit to a primitive natural place to prove and purify oneself in the tradition of Moses, Jesus, or a Native American on a vision quest. These two sets of values — folk and wild, civic and personal — coexisted uneasily in the environmental imagination. But it was this capacity to embrace opposites, its elastic quality, that gave environmental thought its political reach. Environmentalism inspired people across many social and political divisions because it addressed a core concern: the loss of an authentic, harmonious, and intimate relation to nature and the land.

Most histories of the environmental movement focus on national events, beginning, for example, with the Echo Park dam controversy in Dinosaur National Monument in 1954. Addressing events at the national level confirms the movement's broad scope and identifies its prominent political concerns, but this is not the best arena for exploring the connection between environmental ideas and politics. Better evidence of the movement's ideological character is found in state-level politics, closer to the "special places" that were the focus of environmental activism before 1975. The dramatic national events of the late 1950s and the 1960s — the outrage over radioactive traces in human milk, the "cranberry scare" of 1959, the publication of Rachel Carson's *Silent Spring*, the popularization of ecology, and a variety of other concerns — clearly brought a profound change in public consciousness.[2] But the dialectic of popular consciousness and political action began at the local and state level, where citizens first voiced their frustration with blight, pollution, and other modern environmental problems. The role of these formative environmental battles is an important part of the larger story.

Statewide environmental campaigns also offer a unique opportunity to trace the evolution of environmental thinking and assess its role in sustaining the movement over three decades. On the national scene, environmental debates often appear episodic, as the energy of the movement shifts from locale to locale and from battle to battle. Tracing environmental ideas in sequential order at the state and local level shows us how they emerged out of social controversy, and how they spread and changed over time.

Maine and Oregon would make anyone's list of states illustrating America's commitment to environmental values. Just as apparently, they would not be the only states to make such a list. We chose these two because they offer prime examples of natural landscapes that served as compelling sources of regional identity — eastern and western. As iconic rural places adjacent to wealthier and more powerful metropolitan neighbors, Maine and Oregon also represent an interplay basic to environmental thought and politics. Although environmental debates focused on rural landscapes, the movement's core ideas typically radiated out from urban centers. Maine and Oregon show how rural–urban interactions shaped environmental values.

Maine and Oregon are also instructive because of their differences. From 1945 to 1975 Maine's economy lagged behind the nation's, its marginality rooted in its staple-based economy and its downeast location, outside the westerly orientation of national development. Economic marginality gave Maine's brand of environmentalism a unique cast. Oregon's postwar economic growth was far more promising, and thus its basic environmental impulse was cultural rather than economic, part of a longstanding preoccupation with distinguishing the state from its more "worldly" neighbors, Washington and California. These differences highlight the varied conditions under which the movement flourished all across America. Clearly neither Maine nor Oregon can adequately represent the entire spectrum of issues that made up the movement, but the concerns, attitudes, and strategies that surfaced in these two states transcend local circumstances.[3]

Our investigation begins in the years of mingled hopes and fears that followed the Depression and World War II and ends in the mid-1970s, with another period of national reassessment following yet another war. These three decades mark a distinct and significant era in American environmental thought, when a popular environmental imagination coalesced and became a national political ideology. This era coincided with a period of grassroots civic ferment unlike any other in twentieth-century America, punctuated by civil rights campaigns, rising ethnic consciousness, antiwar protest, urban rebellion, feminism, and the rise of a youth-oriented counterculture. The environmental imagination continued to evolve beyond this period of intense civic and personal engagement — indeed, persistence is its hallmark — but this creative and tempestuous era best illustrates its dynamics.

The history this book traces demonstrates the elasticity of the environmental imagination as a political ideology. Chapter 1 identifies the literary pastoral tradition common to postwar Maine, Oregon, and the nation. Tracing this genre of regionalist literature, the chapter explains how an homage to folk and nature resonated in a postwar society seeking freedom, permanence, and authenticity in communion with a natural rural setting. Chapter 2 shows how this literary tradition venerating traditional folk in a timeless landscape came to define the terms of livability for all citizens, as traditional pastoralism galvanized political debate over commercial blight and water pollution in the mid-1950s. Engaged in political dialogue concerning these issues, people in Maine and Oregon began to think about their states — their environments — in new ways.

As the idea of nature became a force in state politics, it incorporated elements of other traditions. Chapter 3 explores the interaction between pastoral aspirations and wilderness ideals in managing rivers, specifically efforts to win protection of Oregon's Rogue and Maine's Allagash as wild and scenic rivers. Chapter 4 describes the high-water mark of state-level environmental engagement in the late 1960s and early 1970s by focusing on two regional scenic icons: Maine's rockbound coast and Oregon's windswept sand beaches. Controversy over oil tanker ports and refineries on the Maine coast generated enormous statewide concern for the state's traditional pastoral seascape. Threats of commercial and highway development along Oregon's beaches provoked a similarly passionate defense of that state's unique coast. These coastal-preservation controversies sparked widespread discussion of each state's natural character. The resulting successful efforts to contain commercial blight and water pollution, preserve wild rivers, and protect scenic coasts were transcendent accomplishments that thrust Oregon and Maine into the forefront of the environmental movement.

New controversies emerging after these initial successes, such as battles to preserve free-flowing rivers from dams, illuminated tensions in the popular commitment to nature and place. Recreationists envisioned themselves roving freely through a pristine natural world; people living closer to rivers clung to older pastoral ideas of a cultivated and managed nature; while ecologists opposed any human use that might disturb the harmonic balance that held together fragile river systems. Negotiations among adherents of these perspectives resulted in the compromises that characterize modern river management. Chapter 5 shows how strategies for using natural corridors on the Willamette and Kennebec rivers helped reconcile these tensions.

In the early 1970s the popular environmental imagination emerged in full form and flower. From its inception in regional pastoral musings about iconic rivers and ocean shores, it expanded to embrace entire regional landscapes in all their ecological complexity. Chapter 6 describes a new genre of "ecotopian" regional literature — futuristic thinking about a world where nature and society harmonized. Ecotopia promised to reconcile the contradictions between the folk landscape and the wilderness, between ecological preservation and recreational use, between urban and rural, common and private. As a first step on the route to ecotopia, Maine, Oregon, and a few other states experimented with comprehensive rural land-use planning. At the height of the Vietnam War, and in the wake of widespread urban unrest all across America, Mainers and Oregonians of all classes embraced the idea of constructing a better, fairer, more humane, and more ecologically harmonious society.

Although the intense search for ecotopian reconciliation receded along with the specter of unrest and social breakdown, the environmental imagination persisted in a variety of new expressions in the mid-1970s, as many Americans experimented with personal rather than political means of expressing their passion for nature. Chapter 7 highlights these emerging trends. The first is a commodified conception of nature and place linked to merchandising of "natural" lifestyles. A second new form entailed an exurban migration that brought a wave of nature and heritage preservation campaigns to the countryside. Finally, many environmentalists, clinging to an older pastoral ideal of authentic folk and their thrifty ways, sought to reinvent environmentalism by living on semisubsistent farms, creating urban gardens, and experimenting with sustainable alternative technologies. Even while many adherents of these divergent expressions of personal choice ultimately embraced consumerism, as a combined cultural phenomenon they affirmed America's allegiance to environmental values.

These and other premises for reforming the countryside, although in some ways compatible, exerted powerful tensions in the movement and in society at large, weakening, at times, the legislative response to the environmental imagination. But they also ensured that the movement would remain dynamic and adaptive. Contradictory though they were, these ideals sustained the environmental movement, and they remain a driving force behind our continuing commitment to protecting our environment.

Richard W. Judd
Christopher S. Beach

Waldo-Hancock Bridge, Bucksport, Maine

CHAPTER 1

FOLK AND NATURE: PASTORAL LANDSCAPES IN THE POSTWAR WORLD

O n a clear fall morning at the close of World War II, Robert P. Tristram Coffin, poet laureate of Maine, flew with a friend in a light plane over his family homestead on the state's fabled rock-bound coast. Lofted above this iconic landscape, Coffin gained perspective on a troubling time of global depression and war. A land as richly textured as Maine's, he thought, would surely buffer its inhabitants against this chaotic world: "how could people living in the midst of such splendor be dull or drab or ordinary?" he reasoned. "They had come naturally by their sparkle." Recoiling from a world torn by economic confusion and social violence, Coffin described a fantasy of Maine island life cut off from the modern world. "We can always blow up the bridges to the mainland ... and return to the free life, the life without fences."

At the same time, regional writer Henry Beston applied a similar vision by retreating to his saltwater farm in Maine to focus on the simple virtues of a life lived close to nature. Here troubling memories of the war dissolved into comforting images of the country home fortified against more familiar forms of adversity.[1] "Strengthened and provisioned like fortresses," he wrote, "our cordwood stacked under cover and our cellars filled, we should not fear the northeaster and the wind running in above the sea." Beston's year on the northern farm culminated in a raucous celebration of nature coming alive in springtime, a "shouting and rejoicing out of puddles and streams, a festival of belief in sheer animal existence." These and other seasonal cycles resonated deep in the souls of his rural neighbors, setting them apart as a pastoral folk enjoying a special bond with nature.[2]

For both these visionaries, home and homestead became panaceas for the multitude of problems Americans experienced after the dark days of 1929. Coffin's romanticized coastal Yankees — those who stayed behind when their forebears moved on to mine richer lands farther west — made peace with a world of thin soils and harsh weather. They adapted to the rhythms of wind, tide, and season, took comfort in simple things, and reduced life to elemental forms: the simple cadence of the farmer's year and the abstract purity of the curve revealed in drifted snow. With souls purged of extraneous desires, they cultivated an intractable individualism that mirrored the randomness of nature's day-to-day processes. Pastoral images like these offered a nostalgic vision of nature and folk interacting in timeless harmony that could salve the ills of modern society.

Speaking similarly of the intimate relation between folk and the land they inhabited, Oregon writer and later senator Richard L. Neuberger explained that those who farmed and ranched the vast interior grasslands of the Pacific Northwest would never succumb to the mass ideologies that seemed to dictate global order in the 1930s and 1940s. Their opinions, he insisted, reflected the "independence, freedom, and uniqueness of the majestic hinterland in which they live."[3]

Neuberger spoke in glowing terms of his home state as a land where "mountain ranges rear three miles in the air, and rivers fall so far they dissolve into plumes of mist. Foaming creeks are alive with trout, and ... even the gaunt desert uplands have a grim and striking beauty." Fellow writer H.L. Davis described forest clearings where wild strawberries were "crowded so close together that you will crunch a handful at every step ... huckleberry swales in the higher mountains where the bushes are bent flat to the ground under the weight of their berries."

Unlike Maine, where unrelenting winters and thin soils shaped a landscape of stark beauty and a people of singular persistence, Oregon nurtured its folk in the lap of lush nature; it was extravagant, cornucopian, monumental.[4] Oregon folk, the sons and daughters of western pioneers, were also more mobile and less traditional than those in Maine, giving the country a "persistent sense of newness, of everything being done for the first time."[5]

But the connection between folk and nature in Oregon was no less significant. Stories of the Oregon trail gave the natural landscape enormous symbolic meaning.[6] Crossing the Cascades, the central mountain spine that separated Oregon's eastern high desert and the broad, lush valleys to the west, settlers entered a parklike world of mixed grassland, woodland, marshland, and forest. Newcomers in a new land, they absorbed the ethos of Eden restored and settled in to transform the western valleys into gardens and grain fields. Like writer E.B. White, who described his saltwater farm in Maine as beginning "close to the house with a rhubarb patch, but ... end[ing] down the bay beyond the outer islands, hand-lining for cod and haddock," Oregon pastoralists represented the landscape as a seamless and indeed timeless blend of human effort and natural process.[7]

Literary representations such as these provide insight into the postwar idealization of folk and their natural environment, and accordingly into the origins of the environmental movement. Microbiologist and nature philosopher Rene Dubos once said that there were only two kinds of landscape so appealing as to elicit the kind of political defense characteristic of late-twentieth-century environmentalism: "one is primeval nature undisturbed by man.... The other is one in which man has toiled and created, through trial and error, a kind of harmony between himself and the physical environment." In Dubos's opinion, it was the latter that proved universally satisfying. "What we long for is rarely nature in the raw; more often it is an atmosphere suited to human limitations.... It did not result from man's conquest of nature. Rather, it is the expression of a subtle process through which the natural environment was humanized, yet retained its own individual genius."[8]

Dubos's biases against wilderness are well known, but the textured and cultivated landscapes he espoused were an essential component of the diverse and complex movement we call environmentalism. Over time, the idea of folk and nature gave way before a more compelling vision of wilderness and ecological balance, but the popular appeal of the environmental movement is rooted in this postwar pastoral ideal, and thus our understanding of the environmental movement must begin with this vision of rural America — nature and folk — as balm for a troubled mid-century world. This core vision of harmony between folk and nature echoes through the movement to the present. It was indeed the pastoral ideal, rooted deeply in American literature and culture, that early on formed the basis of the environmental imagination — the body of popular thought that informed environmental politics.

PASTORALISM AND POLITICS IN MAINE AND OREGON

While neither Maine nor Oregon ranked among the nation's most rural states, in both cases the larger-than-life qualities of the natural landscape and the rich character of the folk — farmers, fishers, loggers — provided fertile ground for the imaginative incorporation of pastoralism into literary representations, and later into politics.

Both states contained an expansive urban-industrial base, but in each instance metropolitan growth was attenuated by a history of regional isolation. Oregon's formidable sea cliffs and the Columbia Bar drove nineteenth-century timber speculators and empire builders north to Puget Sound's more accessible bays, while Maine's northeastward orientation insulated its hinterland frontier from the great migration streams flowing westward across the continent. Thus the urban hubs of both states — Portland, Maine, and Portland, Oregon — retained strong pastoral associations. Nowhere but in Oregon, according to one writer, did urbanites dwell "so close to scenes of breath-taking beauty, ringed round with shining water and snowcapped peaks."[9] Maine's Portland cast itself as gateway to the "Nation's Vacationland," banking on tourism, with its pastoral reflections of mythic Maine, as its economic destiny.

Describing rural landscapes in pastoral terms gave them iconographic meaning and helped prepare the nation for a crusade to save them.[10] Despite the obvious differences in regional identity — Yankee and western — both states epitomized the postwar quest for physical and spiritual freedom in landscapes, as Coffin put it, without fences. In Maine, this liberating experience came through intimate contact with the fog-shrouded coast or the state's thick interior forests. Maine was a place, as F.

Wallace Patch wrote in 1960, "to walk slowly, to feel close to trees and wild life and to know the weather's whims."[11] Oregon's country, by contrast, encouraged the spirit to soar and the mind to contemplate awesome distance. In both cases, however, contact with nature was liberating in contrast to the confining urban world most sojourners left behind.

Pastoral landscapes also embodied a sense of permanence: patterns fixed in the recurring rhythms of nature. E.B. White recalled a local boy's poem about the Narramissic River on the eastern Maine coast that ended with the phrase, "it flows through Orland every day." White found the boy's simple sentiment compelling: "I never cross that mild stream without thinking of his testimonial to the constancy, the dependability of small, familiar rivers." Pastoral landscapes like these were "free from the bonds of history."[12] Here, human activity did not transcend nature but rather blended into it. In this nostalgic construct, past and present were fused into a single vision; folk and nature were suspended, as if in amber, and firmly anchored against the winds of change.

Both states also conveyed a vision of the rural world as authentic, as an asylum for alienated urbanites who had lost touch with the elemental natural forces that had sustained the nation in its youth. Essential forms of work in an open, natural setting grounded the folk in reality, and contemplating the primacy of this relationship gave the urban observer a greater sense of authentic existence.

These pastoral constructs were not unique to Maine and Oregon; nor were they peculiar to postwar literature. America at its founding, according to historian Leo Marx, was to be a "rural landscape, a well-ordered green garden magnified to continental size."[13] The pastoral tradition originated in the Jeffersonian veneration of freehold farming and developed its fine points through the romantic literary tradition and the landscape art of the Hudson River School, whose most prominent advocates, Frederic Church and Thomas Cole, developed a motif of picturesque or sublime nature humanized by some diminished cultural icon.

Pastoralism reached its finest physical expression in the landscape architecture of Alexander Jackson Davis, Andrew Jackson Downing, and Frederick Law Olmsted, romantics who "fused culture and nature into a single representation, usually an idealized, pastoral landscape that mediated the polarities of art and nature, civilization and wilderness." These proponents popularized the pastoral tradition as withdrawal, a "yearning for a simpler, more harmonious style of life," and, as Lawrence Buell adds, appreciation for the "aesthetic pleasure of privileged solitary communion with nature." But at a deeper level, pastoralism expressed a civic aspiration, a vision of beginning civilization anew free of the tensions between culture and nature.[14]

In the popular drama that played out later in the 1960s and 1970s, the city stood as the antithesis of this idyllic world. Conservationist Benton MacKaye saw the city as a glacier, "spreading, unthinking, ruthless." Its tenements, homes, stores, factories, billboards, filling-stations, and eating-stands engulfed the countryside, effacing regional character.[15] Yet traditional literary pastoralism offered examples of creative interface between these two worlds, the rural serving as a spiritual reserve for alienated urban society. Out of the critique of the city, according to historian James L. Machor, came a "vision of environment in which city and country are equally valuable components in an evolving landscape best served when those components operate in harmony."[16]

Pastoralism suggested a physical amalgamation of city and country — nature brought into the city, or an urban society reconstituted in the countryside. Ebenezer Howard explored the latter idea in his influential *Garden Cities of To-Morrow* (1902), suggesting a pattern of rural communities scattered around the metropolis, each incorporating the essentials of city life — industry, transportation, housing, culture — into a pastoral setting. Frank Lloyd Wright's Broadacre City, an urban civilization similarly disgorged into the countryside along high-speed transportation routes, suggested a remedy for the "hopelessly, helplessly, inorganic" twentieth-century urban civilization.[17]

The garden and park movements of the late-nineteenth century exemplified a second alternative: the pastoralization of the metropolis. Frederick Law Olmsted's influential landscape plans were designed to show that the city was "intrinsically connected to the land"; Lewis Mumford stressed urban forms that relied on local resources and local sources of energy. "The design of houses, the use of indigenous materials, the layout of streets, the reliance on local sources of power such as wind and water all reflect a more intimate and functional relation to geographic conditions." Here the city would be "grounded in its geographic environment, in *topos*."[18]

These pastoral associations proved appealing to postwar Americans. Richard Neuberger, who attributed his eventual political success to culti-

vating Rockwellian images of Oregon's pastoral cities, remarked ironically that an explosively mobile postwar generation cherished its provincial-
ism: "although millions ... may be looking for a new address, pride of
place still must be a dominant national trait."[19]

PASTORALISM AND CONSERVATION

Although pastoralism is a hallmark of the U.S. literary tradition, its incorporation into conservation politics is only a mid- to late-twentieth-century development. Pastoralism is fundamentally nostalgic, and as William
Robbins notes, conservation ideology, as it was first formulated in the
Progressive era, "preached virtues that were consistent with the modern-
izing world of industrial capitalism: efficiency, the elimination of waste,
and the development and scientific management of resources." As the
mainstream Progressive conservation movement emphasized the rational
use of nature, the idea of the folk as a landscape feature worthy of protection remained peripheral.

Literary pastoralism did inspire the turn-of-the-century country life
movement. As sociologist Jeffrey Jacob points out, this was a classic political and economic expression of American agrarianism, reflecting anxiety
over the loss of rural identity and rural character as Americans first discovered that theirs was an urban, not an agricultural nation. Progressive
reformers, notably Theodore Roosevelt, reasoned that improving country
life would slow the demographic drift to the city, thus conserving rural
character and values.[20] Pastoral rhetoric also appeared as a rationale
behind the Bureau of Reclamation Act of 1902, promoted as a means of
extending the agrarian republic into arid lands.

In contrast to utilitarian conservationists like Gifford Pinchot, romantic nature-writers like John Muir and John Burroughs stressed the intrinsic value of pristine natural landscapes. These wilderness advocates challenged the idea of rational resource use and drew upon the idea of a
redemptive rural landscape. Despite these parallels, however, wilderness
preservation in the early years of the century drew little from the pastoral
tradition of a lightly humanized rural landscape.[21]

World War I deflected the nation's attention from conservation issues.
Then the need for massive reemployment programs during the Great

Depression, coupled with disasters like the Dust Bowl and floods in the Mississippi Basin, prompted a second wave of conservation politics. New Deal initiatives were marked by numerous programs to create national parks, national forests, and public sanitation programs, and by grand-scale, multiple-purpose dam projects on the Columbia, Tennessee, and Colorado rivers. Unlike Progressive-era conservation, the populist rhetoric behind these projects drew heavily upon pastoral images, but it also enlisted pastoralism in an aggressively manipulative ideology of natural resource use. New Deal conservation rhetoric was static in its images of peaceful, cultivated fields, fenced pasture lands, and quiet country homes, yet dynamic in its goal of conquering rivers and expanding the agricultural domain. Instrumentalist and assertive means could be used to achieve romantic and nostalgic ends.

Like all New Deal programs, the conservation agenda was steeped in the ideology of the folk. In the 1930s, according to Warren Susman, the American intelligentsia "set out to become 'an unlearned class,' to assimilate the culture of the 'people.'" Clifford Odets's plays, John Steinbeck's novels, murals created as part of the Works Progress Administration, the folk celebrations of artist Thomas Hart Benton and songwriter Woody Guthrie, the photographic and literary representations by James Agee and Walker Evans, and numerous other cultural expressions captured "an old and enduring vision within American political culture: the idea of a commonwealth at work."

Drawing on these images, the New Deal framed a centrist version of the Popular Front, recasting the American pageant as a workers' republic peopled by the modern sons and daughters of the yeoman farmer. This democratic humanist tradition — the "grainy photography of croppers and union militants, the Depression–era poetry with its Whitmanesque chanting of the place and river names ... the angry political rhetoric demanding a restoration of the American promise of labor and justice" — distinguished the dam projects, the Civilian Conservation Corps programs, the soil conservation plans, the Tennessee Valley Authority, and other New Deal conservation initiatives from their Progressive predecessors.[22]

Once again war distracted the nation from its conservation concerns. And after World War II the idea of conserving resources receded as Americans enthusiastically embraced the material benefits of their vibrant postwar economy. Global ascendancy and mass-production techniques

revolutionized consumer markets, improved financial security, and created a spectacular boom in suburban building. More prosperous and secure
than ever before, Americans in the 1950s saw around them an environment marred by artificiality, industrial blight, sprawl, lack of open space,
and polluted air and water.[23] To reassert the livability of the postwar
world, they turned first to the suburb and then to the countryside to
renew their commitment to the pastoral ideal.

The nostalgic elements of literary pastoralism soothed the anxieties of
a generation emerging from two decades of depression and war and
offered a psychic release for alienated urban citizens. Postwar pastoralism,
more than any earlier political manifestation — urban parks, country life,
New Deal populism — stemmed from an increasingly negative assessment of America's great cities, a mood epitomized in urban-based *film
noir* productions emphasizing human powerlessness, vulnerability, alienation, and paranoia. Anti-urbanism is an American tradition, but the
Great Depression and World War II gave it a particularly somber cast. The
1930s experience suggested to some that depressions would become
more severe as more Americans left the farm for the city. Urban workers
produced mostly "postponable goods," one economist reasoned, which
were the "first to be sacrificed during depression." Nazism and Fascism
also drove home the terrifying political implications of alienated mass
constituencies.

Coupling these themes to basic changes in mid-century demographics,
in 1946 former *Better Homes and Gardens* editor Elmer T. Peterson published a collection of essays under the revealing title *Cities Are Abnormal*.
Drawing on Depression images, Peterson noted the disparity between
farm and city: "eighty-five urbanites figuratively stand in line on pave-
ments, waiting for the fifteen rural men to feed them."[24] According to
contributing essayist S.C. McConahey, the "root source of most of our
present-day economic, social, and political problems lies in the *unbal-
anced distribution of population as between rural and urban occupations.*"[25]

Consensus from the various occupations represented in Peterson's
anthology — biologist, ecologist, economist, manufacturer, physician,
religious leader, architect — was that the city no longer offered the freedom, the sense of permanence, or the means of authentic existence necessary to sustain the American character. Peterson detected a "vague
uneasiness ... about metropolitan life ... the result of unguided but

sound questioning of things which have heretofore implied material security — concentration of large-scale industry, the apparently stable flow of commerce to river and port towns, the increasing power of those cities."[26]

Confronted with deteriorating neighborhoods, congested streets, and an "atmosphere ... filled with noxious gases, dust, dirt, and smoke," city people lost contact with the "rural landscape which feeds them." Returning each night to a "stifling apartment," the urban worker "loads his stomach with ice-cold drinks and calorie-rich cold food, taking on the heat-producing task of burning up all of these unnecessary calories. Small wonder that he lies in his bed and sweats, half sick and sleepless, until it is nearly time to get up again." Home air-conditioning seemed a distant possibility, but it would "never ... equal, for health and comfort, the shaded house set in a shady lawn well away from other houses on a farm or suburban acreage."[27]

The urban crisis was thus a crisis of character, as the generic replaced the organic in both environment and personal identity. As America moved from postwar euphoria to isolation and conformity in the later 1940s, its literary and cultural critics began a debate about personal autonomy in modern society. Sociologist C. Wright Mills, who expressed his dyspeptic postwar vision in White Collar (1951) and The Power Elite (1959), saw urban America as a vast, homogenous culture ruled by corporate elites and peopled by a middle class that had become rootless, fragmented, and pathologically uncertain of itself. Psychologist and German emigre Erich Fromm profiled the age through its archetype, the authoritarian personality, while cultural critic Lewis Mumford and architect Frank Lloyd Wright highlighted the contrast between organic small-town life and the artificial urban world.[28]

Dystopian perspectives invoke utopian alternatives — new worlds emerging out of the chaos of the old — and the pastoral vision of regional writers like Coffin, Beston, Neuberger, and Davis loomed large on the postwar literary scene as an antidote to the multiple crises of urban America. According to literary historian Lawrence Buell, the pastoral serves a dual purpose, "as ... a dream hostile to the standing order of civilization ... and at the same time a model for the civilization in the process of being built."[29] In this dialectical relationship, the alienation of urban life implied authenticity among rural people. Unlike William H. Whyte's perpetually ineffectual "organization man," Coffin's quietly competent rural folk inscribed their way of life on the land in patterns of work handed down from generation to generation. Oil lamps, lantern-lit tables, mince pies, and hooked rugs on pine floors consummated the quest for authenticity in the postwar world.

THE RISE AND FALL OF SUBURBIA

The pastoral tradition influenced the rapid suburbanization that followed World War II. Suburban growth was a product of political and economic forces such as highway construction, new home-building techniques, urban decay, and the influence of auto, highway, oil, and real-state lobbies. However, as Kenneth Jackson points out, it was a cultural response as well. It was, as Jackson says, a spatial representation of America's love affair with nature.

For postwar Americans the suburb offered the best hope for an urban-rural synthesis. The home in the suburb was at least symbolically a kin to the country cottage. But as the suburb spread outward from the city, the illusion of rural living was lost; a victim of its own success, the suburb effaced its organic connection to the land. Moreover, as historian Adam Rome points out, suburbia presented a whole new variety of environmental problems.[30] Thus it was here, where aspiration met reality in the suburb, that pastoralism moved into the arena of civic activism.

The conflict between nostalgic pastoralism and the expanding suburb was particularly intense in Oregon.[31] Articulated early on by literary critics like H.L. Davis, pastoralism gained popular currency among a self-critical suburban population that saw its own worst qualities mirrored in the rapid growth occurring several hundred miles south in the Los Angeles Basin. Breathtaking natural beauty, a mild climate, productive farmland, social homogeneity, abundant renewable resources, and an exploding suburban population set the scene for Oregon's rapid emergence into the limelight of the environmental movement.

Oregon had linked its conservation ideology to the imperatives of growth at the turn of the century. With roughly half the state's land base under federal control, much of the national debate over resource conservation and use in the Progressive era played out in Oregon's backyard, and federal management in the Cascades national forests became a model for state fish, game, and timber policies.[32] In the 1930s the divisive issue of federal hydropower in the Columbia River basin accentuated the New Deal fusion of populist pastoralism and utilitarian conservation. Richard Neuberger's 1938 *Our Promised Land* described the Pacific Northwest as "an almost limitless region of undeveloped natural resources," a place where "men out of work and men who are hungry and ... bitter and ... homeless" could find jobs if political leaders awoke to the region's vast treasure-trove of resources. Over the coming decade, he pointed out, the Columbia River would be harnessed to provide hydroelectric power and irrigation water on a scale that dwarfed the region's human resources. The question remained: could 3 percent of the nation's people consume nearly 50 percent of the nation's hydroelectric power? Neuberger welcomed the arrival of "eastern slum dwellers and middle western drought refugees"; without them, "most of the one hundred fourteen billion sleeping kilowatt-hours in the Columbia River watershed will be useless."[33]

The Bonneville and Grand Coulee dams, completed in 1938 and 1941, delivered an enormous surge of electric power for war mobilization and postwar industries, launching a spectacular assemblage of new light manufacturing, service, utility, food processing, and consumer non-durable industries. Huge federal irrigation, highway, and defense projects also drew heavy manufacturing, service, and technology industries. The Korean and Cold wars, consumer markets generated by regional in-migration, and a comfortable share of several national growth sectors put Oregon in the forefront of the postwar economy.[34]

Oregonians, like Americans everywhere, were ambivalent about their rapidly urbanizing landscapes. Young, mobile, white, and middle-class, they were drawn to the West not by economic necessity but by personal choice.[35] Rejecting the sprawling eastern metropolis as an urban model, they viewed their new suburbs, like their countryside, through pastoral lenses. Real-estate promoters, according to historian Earl Pomeroy, "cultivated the illusion that the subdivisions they had carved out of pastures and desert had the color of history behind them," and suburbanites incorporated this promotional pastoralism into their vision of the West. They expressed their aspirations for new beginnings in what one historian called "virgin cities," a complement to the quest for virgin land as a measure of regional identity.

However, like suburbanites everywhere, those in Oregon sensed the gap between Arcadian aspirations and the realities of suburban life. Western suburbanites were, as historian Edward Morgan puts it, "comfortably affluent but troubled by a gnawing sense that something was missing from their lives," an impression drawn in part from negative assessments of suburbia like Sloan Wilson's best-selling *The Man in the Gray Flannel Suit* (1955), William H. Whyte, Jr.'s *The Organization Man* (1956), and Paul Goodman's *Growing Up Absurd* (1960). Their environment was "a consciously planned hybrid, a man-made nature that was, at the same time, visibly, even painfully, artificial."[36]

New residents on this expanding suburban frontier began to question the primacy of Oregon's growth-oriented politics.[37] Subdivisions, they discovered, were spreading rapidly into jurisdictions with no building or zoning codes, resulting in a "hodge-podge of shacks, junk yards, and beer taverns" over which municipal or county governments exerted no effective control. Sewerage and fire protection, if any, depended on the parent city.[38] Uncomfortably situated at the interface of two competing land-use systems, suburbanites came in contact with rural practices they found inconsistent with their romantic ideals, and they deluged the state's health officers with complaints about vacant fields used as "dumping grounds," odors from hog farms, accumulations of feces from too many dogs in a yard, mosquitoes breeding in log ponds and abandoned gravel pits, smoke from burning trash or sawmill waste, and pollution in nearby streams and rivers.

As early as the 1930s the Oregon Writers' Project had worried that post-Depression industrial growth would bring "a network of highways clogged with cars and defaced with hot dog stands ... groves littered with tin cans and papers, the hills pock-marked with stumps and ... cities cursed with ... slums."[39] Richard Neuberger, amid his 1938 plea for more people and more industry, noted that "eagles do not soar above smokestacks, salmon do not swim into pollution, and panthers and bears do not roam within sound of whirring machinery." Local writer Stewart Holbrook pondered the implications of a "second New Jersey here among the tall firs and the cattle ranges," and H.L. Davis recoiled from the prospect of "overgrown towns ... leveled forests, and stopped-up creeks." The doubling and trebling of many western Oregon towns, he observed, brought a "swarm of new stucco supermarkets, car and tractor showrooms, chain-saw and logging-truck repair shops, real-estate offices, Assembly of God tabernacles, drive-in movie theatres, country-club residence subdivisions, antique and curio shops."[40] This tradition of pastoral criticism — Neuberger, Holbrook, Davis — gave substance to the suburban malaise.

Oregon old-timers who envisioned their state as a place of quiet towns and rural neighborhoods took these concerns seriously, and their fears were reinforced by the state's recent arrivals, who had fled California's suburban sprawl and now sought to protect Oregon against a similar nightmare. As a symbolic protest against in-migration, Holbrook founded the James G. Blaine Society, honoring the xenophobic Maine senator who coined the phrase "America for Americans." The national media picked up and publicized Holbrook's spoof as a reflection of Oregon's growing alarm at "being invaded by hordes of tourists, most of them from California."[41] Pointing to the concrete and graceless artificiality of suburban southern California, Portland planner Charles DeDeurwaerder warned that the Los Angeles basin "looked like the mid-Willamette River valley once — and just a short forty years ago."[42] How, Oregonians wondered, was the accelerating pace of change to be incorporated into an Edenic landscape most saw as perfect just the way it was?

MAINE CONFRONTS SPRAWL

Maine's pastoral aspirations had a decidedly different cast, woven as they were into a general mood of economic despondency. By indexes like personal income, education, and social welfare spending, Maine ranked among the poorest states in the nation, shut out of America's brilliant postwar economic expansion by declines in its major staples — logging, fishing, shipbuilding, textiles. Chronic low wages, substandard rural living conditions, and impoverished educational institutions exposed the colonialist footing of Maine's economy; here the line between the simple pastoral life and grinding rural poverty was thin indeed.[43] Where Oregon braced against a flood of in-migrants after the war, Maine suffered a net outmigration. While Oregon stood at the center of America's postwar economic expansion, Maine struggled to retain its obsolete heavy industries, offering its natural landscape as a colonized arena available for any sort of economic expansion.

In fact, Maine's economic quiescence was partly a result of its pastoral imagery. Despite concentrations of heavy industry in towns across the state, Maine, in the tourist's eye, was a pastoral landscape, and this nationwide reputation as a land free of industrial infrastructure did little to encourage outside industrial investment. As one magazine editor noted, the Maine way of life immortalized in E.B. White's nostalgic *One Man's Meat* seemed all too discordant with industrial growth:

> More likely those who deplore neon jungles and commercial activity on our main drag swoon with pleasure at seagulls circling over rotting piers, take pictures of deserted sway-back barns, non-operating fish plants, or old, dismantled vessels beached in some gunkhole. Is it the patina of death and disuse that delights their tired eyes? Would those same fish plants operating, those same piers bright and functioning, be vulgar and lamentable? ... To satisfy those who cry out at the vulgar vitality of newness, must Maine go backwards, or at the very least, stand still? ... Humming factories may be one man's poison; but sure as shooting they are Maine's meat."[44]

Adopting a static Arcadian vision, Mainers themselves succumbed to a mood of "quiet decay," as one postwar planner put it. Author Mary Ellen Chase found no Faustian drive among the "folk": a legacy of storms and fogs, winter isolation, and summer drought overwhelmed the people of Maine with a sense of the inevitable, a "patient wisdom" inclined to accept adversity as an "order ... against whose laws only fools will struggle."[45] Immersed in the irrepressible cycles of natural growth and dormancy, rural people learned to submit to larger forces.

These images influenced Maine's postwar economic strategies. Clinging to its small-town political culture, Maine emerged from the war conservative, staunchly Republican, insular, and in some ways antimodernist. In his 1945 inaugural address, Governor Horace Hildreth reaffirmed the state's sovereignty over New Deal–type federalist forces and warned against the inroads of big corporations and big government agencies. Recalling Maine's singular resistance to the New Deal through three presidential election cycles, the Republican governor-elect cautioned that Maine should rely on federal programs for postwar economic readjustment "only as a last resort."

Most of all, Hildreth placed his faith in small, traditional factories and mills in Maine's rural settings, "where industrial and agricultural work may be combined." Like Coffin and White, he saw Maine as a world of small producers insulated from the harsh modernism of the metropolitan core.[46] The Depression demonstrated that U.S. industry had grown too large, members of Hildreth's Development Commission reasoned. In small towns "the worker ... thinks constantly about his job and figures out ways to do it better." Energized by a healthy rural environment, he would advance to the ranks of petty capitalist, mitigating class struggle and buffering the state from the ravages of big government and big business.

Overall, Maine projected a confusing mixture of progrowth sentiment and antimodernist pastoralism. Two generations of declining markets for Maine timber, fish, and farm produce fostered a desperately probusiness political culture in which, as Governnor Hildreth suggested, even the most assertively promotional messages could carry a hint of nostalgia.[47] Postwar planners offered potential investors an economic version of the Maine way of life, where rural living drove workers to greater exertion and superb opportunities for outdoor recreation compensated for low wages.[48]

Maine's vacation industry, one of the few bright spots on its economic horizon, offered another view of this economic pastoralism.[49] Tourism, according to Robert P. Tristram Coffin, was an affirmation of Maine's redemptive character: "we are glad to have people flock here in the summer. And it isn't solely because we want to sell them things.... They are looking for something. Something they have lost or mislaid." Alienated urbanites embarked on an odyssey of self-discovery in Maine, and what they found, they shared with locals perhaps too familiar with the landscape to appreciate it.[50]

Urban tourists and resorters indirectly forced the pastoral ideal into Maine politics. State planners and business leaders knew that scenic resources were nonrenewable, and they anticipated fierce interregional competition for tourist dollars as the northeasterners who traditionally vacationed in Maine became even more mobile.[51] Promoters of Maine's rock-bound coast and its "sturdy, self-reliant breed of Yankee" understood that the state must shield its scenic resources from the tourism industry's own commercial trappings. These leaders formulated a "Keep Maine Scenic" program and local beautification ordinances to distract outsiders' attention from the expanding tourist infrastructure.

When literary critic Bernard DeVoto published an article in *Harper's Weekly* criticizing Maine's cluttered and unsightly roadsides, the

Department of Development canceled state advertising in the magazine.[52] In response to similar complaints and a 1958 federal law providing highway funds to states that controlled billboards, the legislature
passed a measure sponsored by the state Federation of Garden Clubs to
regulate roadside advertising. According to John T. Rowland of *Down East
Magazine,* the "defacement of nature by billboards" was a "folly of the stu-
pidest sort," undermining Maine's most valuable natural resource.
"Do we think these customers are not yearly more aware of the glar-
ing ... despoliation to be seen in drives through once picturesque country
roads?"[53]

As the antibillboard campaign suggests, Maine people understood the
threats to their landscape's integrity.[54] Henry Beston, from his peninsular
vantage on the Maine coast, sensed the urban blight spreading northeastward from Boston along Route 1: a "vulgar confusion of predatory billboards and trashy signs which are destructively allowed to steal and
exploit one of the great landscapes of the nation." Noting a massive transfer of southern Maine real estate to out-of-state resorters, the *Kennebec
Journal* complained in 1956 that "Maine residents approaching the coast
today are apt to feel like they're entering a foreign land."[55]

However, the reaction to blight was less strident in Maine than in
Oregon. This was partly because the wealthy urban resorters who had colonized coastal towns like York, Ogunquit, and Bar Harbor in the previous
century themselves promoted Maine's beautification campaigns. The
Rockefellers, Cabots, Lowells, Dillons, Lamonts, and DuPonts, as a Maine
governor once caustically remarked, made their money polluting other
states and then retreated to Maine for serenity. Landscape preservation —
"saving the view from Mrs. Rockefeller's picture window," as locals put it
— was thus tainted by urban elitism.

Maine's coastal counties also harbored some of the worst cases of rural
poverty in the Northeast, often adjacent to these summer homes, and
among natives the demand for jobs outweighed the quest for tranquility
in an unspoiled landscape.[56] Commercial development along U.S. Route
1 provided jobs and tax revenue, while the locals themselves continued
their way of life unmolested down unmarked byways and secluded penin-
sulas — the place "you can't get to from here."[57]

What's more, to the native Mainer, commercial blight may not have
been all that disconcerting: as E.B. White observed, locals left the major

access highways into Maine a "sorry mess." He was quick to explain, however, that winter "quietly erases" such mistakes, and that even in summer those who knew the "real" Maine were not dismayed by a thin belt of billboards and curio stands: "woods and fields encroach everywhere, creeping to within a few feet of the neon and the court, and the [local resident] ... is always conscious that just behind the garish roadside stand, in its thicket of birch and spruce, stands the delicate and well-proportioned deer."

Mainers were not always sanguine about these changes, however. People living along the coast realized, as an editor put it, that there was "something in the quality of their lives worth saving" — something that kept them fiercely loyal to the traditional Maine, despite its economic backwardness. Clam-diggers and lobster-fishers in Stonington, a remote harbor in Penobscot Bay, applauded the fact that there were "40 miles of crappy road between here and the thundering herd" on U.S. 1.[58]

Mainers' changing perceptions of growth and stability related intimately to the sentiments Coffin and Beston explored in their postwar literature of place. Coastal Mainers intuited a loss of the freedom they associated with the prewar world. "The average Mainer, who had been able to get along because land was cheap and wide open to any, and who wanted to add wild meat to the family larder [or] ... grow his own vegetable garden and raise a pig or some chickens, began to find himself ... squeezed by rising prices and higher property taxes, while his access to wild lands decreased and the cost of feed for his animals made that economy questionable." Traditional pastoral activities — clamming, hunting, berry-picking, fishing, gardening — had enormous symbolic value in Maine, associated with a regional brand of freedom and an earlier day when "local worlds were ... not dependent on a vast impersonal market."[59]

The loss of farmland and open landscapes challenged these nostalgic pastoral ideals. As more and more land fell into the hands of outsiders, Mainers found themselves barred from forage and recreational resources that generations had regarded as an informal commons. Changes in property ownership, new forms of land use, and a host of other vexations impinged on a traditional way of life.

Tourist promoters, real-estate agents, motel and campground operators, and restaurant and service station owners worried about scenery as a nonrenewable resource; summer residents and resorters took an interest in protecting their private pastoral utopias; coastal fishers grew apprehensive about water pollution. Together with those Coffin and Beston identified as the folk, they defined a common interest in protecting Maine landscapes.[60]

ON THE MARGINS OF POPULAR POLITICS

Steeped in the iconography of nature and folk, regional writers gave voice to an anxious popular mood spreading through states like Maine and Oregon in the postwar years.[61] Yet despite the outcries against growth and commercial blight, in the early 1960s Americans were not yet ready to launch a militant defense of the pastoral landscape. In Maine, the sheer abundance of empty land, the relatively static economy, and the seemingly permanent pastoral features like ramshackle fish-houses, abandoned streamside mills, sagging barns and harborside wharfs helped perpetuate the "gentle air of the past" despite the metropolitan forces pressing at the borders.

Inspired by these timeless and graceful ruins, regionalists invoked the power of the folk to remember, and the power of memory to preserve. The mind of the folk and the state's wealth of cultural relics secured the pastoral ideal. Stories, Elizabeth Coatsworth wrote, "become fables," and though the individual is lost, "out of the accumulation of tales remembered ... comes a sense of the entire people, vague, often contradictory, but still real."

Oregon's ruins, unsoftened by weather and time, proved less inspiring; "dead towns" left in the wake of the mining and lumbering frontiers conveyed only a sense of raw clutter, a thin sheen of memory upon which to build an Arcadian vision.[62] Here magnificent natural landscapes and Oregon's newness — its "perpetual youth" — conveyed the hope that the pastoral ideal could be sustained. Just as America was once young and innovative, "so the Pacific Northwest, coming late to that transition ... [had] a chance to develop more rationally than older sections of the country." The city would be tastefully integrated into the gardened landscape.[63]

Sentiments like these delayed the politicization of the pastoral ideal. Nor was it clear how the pastoral landscape could be preserved. In Maine the legislature, Coffin theorized, could impose preservation laws that would "take all the people away from [the coast] ... and the people, mind you, are our greatest destroyers — pull down all their houses. Build scenic highways ... along all the headlands ... make all the coast one great national park."

But was this desirable? Human activity was, after all, as natural to the Maine coast as granite and spruce. What Coffin hoped to preserve was not pristine nature but a precommercial pastoral landscape peopled by self-reliant farmers and fisherfolk living out a traditional way of life in the lap of nature. But could this pastoral landscape be suspended in amber without applying the politics of "mass righteousness," as he put it? Like nineteenth-century romantics contemplating the American frontier, Coffin celebrated the simple life while acknowledging the inevitability of its passing. "In the name of efficiency," he recognized, "the disease comes." Fisherfolk would become hired hands; machines would be imposed between workers and their work; modern technology would strip the sea of its resources; consolidated schools would blot out community identity.[64] The landscape he cherished was, after all, an illusion, a fleeting way of life suspended on the thin threads of memory and custom.

The story of the environmental imagination begins, then, with a pastoral ideal threatened by the glacial out-thrust of the city and its suburbs. Although the construct originated in a nostalgic literary tradition, the evocation of cherished rural values like open space, clean air and waters, natural scenery, and quiet surroundings gradually assumed political significance and progressive implications. But in the early 1960s this oblique challenge to the politics of growth remained latent; for the moment, the standardized suburban tract, with easy access to nature beyond its borders, seemed livable.

However, the decade brought changes as well. Growth in leisure time and mobility and burgeoning dissatisfaction with the physical and cultural landscapes of suburbia, operating in tandem with national events like the publication of Rachel Carson's *Silent Spring* (1962) and the discovery of radioactive isotopes in human tissue, set the stage for a more assertive defense of the pastoral. While this defense took root in different ways in Maine and Oregon, the result was, in both, transformative, redefining political challenges and processes and investing government with the power to preserve the deeply symbolic landscape of folk and nature.[65]

Unlike the expansionist pastoral elements that characterized New Deal conservation, the new images were initially antimodernist, emphasizing, according to one study, a "return to an idealized, organic, agrarian golden age when humanity had not yet despoiled the earth." When the rhetoric of harmony between folk and nature intertwined with the political campaign against blight and pollution, it became a powerful and progressive impetus for environmental reform.[66] A conservative literature of place in Maine, Oregon, and across America provided fertile ground for the growth of the environmental imagination.

Mt. Hood, Oregon

CHAPTER 2

POLITICIZING THE PASTORAL IDEAL: THE CLEAN-WATERS CAMPAIGNS

On the morning of March 15, 1907, the citizens of Lewiston, Maine, awoke to find immense banks of foam billowing out of the mill canals and drifting through their streets. The surging 20-foot drifts delighted the city's children, who tossed the creamy substance into the air and watched it blow away. Older residents turned their thoughts to the huge textile and pulp mills on the river and wondered about the safety of their drinking water. Over the next three decades, these mills dumped approximately 100 million gallons of organic and inorganic waste into the Androscoggin River each day, and the resulting caustic, foul-smelling hydrogen sulphide fumes blackened house paint and afflicted hundreds of valley residents with irritated eyes, sore throats, and nausea. In 1941 the former children of 1907, now disgusted rather than intrigued by the unnatural conditions of the city's most prominent natural feature, gathered in "indignation meetings" and prepared to challenge the industrial powers that made their river a stinking hell.[1]

Oregon faced a similar problem. A 1945 governor's report on the Willamette River, with some 70 percent of the state's population living along its banks, found that heavy waste volumes, low summer stream flows, and elevated water temperatures produced a dead zone near the river's confluence with the Columbia where dissolved oxygen frequently dropped to zero, impairing the chinook salmon run and threatening the health of citizens. In two decades the pollution plug had spread some 30 miles upstream, leaving the river from Eugene to Portland in a dangerously contaminated condition. Impregnated with municipal sewage, the fibers and waste chemicals from the Willamette's paper mills formed deposits on the river bottom and periodically bubbled to the surface as huge slime formations. The river's banks were so tangled with debris that it was difficult to get to the water to sample it.[2]

The solutions proposed to this problem in the 1940s were far from complete. Oregon's officials suggested, along with the inevitable call for more investigation, a system of "stream specialization," designating some rivers for industrial use, some for recreation, and some for fish life. Those who launched Lewiston's clean-waters campaign in 1941 simply aimed to reduce the nuisance to tolerable levels. Lewiston Mayor Edgar St. Hilaire explained his conservation agenda: "we are not interested in a pure river. We [simply] want to get rid of the smell."[3]

By the 1970s, pollution problems drew a very different response. In 1977 Mabel Johnson of Boring, Oregon, complained to Governor Robert Straub that the creek near her home was running a "thick green with a blanket of suds on top." For 20 years she and her husband had battled to save this once-beautiful stream, and for 20 years they had received the same answers from state officials. After her husband died, Mabel persisted, complaining to health officials, sending water samples, and finally writing to the governor: "I'm sorry to trouble you with this, but I'm sure Fred wouldn't want me to give up the battle just because he isn't here to help me." The governor received similarly persistent messages from Margaret E. Benz, who watched the ducks, crawdads, frogs, and trout she carefully cultivated in an old log pond disappear when a nearby city ran a sewer outfall into the stream. Across the state, Mrs. Glen Boehme described the foul-smelling algae blooms in Klamath Lake and pleaded with the governor: "God gave us these beautiful lakes and streams, so why can't we clean them up, so we can enjoy them more [and] ... when visitors come to our beautiful state, we can take pride in showing off the results of our labors?"[4] Rivers, these women insisted, were living systems — habitat for wildlife and centerpieces of the pastoral landscapes they cherished.

The path from accepting rivers as industrial sewers to protecting them as symbols of natural purity illustrates the interaction of pastoral ideals and environmental politics. In the early postwar years, popular perceptions of rivers included little regard for natural values. Valley residents recognized the importance of their urban rivers as "tremendous power plants," turning the wheels of industry, keeping their homes bright, and providing jobs for those who lived along the banks.[5] They saw their rivers as parts of an engineered landscape that included dams, power lines, highways, subdivisions, and other manufactured attributes, a vast system that, as historian William Cronon demonstrates, conveyed the wealth of the countryside to the city. Rivers were viewed as beasts of burden, harnessed to the needs of the urban industrial order. Any relief from pollution was predicated on what Richard Andrews calls a "rudimentary 'cost–benefit balancing' approach" that weighed the damage pollution caused to an injured party against the economic benefits of retaining the polluted river in this mechanistic system. Generally, the courts sided with the system.[6]

This focus on human control of energy and resource flows, as cultural geographer J.B. Jackson argues, was accepted as normal in the mid-twentieth century. Under these conditions, resignation came easily, since few urban people had ever seen their rivers as living components of the landscape. For most, urban rivers were despised features of the city — "like having a damned sewer running through your living room." Residents discarded old tires in rivers, piled brush and rubbish along their banks, and hid the foul waters behind rows of mills or commercial buildings. Asked if the river smell bothered her, one valley woman sighed, "you can get used to anything, I guess."[7]

However, over time, the river came to symbolize a different order: a means of reconnecting nature and the metropolis — of pastoralizing the urban environment. This powerful pastoral vision conveyed romanticized memories of a lost intimacy with nature, experienced not as a place to sojourn but as a way of life. This nostalgic vision, as John Cumbler points out in his history of early New England environmental politics, was essential to the environmental imagination.[8] But in the context of an emerging environmental movement pastoralism underwent a change from a nostalgic construct to a pragmatic and urban vision, a progressive and even modernist form of thinking that acknowledged and indeed embraced the city as part of a new livable landscape.

NATIONAL BEGINNINGS, 1946–1960

Protests over pollution began as a classic, turn-of-the-century Progressive-era urban health campaign, driven in good part by the realization that America had indeed become an urban nation, and that pollution was most extreme precisely in those urban areas were water was needed most for domestic, industrial, and recreational purposes. The campaign was similar to contemporaneous smoke-abatement and sanitary reforms in that its goal was limited to containing the unfortunate by-products of urban industrial growth.[9] However, in the late postwar period the battle for clean waters absorbed the pastoral aspirations of the era to become America's first grassroots environmental crusade, having the essentials of the modern movement in its preoccupation with aesthetics, recreation, ecology, and health.

Nationally, the seriousness of this problem was first evident in cities along the Ohio, Allegheny, Monongahela, Delaware, and Schuylkill rivers, where millions of gallons of industrial waste and raw sewage churning daily into the waters blackened paint and corroded fixtures on buildings along the banks. Mothers grew apprehensive about "water diseases," and treatment costs skyrocketed as engineers "exhausted virtually all of the known methods [of treatment] and ... succeeded largely in producing a brand-new breed of taste which the public does not enjoy." For health reasons, thousands of acres of clam and oyster beds were closed from New York to Maine, and California was forced to bar access to several beaches.[10]

Several federal antipollution bills were voted down or vetoed in the 1930s, but after World War II Congress returned to the issue encouraged by sanitary successes in New Deal and wartime housing projects. In 1945 *American City* ran an article on "America's Dead and Dying Rivers," pointing to the sobering facts of America's mounting crisis: 2.5 billion gallons of raw sewage and 3.75 billion gallons of industrial waste pouring into the nation's waterways every day; the Ohio from Pittsburgh to Cairo gravely overburdened with raw sewage; rising infant mortality from diarrhea and enteritis; and possible connections to infantile paralysis and poliomyelitis. In the next two years Congress debated several bills requiring state permits issued under federal guidelines for new sources of pollution in interstate waters. The 1945–1947 bills stimulated some soul-searching about the despoliation of America's natural heritage, but for the most part testimony focused narrowly on practical health and economic threats: tinkering with the engineered landscape.

None of the bills passed, but they set the stage for America's first modern antipollution law in 1948, a product of intensive lobbying by the Isaak Walton League, the U.S. Fish and Wildlife Service, and various professional engineering associations. The Water Pollution Control Act, passed by the 80th Congress, providing a new Division of Water Pollution Control within the Public Health Service charged with investigating river conditions and encouraging cooperation between federal, state, local, interstate, and private agencies. Irrespective of the 1948 act and subsequent federal measures, states remained the primary underwriters and enforcers of water pollution control policy well into the 1960s. Illinois and Pennsylvania created special water pollution control agencies as early as 1923, and others, including Oregon, followed in the 1930s. But despite this early start, state agencies accomplished little. In 1946 only eight states provided 50 percent or more of their sewered population with any sort of sewage treatment. Maine partially treated only 2 percent of its sewage, ranking it near the bottom nationwide. Oregon's 21 percent fell in the lower third.[11]

As long as pollution was viewed simply as an urban health issue, the remedies were narrowly constructed. Redeeming America's fouled rivers as healthy natural systems, by contrast, would be a vast undertaking involving billions of federal, state, and private dollars and a monumental sense of responsibility for the future. This approach would require a popular imagination that fused love of place with resentment toward those who despoiled it — the catalyst that compelled three Oregon women and thousands of others to write, again and again, to their state representatives.

Environmental solutions, Daniel Botkin writes, "depend on our inner perception of our connection with nature." Americans needed a motive, in short, for improving their relation to nature. This sense of harmony and connectedness — the pastoral vision — emerged as a political force in the late 1960s.[12] The water pollution campaigns of the 1950s and early 1960s began largely as utilitarian concerns, but over time they drew upon pastoral images, transforming this literary construct into political rhetoric and symbol. This, in turn, provided pollution politics with a sense of empathy, or as Sara Ebenreck puts it, an "understanding [of] the flowing interactions with air, water, sunlight, and plant and animal life." As the letters from three rural Oregon women make clear, this empathy had enormous political repercussions.

The environmental imagination — the power of place, memory, and nature in people's lives — emerged out of the traditional pastoral visions that had always provided contrasts with the engineered landscape. In Maine, even in the dark and dirty days of the 1940s, a few old-timers could recall an era when the waters were clean and the rivers were the "meat tubs" of the valley, teeming with salmon, sturgeon, shad, pickerel, trout, striped bass, and smelt — species that had long since disappeared under clouds of poisonous waste.[13] In addition to these fading memories, those living along the industrial rivers knew their headwaters. "We're lucky," an upper Androscoggin Valley resident mused; "You go upstream a little ways, and the river's beautiful." There, the clear, cold waters coursed down winding channels around tiny wooded islands, and the river still served up the pleasures of an Arcadian landscape. Columnist William H. Williamson, standing at the font of the Androscoggin near Lake Umbagog, proclaimed that the river "couldn't have picked a more beautiful spot" to start its 160-mile journey to the sea. Here, too, was a pastoral benchmark for gauging the destruction of the lower river.[14]

In the same way, Oregon's Willamette River began as a collection of clear, pristine streams in the foothills of the Cascades, Calapooyas, and Coast Range, where tree-shaded waters exerted a powerful pull on the imagination of valley residents, highlighting the contrast between natural rivers and those engineered as industrial sewers. Perhaps America's most paradoxically dichotomized river was the Cuyahoga, flowing out of east-central Ohio to Lake Erie. Made famous in 1969 when its polluted waters near Cleveland actually burst into flame, upriver sections were pure enough to be consecrated as a part of a state wild and scenic river system just a few years later.[15] In the emergent environmental imagination, the purity of these upper rivers symbolized a golden age of harmony with nature lost when America engineered the countryside, but now perhaps recoverable as part of a more livable landscape.

EARLY CITIZEN PROTEST AND ADMINISTRATIVE RESPONSE IN MAINE, 1941–1960

Maine and Oregon illustrate the importance of pastoral ideals in setting a moral tone for antipollution politics. In Maine, concern about water pollution began at the turn of the century with a series of typhoid outbreaks along the Kennebec and Penobscot rivers. But the first official statewide recognition of the problem came in 1929, when Governor William Tudor Gardiner, acknowledging local protests, asked the paper industry to conduct a survey of the state's rivers. The study, funded by the S.D. Warren Paper Co., found ample evidence of pollution but defined the problem in ways favorable to the status quo: it used unrealistically low standards for defining nuisance conditions, obscured industry responsibility by underscoring municipal pollution, and assumed industry's right to the rivers' entire quotient of dissolved oxygen beyond the minimum necessary to oxidize organic materials and abate extreme nuisance conditions. As the Warren report framed it, the solution was simply a matter of allocating the river's usable supply of dissolved oxygen more efficiently among the mills and factories that lined its banks.[16] This belief remained at the core of the antipollution campaign into the 1960s.

Citizens living along the Androscoggin River, the state's filthiest, were the first to challenge the principle that Maine's waters were owned by industry. As factories geared up for war in 1941, the fumes exuding from the Androscoggin became intolerable. Depositions taken a year later by Auburn attorney Seth May and state Attorney General Frank Cowan reveal the severity of the conditions that finally aroused citizens to action. Rumford drug store manager Leo R. Good related that he "came across the bridge one morning and threw up." Customers ordering ice cream at his store left without touching it. Others complained that children awoke in the night nauseous and vomiting, and physicians worried about health conditions throughout the valley. As machinist Cleas Lacombe reported, "everything I eat seems to ... taste the same as the river smells."[17]

Citizens expressed their concerns primarily in economic terms: "[w]hat business or professional man ... is so detached that the stench does not lessen his personal efficiency? What clerk, after days and weeks of this punishment, can be expected to be at his, or her, best? ... What about pride in community when it literally smells to high heaven?" Working through "action clubs," town merchants and proprietors organized protests, petitioned the state legislature, and met with Governor Sumner Sewell to urge a new survey of the river. Odors, caustic fumes, and raw sewage, they pointed out, threatened the health of citizens, devalued commercial property, drove away customers, and forced owners to repaint their peeling and discolored buildings yearly.[18]

Given the power of big industry — paper and textiles in particular — the concerns of Lewiston's merchants and real-estate owners failed to move officials in Augusta. The state Bureau of Health lectured that the only solution "would be to have all the industries on the ... river cease business and move away, and have all the people follow them."[19] The legislature established a Sanitary Water Board (SWB) to investigate the Androscoggin and other industrialized rivers but gave the new agency only $400 in funding. This halting response reflected the conflicted mood among citizens along the river.[20] Those living upriver viewed water quality as a local issue and maintained a high tolerance for "aroma riverensis," as a mill-town journalist called it; downriver protests continued to be circumscribed by the Warren report's narrow definition of nuisance and by Maine's continued economic desperation. Although business owners hoped to protect their own investments from the scourge of polluted waters, they understood that the huge upriver mills were the core of the valley's economy, providing jobs and giving value to the region's woodlands. In September 1941, as the odors abated with colder weather, a newspaper editor asked the inevitable question: "supposing it appears that a considerable sum of money must be expended to care for the situation.... What then?" The question supplied its own answer.[21]

Corporate publicists countered rising popular resentment by arguing that their discharges were "practically sterile"; odors resulted only when fibers and chemicals were "inoculated" with municipal waste. Although 90 percent of the pollution load came from the mills, they insisted that sewage created the nuisance, and for the next two decades they would demand municipal abatement before beginning their own cleanup.

Corporate intransigence was firmly fixed; as late as 1965 a paper company official argued that there was "no such thing as pollution.... The only way you can pollute a river is to change its character ... and ... by this definition if we start to clean them up, we are polluting them because we do change the character."[22]

In response to continuing complaints from Androscoggin Valley citizens, the state began searching for a conflict-free solution to the pollution crisis. Administrators ordered companies to dump chemicals into the river to hide the smell, to build storage lagoons at the mills, and to collect sulfite wastes and spill them only when fall rains and cool temperatures increased the river's carrying capacity. After comparing production figures with past citizen complaints, mill owners proposed a formula allowing three tons of finished paper production for every million cubic feet of river flow above Lewiston. The consulting engineers expressed misgivings, but the court agreed to the formula. Thus was negotiated what Industrial Development Commissioner Harold Schnurle described as the "exact science" of water use.[23]

Like agencies in several other northern states, the Sanitary Water Board began classifying each river according to its "highest use" as a basis for regulating new discharges. But the law grandfathered all existing pollution sources and prohibited new discharges only if they increased pollution "in a manner inconsistent with the public interest." Companies could overturn a ruling simply by proving that pollution was "necessary in the prosecution of ... business." More significantly, the major industrial rivers were exempted altogether. Given its weak legislative mandate, the agency hastened to cooperate with offending companies. Board members met frequently as guests of the mill managers and adopted the millowners' perspective on the technical feasibility of water treatment.[24]

Changes in state politics in the mid-1950s provided a firmer grounding for the attack on pollution. During these years, a subtle antimonopoly mood, fueled by postwar labor militancy, brought to power a generation of Democrats and liberal Republicans willing to challenge corporate control over natural resources. Resurgent liberalism all across the northern states was driven by a constituency that included skilled and semiskilled white workers moving into new suburban homes and entering the culture of abundance. The national mood was growth-oriented, but in state politics it contained an undercurrent of skepticism about corporate power and a growing interest in quality of life.[25] In Maine, this populist mood crystalized in 1954 when voters rejected a century-long legacy of Republican rule and elected Democrat Edmund S. Muskie governor. Because Maine's September elections served as a barometer for the national vote in November, Muskie's upset caused a nationwide stir. Commentators pondered whether the Big Three — paper, hydropower, and textiles — could retain "sovereign control" over Maine politics.[26]

Several factors propelled Muskie's victory. The Republican Party was distracted by a breach between the Old Guard and a younger liberal wing, and Muskie personified the youthful, grassroots energy and populist outlook voters in this depressed state were looking for. With the economy soft and labor on the move, Muskie, a Polish Catholic raised in a central Maine mill town, received solid union backing, particularly from the United Textile Workers and their Catholic Franco-American rank and file. Along with labor, Muskie attracted farmers, small shop and factory owners, and professionals frustrated by the state's poor economic performance. In Maine and in other states, this Democratic resurgence amounted to a "popular resentment against the special interests" and the political organizations that supported them.[27]

Maine's revolt against the "special interests," however, remained incipient. The small-business element in Muskie's constituency dictated moderation, and the new governor faced an overwhelmingly Republican legislature. Thus Muskie turned the populist surge in the direction of encouraging small-business growth. Aware that his state was losing some 2,000 textile jobs yearly, he offered a new development offensive, including improvements in schools, roads, and research capability and fuller utilization of Maine's forests, rivers, and wildlife. Although Muskie was unable to solve the state's thorny economic problems, his politics of growth won him a second term.

In 1958, Waterville lawyer Jerome G. Daviau published a powerful political exposé detailing the historic abuse of Maine's rivers and the "the art of public deception" that sustained the paper companies' monopoly over the natural landscape. Daviau listed, in lurid detail, scandal after scandal perpetrated by the legislative "suitcase brigade," compiling a case that the industry had a "vested interest" in keeping Maine underdeveloped. *Maine's Life Blood* was the earliest of several populist tracts written about Maine that added specificity and direction to the vague resentments first expressed in the 1954 election.[28]

Still, the 1950s were not propitious years for moving beyond the ineffective early efforts at pollution control. Americans were in a transitional age, fascinated by the nation's capacity for material wealth, driven by mass advertising designed to promote consumer spending, and as yet only vaguely aware of the need to control the by-products of their affluence. Production and advertising were geared to the idea of disposability, creating a society of "waste makers," in Vance Packard's term, and pushing regard for landscape and water quality to the margins of public consciousness.

In Maine and elsewhere, political obstacles were numerous. In an era that subscribed to the tradition of citizen–legislators, the idea of giving a small government regulatory body veto power over corporate activity was troubling, and a brace of corporate lobbyists — often known as the Third House in a state legislature — found it easy to strip pollution bills of any meaningful content.[29] But corporations were not alone in their opposition. No community was willing to invest in pollution control until those upstream agreed to shoulder similar costs, and there was little agreement about where authority over pollution control should lodge. Business owners insisted on state sovereignty, while conservationists felt that federal laws were necessary to level the regulatory playing field. The general attitude, as one consulting engineer put it, was one of "bewilderment"; some felt, in fact, that industrial waste benefitted cities by killing harmful bacteria in raw sewage. Maine, as one legislator argued, still had plenty of unpolluted water, and the Androscoggin had not yet caused "a single death."[30]

For Maine, the most significant barrier to clean-waters legislation was the pressing issue of industrial growth. Sensitive to southern competition in textiles and newsprint paper, legislators listened when mill owners insisted that their expansion plans would depend on the shape of Maine's pollution laws. How could Maine invite new industries to the state, a legislator asked, when "under [proposed regulatory] laws they would be a public nuisance, an outlaw, as soon as they started to turn the wheels?"[31] Nor was it easy to mobilize community leaders. When a tannery owner planning to relocate to an upland mill town demanded a dumping permit, a local Chamber of Commerce official wrote to the governor pointing out that he need not worry about ruining a pristine river. Sewage-impregnated industrial detergents already created "such foam at our falls ... that ... bubbles flow all over town." No one in town, the official explained, was "worried about a few [more] suds in an industrial river."[32]

The tannery received its permit, but in the shifting official rhetoric of the mid-1950s it is possible to distinguish a subtle change in such attitudes, as public officials realized that polluted rivers were as much an impediment to growth as were regulations on pollution. Prospects for allocating dissolved oxygen among a growing number of industries were dimming, a Muskie administration official explained:

> Every industry ... would dearly like clean water to use, but by [implicit] agreement among them, each mill ... spends money to purify the water it uses and adds its waste to the pollution of our streams and rivers, when it would cost no more to build disposal plants to keep the waste out of the water. Not only is the beauty of the country being despoiled, the fishlife being killed or driven out, and the health of our citizens being endangered, but the problem of attracting any industry needing clean water in its process to the State of Maine is obviously doubled if that industry must make arrangements for a water purification system before it can start operation.[33]

Thus Governor Muskie's growth initiative included promoting and to some degree protecting Maine's "excellent water systems," and mill towns along the Androscoggin, struggling to redefine their future in a changing economy, adopted this new development-oriented antipollution message. Earlier business leaders had reasoned that since the rivers were "already contaminated," new industries would find them attractive as convenient sewers; by the mid-1950s these same grimy waters seemed to chain the river cities to their dubious heritage as nineteenth-century mill towns. The "filthy and foul river," Lewiston Mayor Emile Jacques declared, had "tragic consequence on the social and economic life of the entire area."[34]

While mill-town officials were reassessing the use of rivers as sewers, Maine's burgeoning tourist industry entered the equation, and here conservation rhetoric began to absorb the pastoral images captured by writers like Robert P. Tristram Coffin and Henry Beston. The industry became aggressively promotional in the mid-1950s. A flurry of state-sponsored travel guides updated Maine's pastoral images for the auto age, and a lavishly illustrated periodical, *Down East Magazine*, first published in 1954, helped elevate Maine's covered bridges, rock-bound coves, and lobster landings as nationally recognized icons. Articles celebrating fly-tying, lighthouses, one-room school houses, state fairs, and town meetings gave substance and texture to the ellusive "Maine character."[35]

Glossing the image of Maine as a pastoral paradise, tourism set the tone for a more assertive effort to clean up the state's rivers, lakes, and ocean beaches. Recreation, now Maine's second-largest industry, reached a critical juncture when the nation's new interstate highways linked the Northeast to the rest of America: the roads were "just as likely ... [to] take ... people away from Maine ... as they were to bring them into the state." A traveler from Philadelphia complained to Governor Muskie about the repulsive odors along the Androscoggin River that spoiled his family's vacation; Muskie would have to do better if Maine was to "continue to advertise itself as Vacationland." Bucksport Representative Frank Pierce fired off a pithy memo to Muskie after the latter's election: "good article in *SE Post* on water pollution, p. 19; Suggest you read it — next session we're going to *have* to do something about it."[36] Pastoral promotions melded with industrial promotions and began to steer Maine politics in new directions.

Maine was not alone. The links between pastoral ideas and political rhetoric were forged by an expanding national network of professional, quasi-governmental, and grassroots organizations concerned, for various reasons, about the crisis of water pollution. In Congress, hearings on water pollution attracted national conservation and engineering organizations such as the American Society of Civil Engineers, the American Public Health Association, the American Water Works Association, the Federation of Sewage Works Associations, the Izaak Walton League, the American Wildlife Institute, and, because pollution was an urban concern, organized labor.[37] But since most early antipollution legislation was enacted and enforced at the state level, local and regional organizations were far more significant in the first stages of the clean-waters movement. With low administrative requirements and few standing commitments to members, these organizations responded quickly to the crisis.

Coalitions of fish and game clubs, parent–teacher associations, statewide Audubon societies, and a variety of women's groups working through their conservation committees, including garden and literary clubs and local chapters of the League of Women Voters, the National

Federation of Women's Clubs, and the Daughters of the American Revolution, provided enthusiasm and innovation that eventually infused national outdoor recreation and conservation organizations. Here women and men concerned about pollution could find sympathetic neighbors grouped together in familiar associational settings. The organizations were small and informal, but in the aggregate they constituted a powerful community force.[38]

Since the 1980s, grassroots environmental politics has been associated primarily with local "not-in-my-backyard," or NIMBY, campaigns, with direct-action organizations, or with environmental justice.[39] The grassroots groups that initiated the environmental movement in the 1950s shared this local perspective, but they operated in a different context. First, unlike some more recent neighborhood environmental threats, polluted rivers were both local and trans-local: they invoked a strong sense of place yet they were regionally significant as well, because rivers joined several towns along a watershed. Grassroots organizing around water pollution encouraged regional coalition-building, and this in turn led to broader visions that encompassed the concerns of a diverse regional constituency.

Second, citizen–activists in the 1950s faced a more open and responsive state-level political leadership. Earlier activists believed that state and federal governments were capable of providing solutions, and they expected their efforts to produce lasting social change. Thus they held a civic vision that wedded personal quality-of-life concerns to the common welfare. The links among localities and between personal values and collective responsibility distinguished postwar politics from the movements of the 1980s, and served as the basis for the era's challenge to large-scale social processes.[40]

Maine's early grassroots campaigns reflected these organizing conditions. In 1953 the Jaycee Wives of the Kennebec and Androscoggin valley towns began holding joint meetings with local PTAs to "arouse public opinion" about waterborne diseases that threatened their children's health. "Whether more substantial progress will be realized at the next [legislative] session will depend largely on the attendance at such meetings as tonight's," a Jaycee host urged. In 1955 the Maine Federation of Women's Clubs began campaigning for water pollution controls in response to a national convention resolution proclaiming America's waterways vital to human existence.[41]

Local volunteer organizations like these pressured the state legislature and highlighted the issue of pollution, but with small staffs and no professional lobbyists, their forays into politics were limited to writing editorials and making an occasional appearance before legislative committees. However, in the early 1960s these informal groups began to congeal into statewide conservation organizations. Maine's transition began in 1953 when conservationists, concerned mothers, and outdoor-sports enthusiasts from the Androscoggin Valley and beyond launched Citizens for Conservation and Pollution Control (CCPC), Maine's first statewide antipollution organization. The CCPC issued broadsides and press releases drawing attention to river conditions and introduced legislative measures to remove exemptions to pollution laws.[42] Organized efforts along the Kennebec River started with William H. Pynchon, whose interest in conservation dated from an appearance at a national water-pollution conference in Washington, DC, in 1954. In 1956 Pynchon moved from Massachusetts to Maine, and in 1960 he helped found and became president of the Maine Water Improvement Council, based in Phippsburg, a resort town near the mouth of the heavily polluted river. As head of the organization, Pynchon toured the state, speaking to civic groups and drawing together a constituency composed of concerned mothers, outdoor enthusiasts, and tourist-oriented business owners. Pynchon stressed the issue of pastoral pleasures. In Maine, he felt, there was a "satisfaction, spiritually and physically, in knowing the natural world ... a wealth of opportunity to get outdoors, do things, and get off the beaten path."[43] The CCPC, the League of Women Voters, the Maine Fish and Game Association, the state Grange, and a variety of other recreation, women's, civic, and service clubs, each interested in clean waters for their own reasons, began voicing a common theme in editorials and before the legislature.

In the beginning, these groups drew only limited legislative attention, specifically, the allegiance of a small legislative coalition headed by liberal Republican Ezra James Briggs, a maverick representative from northern Maine and vice president of the Maine Fish and Game Association. As the legislature's incipient environmental conscience, Briggs became a conduit for ideas emerging from the local and state civic organizations. Armed with photographs of rivers clogged with decaying fish and garbage, Briggs introduced clean-waters bills in 1955 and 1957 with backing from both the CCPC and his own Maine Fish and Game Association.[44] Playing upon pastoral themes and the growing resentment toward those who polluted public waters, he urged the legislature to action: "we have watched many beautiful God-given waters deteriorate from their natural beauty to a degree of filth that a self-respecting rat would scarcely tolerate.... This condition was more often than not produced by a selfish few and could have, in all cases, been at least partially prevented were it not for the loopholes ... [in] existing laws."

Briggs's bills received widespread press coverage, marking a new link between pastoral sensibilities and political action. Although they died in committee when industry lobbyists pointed out that they would "spell the end of industrial growth and development in Maine," Briggs's courage, persistence, humor, and eloquence inspired a new generation of legislators to take up similar proposals.[45] In an era when at least some lawmakers could still be considered grassroots activists, this was an index of changing popular sentiment in Maine.

Although the 1955–1957 clean-waters campaign drowned in a chorus of pro-industry arguments, it signaled some important changes. First, by mobilizing considerable popular support, clean-waters advocates highlighted pollution as one of the foremost issues in Maine public life. Second, they targeted industry as the source of pollution, linking their cause to anticorporate sentiments in this colonized state. And most important, the small but growing coalition helped shift the focus of the pollution debate from economic development to quality of life. A legislative representative from the Penobscot Valley pointed to the irony of "living on one of the most beautiful rivers in our Country" yet driving 25 miles to find a place where children could swim.[46] While Maine political leaders continued to emphasize economic themes, grassroots supporters were beginning to rally, as James Briggs put it, to the "tangible and intangible aesthetic benefits" of pure, natural rivers as part of a broader, more livable landscape. The constituency was small, but the campaign was in the hands of citizens willing to challenge a view of conservation that simply assumed industry owned Maine's waters.[47]

In 1959 this inchoate movement was given firmer direction by formation of the Natural Resources Council of Maine (NRCM), a statewide conservation umbrella organization. With Bar Harbor architect Robert Patterson and James Briggs among its leading activists, the NRCM became a sounding board for Maine's grassroots environmental politics. Although pollution was not among its early priorities, over the next decade the NRCM, working with its member organizations, took up this and other environmental issues, becoming a "thorn in the side of every state agency or industrial interest whose programs or policies relate to resource conservation." The NRCM provided technical, legal, and political advice, while its affiliated women's, outdoor, and civic clubs contributed the grassroots clout of an aroused citizenry.[48] As this and other statewide organizations brought their powers to bear on the pollution question, legislators began to respond.

PROTEST AND RESPONSE IN OREGON, 1941–1960

As in Maine, pollution control in Oregon began as a health and economic issue, but here concern quickly gravitated to quality-of-life considerations as pastoralism emerged as a dominant theme in Oregon politics. Threats to the salmon fishery, perhaps the state's most distinctive emblem of natural well-being, spurred trade, sports, and civic groups, led by the Issac Walton League, to lobby for corrective measures, and in 1938 Oregon created an Oregon Sanitary Authority (OSA) within the Board of Health. Like Maine's Sanitary Water Board, the OSA focused its early efforts on municipalities, which were to "set the example [for industry] by solving their share of the stream pollution problem first." In 1949 the OSA ordered paper mills to build waste-storage lagoons, but with the agency's statutory authority limited to "voluntary compliance," industry found it cheaper to procrastinate. According to the Izaak Walton League's Stream Purification Committee, company officials were openly contemptuous of antipollution laws, and as in Maine, the OSA deferred to the slightest objection by company engineers, even when effective technologies were in use elsewhere.[49]

In Oregon, too, exploitation of the state's natural resources brought popular resentment about "powerful monopolies [that] ... guard the Northwest as their private domain." Like Maine and several other northern states, Oregon in the early 1950s was dominated by conservative Republicans reluctant to challenge corporations on conservation issues.

Here the catalyst for revolt was the question of public versus private hydro-electric power development, a live issue since the Progressive era, particularly as it related to the Columbia Basin dam projects. "Populist sentiment," as journalist Brent Walth points out, "often fell to the public power advocates," with organizations like the state Grange supporting the federal projects. In 1952, President Dwight Eisenhower moved to reverse the New Deal policy on federal hydropower, and with Oregon's ex-governor Douglas McKay as Eisenhower's interior secretary, the issue cut to the heart of Oregon politics.[50]

Invigorated by the Republicans' unpopular stand on hydropower, Democrats went on the offensive, and in 1954 they won a conspicuous statewide victory that swept writer and longtime public power proponent Richard L. Neuberger into the Senate, where he joined Wayne L. Morse, a fellow liberal. Morse had been elected to the Senate on a Republican ticket in 1944 and 1950 but deserted the party in 1952, principally in protest over Eisenhower's natural-resources policy. In 1956 Morse ran as a Democrat, easily defeating McKay, who left the cabinet to run against him. As in Maine, the new Democratic vitality was a result of populist politics — the dam issue in this case — and a resurgent labor vote. Oregon's industrial base was expanding, and the working-class families flocking to its cities shifted the political base of state politics.[51]

Working with business leaders in Oregon's growth industries, liberal Republicans modernized their own organization and managed to head off a Democratic gubernatorial victory in 1958 by electing Mark Hatfield. A maverick like Morse, Hatfield deemphasized partisan issues and ruled through bipartisan coalition.[52] Like Muskie, Hatfield shifted the populist mood in the direction of economic growth. By the early 1960s Oregon's economy was slipping and seasonal slowdowns and mechanization in agriculture and lumber production left rural Oregon with high unemployment. Hatfield launched his development program with a question that set Oregon's pastoral aspirations and its commitment to liberal developmentalism at opposite poles: "should Oregon, the Beaver State, be as eager and industrious as its nickname implies, or should the state settle back in its scenic splendor and let the world pass it by?" He promised funds to "augment local efforts to eliminate air and water pollution," but, like Governor Muskie, he asked little of the state's antipollution agency, especially where new regulations mitigated against industrial growth.[53]

Oregon's political leaders managed the volatile mix of populist disaffection, pastoralism, and growth politics by emphasizing the latter, and Oregonians accepted this emphasis on growth into the early 1960s, but with growing reservations. As they sensed the limitations of land and water use — the cut-over forests, disappearing wildfowl habitat, and rivers lost to fish and fishing — they began questioning demographic and industrial growth.[54] Noting the completion of Oregon's new superhighway system, writer Stewart Holbrook mused in 1963:

> Figuratively, these highways lead into a future for Oregon that will assuredly be much faster paced.... The highways lengthen and straighten, the cities expand, the dams multiply, and the industries grow. The older resident is amazed that all this came out of the wilderness in a few short decades. But, like most Oregonians, he also knows that if care is not taken, if the balance is not preserved between natural and man-made things, these civilized wonders can, in a few short years, destroy ... many of the things that make the incomparably scenic Oregon country so cherished by its inhabitants and its visitors, and in the end create another and no less harsh kind of wilderness.[55]

As governor, Mark Hatfield found it increasingly difficult to bridge the gap between growth politics and pastoralist aspirations. In 1960 he negotiated with oil companies for offshore exploration, triggering a heated discussion about the future of the Oregon coast. New dam projects, particularly the Pelton Dam in eastern Oregon, similarly animated the editorial columns. Hatfield praised the expansion of a Weyerhaeuser paper mill at Springfield, but a year later his sanitary officers threatened legal action against the firm for polluting the Willamette River. "Taken together," a journalist pointed out, "the incidents underline a conflict of interest and purpose seldom so clearly visible."[56] Hatfield's eight-year tenure as governor brought Oregon some 180,000 new jobs, but this heady growth triggered a number of problems that set the scene for a mandate on Oregon's environment at the end of his term.[57] Where Maine officials began to view filthy rivers as a barrier to economic growth, Oregonians saw economic growth as a barrier to cleaner rivers.

Growth assumed its most disquieting form in the burgeoning suburbs of the Willamette Valley, where population increased 69 percent between 1940 and 1960. Explosive development on the fringes of cities like Portland, Salem, and Eugene resulted in "the virtual elimination of city limits for practical ... purposes." Here, and across the nation, as historian Adam

Rome has written, developers overextended sanitary facilities, built over unsuitable soils and shallow aquifers, and taxed municipal wells to their capacity. In Oregon, some developments simply doused sewage with chlorine and pumped it into ditches along the county roads. Complaints arising from the suburban settlements along the Willamette suggest a changing political mood in the densely populated sections of western Oregon.[58]

Like Maine, Oregon made little concrete progress against pollution in the early 1950s. But by the end of the decade the state's rapidly expanding and water-hungry economy generated a broad coalition of antipollution activists. Suburban residents, sensitive to the need for better sanitary engineering as they pushed out onto the valley floor, turned to the river problem. Farmers with high-value orchard and grain crops required cleaner irrigation water, as did food processors and those dependent on Oregon's recreation and tourist industry. Statewide fish and wildlife organizations, which had been fighting to protect wetland habitat since the 1920s, molded these health, business, and suburban concerns into a movement.

By the late 1950s Oregon paper mills faced larger and more adamant crowds at public hearings when they applied for dumping permits. Under closer press scrutiny and mounting public pressure, the Oregon Sanitary Authority became more assertive, pushing mills to experiment with better lagoons and storage tanks, ammonia-based processing, spray drying, burning, and evaporating. When a 1958 survey indicated that primary treatment was insufficient to restore oxygen levels, the authority instructed cities to build secondary treatment plants.[59] The program was expensive, but the cities reduced their discharges by 50 percent at a time when their populations were doubling.

Oregon's counterpart to the Natural Resources Council of Maine, the Oregon Environmental Council (OEC), was not organized until 1968, perhaps because the veteran Izaak Walton League and the National Wildlife Federation filled the need for an umbrella organization in the Beaver State. Like the NRCM, the OEC was a coalition of grassroots civic, conservation, planning, and fish and game organizations, and like the NRCM, it hired a lobbyist and maintained a telephone tree for mobilizing citizens and organizations. The OEC's several committees included environmental planning, natural resources, recreation, environmental education, and pollution control.

In sum, shifting from "soft" public education to "hard" legislative lobbying, the conservation establishments in both Maine and Oregon mounted aggressive campaigns that brought river pollution before the public and the legislature.[60] Maine's early antipollution legislation grew from a search for new ways to commodify the landscape as a pastoral playground; Oregon's campaign was rooted in the pastoral dreams of its burgeoning Willamette Valley suburbs. In both cases, the deepening crisis along rivers threatened an emerging regional identity dependent on visions of the countryside as spiritual refuge. Maine's economic role as a pastoral vacationland for the Northeast dictated a serious effort to control pollution of its lakes and rivers, whereas Oregon's endeavor to reinvent the suburb as a landscape of livability mandated a stronger antipollution effort. Blurring the boundary between city and countryside, citizens in Maine and Oregon began a campaign to renew the rivers that linked these two worlds.[61]

CRISIS IN EDEN

As citizen groups became more organized, they learned new rhetorical strategies, many of which served as guides for subsequent environmental politics. It was here, for instance, that activists learned to parlay exponential projections of environmental disaster into a potent political tool. They were also able to assert public claim to rivers and present pollution as an abuse of a common birthright by private, self-aggrandizing despoilers — the corporations. Over time, environmentalists managed to isolate these despoilers from their economic context, casting the clean-waters campaign as a battle between good and evil.

Certainly water pollution in the 1960s offered grist for a sense of crisis. Due to pollution from sewage outfalls, only 35 of the 775 miles of beaches around New York City were safe for bathing, and Lake Erie, according to U.S. Rep. John D. Dingell of Michigan, was so polluted that a bucket of lake water in the middle of a public square would be considered a health hazard. In congressional hearings a feeling of foreboding prevailed. "The waves of pollution shocks will become somewhat additive — there will be little time for the stream to recover.... Repeated re-use of waters will become the rule — not the exception."

The federal Clean Water Acts of 1960 and 1966 raised expectations, but they accomplished little, and for an increasingly affluent nation the situation seemed all the more tragic.[62] Massive fish kills, ever more dire health advisories, and the loss of waterways to swimming, fishing, and boating focused public attention on America's pollution crisis. In the late 1960s algae blooms from phosphate pollution threatened to choke off all life in the lower Great Lakes, and in June 1969 Cleveland's Cuyahoga River, laden with toxic compounds and organic pollutants, burst into flame when ignited by a spill of hot metal slag. That year the nation witnessed the drama of a massive oil spill off the Santa Barbara Coast that kept pollution in the headlines for weeks.[63]

In Oregon, wide seasonal fluctuations in stream flow accentuated the sense of crisis. Each summer brought renewed concern as stream flow diminished and the pollution load overtaxed the Willamette's supply of dissolved oxygen. These fears came to a head in 1961 when the state legislature debated Senate Bill 36, the first major change in Oregon's water pollution statutes since creation of the OSA. Under activist scrutiny in legislative hearings, the bill drew extensive media attention. Along with the Izaak Walton League (IWL), the National Wildlife Federation (NWF), and the Audubon Society, local volunteers joined the lobbying effort, among them a group of resort owners in Newport, a coastal resort town at the end of an eight-mile-long effluent pipeline from the Georgia–Pacific mill in Toledo. Motel owners testified that fumes from mill discharges discolored buildings, ruined shrubbery, and "chased the tourists away." As evidence, they brought jars of effluent to the hearings, and when one slipped to the floor and broke, the room had to be cleared, cleaned, and aired out. Upon returning, the sobered representatives agreed to clear the way for passage.[64] Senate Bill 36, an easy victory for these volunteer activists, popularized the pollution crisis in Oregon and helped define the public claim to pure water. "No one ... has any right to use the state's water unless it is returned to the stream as clean as when he got it," state Senator Robert Straub of Eugene insisted.[65]

A series of events following the battle over Senate Bill 36 accentuated the sense of crisis. In 1962, at an "opportune moment," as one commentator put it, newscaster Tom McCall and KGW-TV produced an hour-long documentary film titled *Pollution in Paradise*. The film, as the title suggests, accented the contrast between Oregon's pastoral identity and polluted streams ruined by paper company giants. "No part of America still retains more of Nature's original works than the state of Oregon, a paradise for those who treasure the unspoiled in sight, in smell, in sound," the documentary announced. "But who are these foul strangers in Oregon's paradise?"

With the film airing in Portland and throughout Oregon, antipollution sentiment "snowballed." When Multnomah County Senator Ted Hallock prepared another antipollution bill for the next legislative session, it drew impressive popular support.[66] Citizens testifying during committee hearings spoke of the river as a commons: "if the waters of our stream systems belong to the State, meaning the people," John S. Wilson of the Columbia River Sportsman's Council reasoned, "then in our opinion the users should be responsible for the changes in water quality that they create."[67]

Low water in the summer of 1965, widely reported in the media, added to the sense of crisis. When dissolved oxygen in the lower Willamette dropped to 1.6 parts per million, the Bonneville Power Administration and the U.S. Army Corps of Engineers were forced to release water from upstream reservoirs to prevent fish kills and a major health hazard. The releases came with a warning from Interior Secretary Stewart Udall that the dams were not meant to be "pollution control devices" for the valley's pulp and paper mills. The public responded with a surge of angry comments in the media and before legislative committees.[68]

Events like these melded with a more pervasive sense of crisis in U.S. society. Unprecedented prosperity after years of depression and wartime scarcity left Americans in a supremely optimistic mood that, paradoxically, masked deep underlying fears. From the New Frontier through the Great Society, national politics "brimmed with hopeful rhetoric" yet reacted to an economic and social system burdened by social injustice and corporate arrogance. America's most idealistic programs — the Peace Corps, the Alliance for Progress, the War on Poverty — contrasted with the grim persistence of the war in Vietnam, race and gender inequality, and a seemingly unresponsive federal government. All across America, deteriorating downtowns, inner-city poverty, neighborhood and school segregation, freeway congestion, and urban riots gave the impression that social problems were out of control.

In his widely read 1963 book, *The Quiet Crisis*, Stewart Udall noted that "America today stands poised on a pinnacle of wealth and power, yet we live in a land of vanishing beauty, of increasing ugliness, of shrinking open space, and of an overall environment that is diminished daily by pollution and noise and blight." These contrasting themes — urban despair and the optimistic visage of America as a redeemer nation — underscored the sense of urgency fueling pollution politics.[69]

Like journalists, politicians, activists, and environmental organizations everywhere, those in Maine and Oregon funneled this popular frustration toward easily identifiable corporate polluters. They were quick to identify a culpable despoiler behind each issue, linking the clean-waters campaign with a broader challenge to corporate behavior that included the military-industrial complex, automobile safety, commercial media, and a host of other contested public issues. Rachel Carson's *Silent Spring* (1962), an exposé of the stream of synthetic chemicals and pesticides that threatened entire ecosystems and the humans that depended on them, was only the best-known of these environmental challenges to corporate America. Fascinated by futuristic advances that altered daily life, Americans became avid readers of popular science in the 1960s. Though most of what they read celebrated the successes of the new scientific age, some reports drew attention to the problems associated with developments like nuclear energy and synthetic chemicals. Thus, while Rachel Carson was preparing the book that would shatter American complacency regarding the environment, clean-waters activists were stalking corporations along a similar front.

Congressional testimony drew attention to new industrial chemicals and metals in the waste stream, including highly toxic pesticides, synthetic detergents, and other compounds virtually nonexistent only a decade before. Resistant to all known purification practices, this new generation of industrial wastes gave pollution politics, initially focused on municipal sewage, a stronger anticorporate bent.[70] Indirectly, federal waste-treatment funding fed this anticorporate outlook. In 1956 Congress offered grants for municipal treatment facilities on a 30 percent matching basis, a promise belatedly fulfilled under the Federal Water Quality Act of 1965 and the Clean Water Restoration Act of 1966 — both largely the work of Senator Edmund Muskie. Because federal law dealt mainly with municipal treatment, it had little direct effect on corporate compliance. But for years industrial lobbyists had argued that municipal sewage treatment should precede industrial programs, despite having evidence that municipalities were fiscally unprepared to

"America today stands poised on a pinnacle of wealth and power, yet we live in a land of vanishing beauty, of increasing ugliness, of shrinking open space, and of an overall environment that is diminished daily by pollution and noise and blight."
—Stewart Udall, Secretary of the Interior
under President John F. Kennedy

meet these goals. The federal matching funds put city officials behind the cleanup program, and industrialists lost a critical ally in municipal administrators.[71]

National trends and local crises put Oregonians in a fighting mood. Pulp and paper producers were responsible for around 90 percent of the pollution load on the Willamette River, and when corporations were drawing political fire for a wide variety of reasons, the public was easily convinced that this was unacceptable. Politicians, journalists, activists, and organizations, tapping a popular legacy of western resentment toward eastern corporations, identified the valley's pulp and paper mills as agents of these crises. Despite a decade of cajoling, mill discharges had not diminished; those who were paying taxes for municipal treatment plants, a news editor summarized in 1965, had "just about had it."[72]

When Senator Ted Hallock conducted a new round of public hearings for an Omnibus Water Pollution bill in 1966, citizen testimony focused heavily on the pulp and paper mills. In Eugene, activists denounced the nearby Weyerhaeuser mill, which despite its advanced abatement equipment had become a major source of pollution. Mill owners were clearly on the defensive: "I know we are a target here today," one confided. "We did have a problem. There is no question about that, and we are on the way to solving it, I believe." Fish Commissioner James B. Haas demanded that polluters be made "financially liable" for injury to fish, wildlife, and their habitat, and Robert Straub, now state treasurer, pointed out that criminal prosecution would put pollution "into proper focus."[73] Caught off guard by the vigor of the antipollution debate, OSA officials commented that "a year ago we were ahead of the people and couldn't go any faster.... Now we apparently are behind." According to OSA Chair Harold Wendel, the agency would "knock [their] ... hats off this year when we present our budget."[74]

For their part, Maine clean-waters activists, inspired by the state's reputation as a paper-company colony, had targeted corporate polluters since the 1950s. Early broadsides from Citizens for Conservation and Pollution Control castigated the "crooked, rotten, polluted cesspool of selfishness known as the inner circle that runs Maine." CCPC Director Norman Tufts urged clean-waters advocates past their confrontation with public officials, whom he thought were "set up as punching bags by the privileged interests of this state." Direct attacks on the mill owners would "force into the open the first team ... who should be made to answer directly to our citizens." At one memorable CCPC meeting, protesters engaged in a three-hour verbal battle with Fish and Game Commissioner Roland Cobb and Industrial Development Commissioner Harold Schnurle, charging that both were conspiring with corporations to gut antipollution codes.[75] Emboldened by the increasingly shrill rhetoric of the CCPC and other groups, in 1963 Lewiston citizens once again began publicizing their pollution plight. When local spokesman J. Dennis Bruno denounced mill-owners' persistence in treating pollution simply as a political or cosmetic question, the outspoken activist was "swamped with telephone calls ... rallying to his side of the dispute."[76]

Maine's most prominent pollution crisis — extreme degradation of a small trout brook flowing from central Aroostook County over the border to New Brunswick — accented this anticorporate bias. In the early 1960s the New Jersey owner of a northern Maine potato processing plant proposed building a sugar-beet refinery near Prestile Stream to help diversify northern Maine's single-crop potato economy. Maine's top Democrats endorsed the plan, and state, county, and federal officials backed the owner, Fred H. Vahlsing, Jr., with grants and loan guarantees. As the beet factory neared completion, Vahlsing asked the legislature to downgrade the Prestile from B to D — the lowest classification on the books — even though the firm claimed the plant would produce no pollution. In view of the plant's importance, Governor John H. Reed and Senator Muskie personally pleaded on Vahlsing's behalf before the legislature. Over vigorous objections from the League of Women Voters and downstream citizens, the legislature approved the change. The declassification in the mid-1960s triggered a storm of protest from farmers, women's club members, newspaper editors, chamber of commerce officials, PTA and civic club officers, and fish and game enthusiasts in northern Maine.[77]

The incident gave Maine's environmental organizations a perfect venue for testing political strategies. The rapid degradation of the stream — actually from production increases at Vahlsing's potato factory, not from his new beet plant — generated a sense of crisis in towns along its banks. The trout stream had been used by generations of locals as a common resource; the despoiler was from "away" — New Jersey, no less — and the stream conditions drew together a community of interest that spanned the international border. Northern Maine citizens, having shown a notable lack of interest in river classifications a decade earlier, circulated petitions to restore the B rating, which would have made the stream suitable for swimming and for fish habitat, and the people of nearby Centreville, New Brunswick, bulldozed a dam across the stream to hold the fetid waters on the American side of the border. When two more Aroostook potato processors sought discharge licenses, the hearings drew "tremendous opposition" from both sides of the border.[78]

By the time Maine's attorney general swung into action in 1969, the potato processor's loans had gone as sour as the Prestile, and the political stench elevated the incident into a statewide concern. The complicity of Maine's top officials, the unanswered questions, and the now-nationwide publicity outraged

Maine citizens and energized its environmental community.[79] Stinging from the national media attention, Reed's successor, Democratic Governor Kenneth Curtis, proclaimed an "all-out war" on pollution, and Senator Margaret Chase Smith urged her fellow Maine Republicans to add a "major plank" on pollution to their state platform. With an indirect reference to Muskie's role in the affair, she called for a "firmer ... posture toward ... out-of-state industrialists [who promise] ... economic benefits at the expense of ... our natural resources." Muskie himself, now a contender for the Democratic presidential nomination, denounced Maine's "rigid, outmoded, and inadequate water classification system."[80]

Maine's clean-waters campaign was further publicized in 1968 when two Yale graduates, John N. Cole and Peter W. Cox, founded the *Maine Times*, an investigative weekly tabloid aimed primarily at environmental issues. The journal quickly developed an aggressive style of reporting that within a year commanded a subscription list of 10,000 readers. Its iconoclastic approach raised the level of environmental coverage in journals throughout Maine.[81] Following on the heels of this media sea-change, in 1970 Ralph Nader's Center for the Study of Responsive Law sent interns to Maine to investigate the paper industry. Two years later, 28-year-old attorney William Osborn published the resulting report as *The Paper Plantation*, a detailed affirmation of Maine's colonial status and the implications for its rivers and other natural resources. According to journalist Bob Cummings, it had been an "open secret" in Maine that industry lobbying kept antipollution efforts underfunded and ineffective. Osborn's conclusions — that Maine was a "water wasteland with its great rivers sullenly serving as private sewers for the mills" — splashed this "open secret" across the front pages of the Maine press.[82]

Paper Plantation focused the clean-waters campaign on a highly visible target at a time when corporations across America were drawing fire. For those who failed to read the report, newspaper coverage was detailed and surprisingly sympathetic. Journalists quoted Osborn's most famous line — Maine was a "land of seven giant pulp and paper companies imposing a one crop economy with a one crop politics which exploits the water, air, soil and people of a beautiful state" — as they went on to embellish his points. In this new climate, Maine's political leaders also expressed sympathy with the report. Senate Majority Leader Richard N. Berry claimed that the study made "a major contribution toward exposing the crap these people ... spread," and Governor Kenneth Curtis labeled some of Osborn's recommendations "good ideas." Momentarily at least, the Nader report loosened the corporate grip on the legislature, influencing the outcome of several key environmental issues.[83]

POLLUTION, GROWTH, AND THE PASTORAL IDEAL

Public awareness of the pollution problem was conditioned by a sense of crisis and by the dramatic contrast between common good and corporate evil. It also expressed a growing appreciation for rivers as symbols of pastoral identity. The pastoral entered political discourse as a counterpoint to the emphasis on economic growth that had guided state politics since the end of World War II. Nationally, the balance between growth and livability began to shift in the late 1950s when liberal intellectuals like Arthur M. Schlesinger, Jr. and John Kenneth Galbraith sought to chart a path away from the New Deal emphasis on expansion, jobs, and wages. In his influential book *The Affluent Society* (1958), Galbraith urged Americans to consider not only higher production but also the "uses to which [that] production was put." Similarly, Laurance S. Rockefeller asked "whether our affluence will create only bigness or whether it will create greatness as well." Rising concern over urban decay, poverty, racism, and corporate power popularized these admonitions, and with the economy seemingly stabilized at a high level of production and consumption, Americans could turn to quality-of-life issues.[84]

The reassessment of growth was noticeable in state politics. By the late 1960s Oregon's population was once again surging, and demographic projections suggested that even a vigorous pollution control program would be insufficient if the population and number of polluting industries continued to rise. Deteriorating air quality also contributed to Oregon's jaundiced view of growth. Emissions from lumber mills, paper mills, vehicles, and fires to dispose of logging slash and seed-grass residue produced a thickening haze during summer and fall. Following a League of Women Voters campaign, in 1951 Oregon became the first state in the nation to pass air-quality legislation, but with the economy expanding, air quality continued to deteriorate. It was not without cause that Eugene residents decided in the mid-1960s to encourage "only those industries ...

that are capable of a high degree of control over the emission of air pollutants."[85] The "old question" when a prospective business was mentioned "was 'how many jobs'; Now it's 'how much pollution?'" Giving voice to the new mood in the mid-1960s, Governor Tom McCall inveighed against the "human tidal wave of migration" threatening his treasured pastoral landscape. With his entreaty — "Please come visit us in Oregon again and again. But for heaven's sake, don't come here to live" — McCall became a national spokesman for the idea of limits to growth.[86]

A textile recession buffered Maine from such growth pains into the 1960s, but late in the decade a land boom along the southern and central coast generated fears of congestion and loss of open space. A 1970 article by news commentator Bob Cummings titled "The Late, Great State of Maine" revealed that only 1.4 percent of Maine's convoluted 3,000-mile coast was in public hands. Cummings called for a moratorium on development while the state considered zoning, easements, purchases, and other measures to protect its most valued natural sites. Jean Anne Pollard's *Polluted Paradise*, published in 1972, warned Maine people against a horde of developers moving north "like a plague of locusts." Echoing McCall's admonitions against growth, a journalist described Maine as a "land of sprouting vacation homes, of mountainsides being carved into ski resorts, and of condominium clusters building up like barnacles along lake and ocean frontages."[87]

While the land boom gave pause to the advocates of growth, it also demonstrated the state's importance as an Arcadian refuge. Maine underwent a "shock of recognition," as *The New York Times* put it, when its citizens found that thousands of mobile northeasterners considered its wilderness and its "bleak, rocky coast" valuable recreational and aesthetic assets. The state had not yet fully recovered from the shock of textile mill closings, but civic leaders were beginning to see another side to this sagging industrial economy. Acutely aware of the metropolitan pressures to the south, they also knew they had time to plan out a new role for urban waterways and other precious landscape features as Maine moved into the economic mainstream. Indeed, the entire Northeast looked to Maine to see how the battle between corporations and conservationists would be played out, and this concern left a deep impression on the state's politics.[88]

In Oregon, too, fears of overdevelopment helped sharpen pastoral images of the state. In a long article in the Portland *Oregonian* in 1969, planner Charles DeDeurwaerder pointed out that cheerless urban architecture and runaway highway development undermined personal identification with Oregon's cities. But he envisioned a new era in which greenbelts, parks, and conservation easements would restore a sense of community and make the city a "gracious place again." DeDeurwaerder saw the Willamette Basin as a single sanitary district that would collect, treat, and recycle effluents. Sewage would become fertilizer for Oregon's forests, and reclaimed water would irrigate farms or evaporate before returning to earth as "clean atmospheric moisture."[89]

In the 1950s Portland residents had shielded themselves from the river's rank waters with freeways, scrap yards, port facilities, and heavy industry; now the recovering river beckoned as a "natural beauty spot" around which to reweave the urban fabric. Seeking to restore a "harmonious visual and physical relationship between nature and the urban environment," planners throughout the valley imagined an integrated, livable landscape. Cleaner rivers would knit together city and country as floodplains and waterfronts became postindustrial playgrounds, providing green hearts for otherwise heartless cities. Dystopic fears of sprawling urbanism gave rise to utopian visions of the river as the core of the postindustrial city.[90]

The pastoral images recruited by clean-waters advocates helped shift the campaign's rhetoric from health and the economy to classic environmental concerns, such as those that drove Mabel Johnson, Margaret Benz, and Mrs. Glen Boehme to write their representatives on behalf of the streams and lakes they loved. In Oregon, Maine, and across the country newspaper and journal articles revealed the gap between polluted waters and pastoral aspirations by contrasting a river's pristine headwaters with the engineered landscapes below. These journey-to-the-sea narratives typically began with canoeists entering a headwaters stream cloaked in the tranquil beauty of the surrounding wilderness. At some point, the river came to a halt behind a dam. Swallowed up by an industrial complex, its waters reemerged below, dead, rancid, and channeled between brick buildings and banks of dried mud and refuse. Kenneth Thompson's classic story of a trip down New York's Hudson River, for example, included photos with captions like "puking paper plants," "percolating purple pools," and "open sewers and orange carp." These contrasts — the pristine stream and the industrial sewer — accented the disjuncture between town and country, city and nature. The prospect of reuniting these opposites appealed to the public, drawing together a constituency that saw the river not as an engineered resource but as a means of bringing nature back into the city.[91]

Pastoral rhetoric like this changed the way citizens viewed their rivers. Those living downstream from the Vahlsing potato plant asserted their right "to live comfortably and in good health" along the banks of the Prestile, and those below the Weyerhaeuser plant near Eugene testified that they had moved to the area "because they liked the pleasant rural atmosphere."[92] Maine State Senator Bennett D. Katz proposed a five-mile greenbelt along the Kennebec River in Augusta, featuring scenic outlooks, picnic areas, trails, boat launches, a parade ground, and a museum. Although Katz's plan languished for lack of funds, a *Kennebec Journal* editor observed with satisfaction that public land along the recovering river was being used as "a lunch spot for Augusta office workers and a picnic area for families." Boat launches, marinas, and rentals appeared on the river, and new restaurants and homes sported decks looking out across the waters. Officials planned for the return of a "bustling riverfront" by reinstituting the office of harbor master. Some day, a Kennebec Valley woman sighed, her children's children would enjoy the rivers "the way we used to."[93] The imaginative leap from engineered river to postindustrial amenity had been made, even if it was yet to be implemented.

Infused with the rich imagery of pastoralism, the antipollution campaign was changing. While national political leaders continued to view rivers in the context of engineered landscapes, activists in Maine, Oregon, and other states now argued their aesthetic and biological importance. In Oregon, anglers, swimmers, boaters, vacationers, resort owners, and housewives testified to the foregone pleasures of pure waters. Exurban and suburban residents, fanning out along the Willamette tributaries, hoped to see the countryside "remain as unspoiled as it now is." Newcomers drawn by the promise of postindustrial livability, and established citizens in cities that had already funded secondary and even tertiary sewage treatment, demanded a halt to pollution "no matter whose toes get stepped on."[94] The multiple facets of the environmental imagination — health, outdoor recreation, business climate, fisheries, livability — drew together a constituency that bridged many geographic, social, and political divides.

THE ENVIRONMENTAL IMAGINATION SUCCEEDS

The 1972 amendments to the Federal Water Pollution Control Act set national standards, required permits for any pollutant discharges, and stipulated that all permits were temporary pending improvements in prevention technology. Waters were to be "fishable and swimmable" by 1983 and sustain zero pollution discharge by 1985 — ambitious goals that were later pushed back or in some cases abandoned.[95] The act signaled a dramatic change in federal responsibility for the nation's rivers, and state politics over the previous decade had laid the groundwork for this path-breaking legislation. Guided by successful rhetorical strategies for mobilizing public opinion and by a pastoral imagination, activists, journalists, and conservation-minded political leaders put in place the administrative apparatus that made implementation of the 1972 federal act possible. The route to effective pollution control in Maine was typical, although shorter than the path taken in most states.

In 1961 Maine's Water Improvement Commission (WIC) — previously the Sanitary Water Board — proposed its first standard for an industrialized river: it awarded a "D" rating to the lower Kennebec, which would require only primary treatment to remove solids over a generous 15-year compliance schedule. Even though news commentators considered the resulting legislation "about as lenient ... as an anti-pollution bill could be," state officials were openly pessimistic about passage, and William Pynchon's Water Improvement Council braced for the debate by gathering support from a broad spectrum of womens', civic, yachting, and recreational groups along the lower Kennebec. The council's activity convinced the legislature that there was "an impressive weight of public opinion favoring up-grading the Kennebec," and Maine's first major cleanup law passed by a substantial margin.[96] When the WIC took up classifications for the Penobscot — a river "too thick to drink and too thin to plow," as one resident put it — state officials were prepared to ask for more: "we knew attitudes had changed." The Penobscot was rated "C": suitable for recreational boating and fishing and for wildlife and fish habitat. With the Androscoggin similarly classified in 1967, Maine moved ahead of the nation on water-quality legislation.[97]

Maine legislators introduced more than 60 conservation bills during the 1969 session, and nearly half eventually passed, placing Maine among the leaders in statewide environmental planning and regulation. In 1970 Governor Curtis unveiled a two-year, $60 million attack on municipal water pollution and appointed Colby College geology professor Donaldson Koons to chair the state's new Department of Environmental Protection (DEP). A good match for the agency's new activist posture, Koons, according to one journalist, "almost singlehandedly ... lifted the Commission from its role as a department of ... legendary impotence to the level of a genuinely aggressive regulatory agency."[98]

The attorney general's office also assumed a more aggressive stand on enforcement. And federal matching funds for water treatment facilities released in March 1970 triggered a rush to complete the municipal phase of compliance schedules. With some incredulity, the *Kennebec Journal* reported that the "strident outcry for clean air and water has prompted a re-examination of economic development efforts ... in the state with the lowest per capita income in the Northeast."[99]

By the time the federal Clean Water Act passed in 1972, Maine's waterways had been classified, and water pollution control became a matter of gradual improvement under ever-tighter restrictions. Mill owners responded with more effective recycling and waste-reduction technologies, primary and secondary treatment plants, settling basins, save-all screens, foam booms, oxidation towers, new kraft processing techniques, sludge disposal, and other means of abating air and water pollution, reducing waste flow by 50 to 90 percent. In 1974 Maine's paper mills were still the dirtiest in the nation, but by 1981 they met state schedules well in advance of federal standards. These accomplishments came "not because industry has suddenly relented ... but because suddenly a lot of ordinary Maine people are concerned about water pollution."[100]

Oregon's accomplishments under Governor Tom McCall were similarly dramatic. With his environmental credentials established by the 1962 documentary and a vigorous antipollution stand in legislative committee hearings, in 1966 McCall ran against Democratic State Treasurer Robert Straub for the governor's seat. Straub, who also enjoyed a reputation as an environmentalist, lost heavily, but he won a second term as state treasur-

er in the 1968 off-year election and went on to succeed McCall as governor in 1974.[101]

In Tom McCall, Oregonians gained a clear champion of the environment who pursued his goals with considerable rhetorical flair. Nevertheless,
stressing McCall's agency in quickening the environmental pulse in Oregon
means overlooking popular grounding for the movement. As the party of
power in the late 1960s, Republicans took the lead in environmental issues,
but in Oregon's system of weak parties and strong personalities, this environmental focus was a bipartisan matter. Each party had something to gain
from environmental advocacy. For Democrats, environmental regulation
was a natural extension of the Kennedy–Johnson emphasis on federal welfare and balanced growth; for Republicans, state-level environmental politics was a means of building a strong bureaucratic counterweight to the
massive federal presence in Oregon.[102] Given the expansive nature of the
environmental imagination, these issues appealed across party lines, and
Republicans and Democrats vied for the popular environmental vote.

The pull of the burgeoning environmental constituency on Oregon's
dominant party is best illustrated by Robert Packwood's rise to the Senate
in 1968. In 1965 the aggressive state legislator helped organize the
Dorchester Conference, a forum for young Republicans seeking to build a
new suburban constituency that would restore the party's fortunes after
Barry Goldwater's disastrous 1964 campaign. Over the next few years
Packwood became known for his driving ambition and his attempt to
capture the politically uncharted "broad middle ground" waiting in the
suburban West. In 1968 Packwood defeated Senator Wayne Morse in a
national election year strained by urban riots, political assassination, and
massive demonstrations against the Vietnam War.

Morse's antiwar stand alienated hawkish fellow Democrats, while liberal Republicans abandoned the senator because he failed to support fellow dove Mark Hatfield in his successful 1966 senatorial race against
Democratic hawk Robert Duncan. Morse looked for common ground in
the fact that he had sponsored virtually every public power project completed in Oregon during his term, but by 1968 the appeal of this prodevelopment strategy was fading. Democrats, disheartened by a primary
fight between Hubert Humphrey, Eugene McCarthy, and Robert Kennedy,
fared poorly in the Oregon general election, and Packwood won in a close
vote emphasizing his own liberal credentials and Morse's age.

Yet the campaign, directed largely at Morse's failings, left Packwood without a stand on issues.[103] Having defeated the legendary statesman, Oregon's junior senator canvassed voters and found a remarkable 93 percent of respondents favoring environmental quality over economic growth. Sensing a "groundswell," Packwood adopted this standard as "his leading cause." Oregon voters set the agenda for Republican Party leadership, and McCall responded to these popular pressures with an ambitious environmental administration.[104]

Still, the popular governor sometimes found it difficult to meet rising expectations. Shortly after his election, American Can Company requested a permit to build a paper mill on the banks of the Willamette. After "painstaking scrutiny," McCall backed the OSA's decision to issue the permit, but as an editor mused, "a lot of supersensitive noses" would be "sniffing the air" when the mill went on line.[105] McCall's pollution policies also drew criticism from federal administrators. Federal funds supported about one-third of Oregon's municipal pollution program, and federal officials in the Democratic Johnson administration were not averse to interceding when state efforts fell short of federal standards. In February 1967 the Federal Water Pollution Control Administration issued a report that found Oregon's program inadequate. Only three of the valley's five paper mills had managed to reduce waste discharges, and the OSA, the report concluded, was "understaffed, underpaid, and overabused." Oregon, a state health official brooded, had "forsaken its leadership in pollution control."[106]

Officials in Salem scrambled to restore the legitimacy of their own — as opposed to federal — policies. That spring, in a burst of effort, the legislature replaced the OSA with a Department of Environmental Quality (DEQ), increased its staff, developed a permit system for controlling discharges into all Oregon waters, and created legal procedures subjecting polluters to court order. The legislature passed even higher standards in 1969, bringing Oregon acclaim as a national leader in pollution control. In May 1972, following another morale-boosting legislative session, L.B. Day, director of the DEQ, shut down Boise Cascade's Salem paper mill temporarily when it lagged behind other mills in meeting Oregon's new standards. The mill remained idle for 12 days, and when it reopened its air and water emissions met state and federal standards. By 1973, a year after passage of the 1972 Clean Water Act, all 600 industrial plants and 20 municipalities on the Willamette had secondary treatment plants.

The regulatory initiative fell short of restoring the rivers to their natural conditions. Both the Androscoggin and the Willamette continued to bear significant pollution loads, and the primary source of the problem remained the paper mills, which regularly violated state and federal antipollution laws. Residual organic sediments continued to consume dissolved oxygen, and dams and water releases destroyed spawning grounds and altered riparian habitat. New fish-ways and massive plantings revived salmon runs on the Willamette and Maine's Penobscot River, but dams, low oxygen, and nonpoint source pollution like failing septic systems, agricultural runoff, landfill seepage, and siltation from timber operations frustrated these efforts. Reintroducing brown trout in the Androscoggin above Lewiston required a $2.3 million liquid-oxygen pump to boost oxygen levels, and the trout had to be restocked yearly because they lacked natural spawning beds. Brown trout planted in the Kennebec River naturalized more completely, but the discovery of mercury and dioxin in the waters prompted warnings that they could be consumed only in limited quantities.[107]

As rivers began to conform to the popular canons of livability, and as pollution issues became more complex and enmeshed in technical jargon, media sensitivity waned. By the early 1980s the nation was still far from its goal of zero water pollution discharge, but with no "gushing pipe" to keep the public focused on specific polluters, clean waters dropped from the list of demands that rallied popular environmental support. Still, as the letters penned from 1975 to 1977 by Mabel Johnson, Margaret Benz, and Mrs. Glen Boehme testify, rivers that fell below standards of livability could still rally local support. On a rainy May afternoon in 1977 Harry F. Rivelli, whose family had lived near Portland for 60 years, sat down to write Governor Straub about storm runoff into a nearby "crystal clear creek." After hearing "so much talk these days about clean water," Rivelli was disheartened to see even nonpoint source pollution in his little stream: "as I sit in the house and see the dirty water from the street drain into the creek ... it makes me very mad."[108] The people of Lewiston protested in 1941, as Mayor St. Hilaire put it, to "get rid of the smell." By the 1970s the demand for pure water, backed by state and federal regulations, had grown more exacting: "anything that inhibits full use of the rivers, which prevents the public from reclaiming its lost domain, is a problem, whatever label one uses."[109]

In the aggregate, complaints like those from Mabel Johnson, Margaret Benz, and Mrs. Boehme accomplished a great deal, and the success stories in Maine and Oregon were repeated across the nation. The lower Cuyahoga, once a fire hazard, sported night clubs, marinas, restaurants, and pleasure boats by the early 1980s; charter fishing boats appeared in New York's dreaded East River; and the Potomac, infamous for its stench, became a major recreational resource for Capitol-area residents. After a century of decline, the shores of the lower Great Lakes reopened to swimming and other forms of recreation.[110] The Connecticut River was perhaps the East's greatest success story. In 1970 a group of students and their instructors canoed the river and documented its reputation as "the nation's best landscaped sewer." Eleven years later the Connecticut River Watershed Council traced the same route while drawing a far more optimistic picture: the river was safe for swimming and fishing, salmon were returning in limited numbers, local governments were piecing together a protective corridor, and Hartford and Springfield boasted plans for riverside parks, greenbelts, walkways, nature trails, playgrounds, picnic areas, shops, and restaurants — repastoralizing the river.[111]

Out West, journalist Paul Schneider described the Boise River in Idaho in similar terms, as a way of bringing nature into the heart of the state's capital city. Disparaged for years, the river had become "a symbol of civic pride and renewal." Boise's mayor kept a kayak in his office near the river, and an appreciative citizenry made heavy use of the foot trails, bike paths, parks, and soccer fields along its banks. Beaver were at work in a park near the capitol building. Summarizing the results of America's two decades of clean-waters agitation, Schneider admitted that pollution had not been eliminated, "but then, neither has the Civil Rights Act eliminated a legacy of racism." What had changed in both cases, he concluded, was public pressure preventing a return to the old ways. "Citizens have noticed the improvement ... and they are not interested in returning to a polluter's free-for-all." Expectations had changed, "giving the public the idea that progress toward clean water is possible. Not inevitable, not easy, and not cheap. But possible. And right."[112]

The clean-waters campaign played a significant role in the evolution of environmental thought and environmental politics. The shifting images of the river, from a lowly component of the engineered landscape to a symbol of pastoral harmony in the livable city, gave force and direction to the politics of pollution, and the fight for cleaner rivers in turn sparked the environmental imagination. Urban rivers fulfilled a crucial symbolic function by bringing timeless nature into the heart of the inhabited landscape — the pastoral city. This, then, was the first stage in the evolution of the environmental imagination: harmonizing nature and the city by altering the city's relation to its rivers. Another phase of this movement — protecting the wildness of America's rural rivers — coalesced during the same period. Where the clean-waters campaign rested on a traditional pastoral idea of harmony between nature and society, the effort to preserve wild rivers introduced a different tradition: freedom, permanence, and authenticity actualized through solitary recreation in an uninhabited natural landscape. This, too, had enormous potential for shaping both the environmental imagination and environmental politics.

Allagash, Maine

"*Objects come into focus and then quickly fade away, as in a land of fantasy.*"
—William O. Douglas, Supreme Court Justice

CHAPTER 3

RIVERS, WILDNESS, AND REDEMPTIVE PLAY: PRESERVING THE ROGUE AND THE ALLAGASH

W hile activists in Oregon and Maine as well as across the country were addressing the pollution crisis in urban waters, others began defending rural rivers from dams and other forms of engineering. Groups such as the Izaak Walton League, the National Wildlife Federation, the Sierra Club, and the Wilderness Society drew popular support as they focused public concern on the nation's remaining free-flowing rivers.[1] These campaigns were inspired by a different form of appreciation for water resources and directed by a different form of activism. Like efforts to create the pastoral city, wild-river preservation was predicated on a sense of well-being gained from communion with nature, and like the clean-waters campaigns, wild-river preservation emphasized the sense of freedom, authenticity, and permanence derived from nature. But here these ideas were modernized to accommodate a new set of recreational and wilderness values. Where the pastoral landscape was utilitarian, social, and humanized, the wilderness landscape was spiritual, personal, and purely natural. The story of wild-river preservation represents the environmental imagination in transition from an older pastoral inspiration that stressed the harmonious interaction of folk, work, and nature to a newer wilderness concept that saw nature in nonutilitarian terms, as a place entirely separate from human artifice.

The popularity of river-oriented sports like fishing, canoeing, and rafting reflected a rising interest in primitive landscapes all across America. "Barring love and war," wilderness advocate Aldo Leopold once said, "few enterprises are undertaken with such abandon ... or with so paradoxical a mixture of appetite and altruism as ... outdoor recreation."[2] Americans in the 1960s enjoyed greater access to national parks and forests, a broader array of outdoor recreational opportunities, and a rising media emphasis on wildlife, all of which fostered a rebirth of the romantic appreciation of wilderness. The ways in which these new values and activities were incorporated into the environmental imagination are best revealed in the fight to save America's wild rivers.

The wilderness movement began in the 1920s with Aldo Leopold's management plan for portions of New Mexico's Gila National Forest and Arthur Carhart's design for a small section of Colorado's White River National Forest, and it culminated in the 1964 federal Wilderness Act. This long battle to protect untrammeled nature in the western federal domain mobilized national conservation organizations and altered popular ideas about the liberating effect of wildness on the human spirit.

Advocates rationalized wilderness preservation on therapeutic grounds against a background of urban malaise. As Max Oelschlaeger notes in discussing the Romantic philosopher Jean-Jacques Rousseau, people who imagined they were living in a corrupted human estate idealized wilderness as an "oasis free of the ills of civilization, a retreat to which the harried and battered, the suppressed or oppressed, might turn for relief."[3] In a popular magazine article, Senator John Saylor of Pennsylvania spoke of rivers as a medium for "redemptive outdoor activity" in a setting far from the "grinds and the strains of large cities." Washington Senator Henry Jackson likewise valued free-flowing rivers as a "bulwark against the strains and stresses of our fast-moving, complex civilization."[4] Viewed through lenses both pastoral and wild, even the heavily used rivers of the East — the Shenandoah, Ausable, French Broad, Cheat, Cache, and others — evoked a kind of "undisciplined beauty" that resonated with the spirit of true redemptive nature.[5] Many postwar urbanites and suburbanites out of touch with the natural world and their own inner selves found that redemptive recreation in fluid landscapes like these transformed them into avid nature preservers.

Unlike the traditional pastoral landscape, wilderness recreation brought to mind a pure nature, without the pastoral attachment to the folk. Recreationists immersed themselves in natural settings, just as the folk immersed themselves in the traditional pastoral landscape, but recreationists also idealized their activity as a way of distancing themselves from the broader masses of humanity, and the intensely personal implications of this activity implied an exclusionary bias. Writing in the *Sierra Club Bulletin*, Galen Rowell celebrated the spectacular wildness he discovered in northern Canada, but he found himself unwilling to share this experience with others: "Where is it? In a remote mountain range of Canada's Northwest Territories. Exactly where is it? I will not say in the presence of 140,000 readers who might take it into their heads to visit this fragile region."[6] Wilderness images embodied an absolute distinction between the artificial and the natural and favored personal redemption over social reform. In this sense, the fight for wilderness was far more exacting in its demands upon the rural landscape, and far less effective in uniting a diverse constituency around a goal of nature preservation.

River preservation shared some of these wilderness values. The campaign to preserve wild rivers began as a public reaction to the federal dam building mandated by the Bureau of Reclamation Act of 1902. In the years between World War I and World War II, state affiliates of the Izaak Walton League (IWL) and the National Wildlife Federation (NWF) challenged several U.S. Bureau of Reclamation projects, but this was, at least initially, a debate over how best to "improve" the river as a water-delivery system that also offered fish and game habitat. In the postwar period a darkening view of the city and the growing popularity of outdoor recreation put a premium on free-flowing water, and this new symbolic freight precluded the heavy engineering that dams entailed.

A key early victory for preservationists was the defeat of a proposed dam at Echo Park on the Green River in Dinosaur National Monument near the Utah–Colorado border, a campaign now viewed as a crucible for modern environmentalism. Preservationists, led by David Brower of the Sierra Club and Howard Zahniser of the Wilderness Society, launched a massive direct-mail campaign and kept the dam in the news for half a decade. As historian Mark Harvey points out, this mid-1950s controversy crystalized a vision of wilderness for the average citizen, "striking a chord with increasing numbers of Americans who were ambivalent about ... postwar prosperity and the population boom."[7]

Victory at Echo Park brought no relief from new dam proposals. The federal program was "awesomely uncomplicated," a *Field and Stream* writer observed: "they want it all." By this time artificial reservoirs represented a shoreline longer than that of the continental United States, and federal agencies proposed to more than double the storage capacity of dams. As the war between builders and preservers moved east in bitter fights over the Little Tennessee, Buffalo, White, and Potomac rivers, preservationists couched their defense in the rhetoric of permanence, authenticity, and human liberation actualized through contact with pristine nature.[8] Although often burdened with a history of use and engineering, rivers under threat by dams became, by definition, wild. Characterizing natural rivers as the antithesis to the artificial city, preservationists came to view them as pristine, stable, and essentially nonhuman.

These campaigns came to a head with passage of the federal Wild and Scenic Rivers Act in 1968, four years after passage of the Wilderness Act. Although the act itself became a national issue, state politicians and local citizen groups led the fight to save the rivers included in the system. Locals lobbied for and against designation, and in most cases local politics determined management strategies for river portions subsequently designated as wild or scenic. Federal status merely confirmed a state-level consensus about recreation, wildlife, dams, and environmental management of the river.[9]

These place-specific battles involved a complex search to balance two sets of landscape values. Classic conservationists viewed rivers as sources of energy, as transportation routes, and foundations for agricultural expansion, all realized through construction of multiple-purpose dams, which could prevent floods, stabilize riparian lands, facilitate navigation, provide clean power, irrigate the nation's food and fiber crops, sometimes improve fishing, and create flat-water and whitewater recreational opportunities. These multiple benefits, for the most part, assumed traditional pastoral ends — small, irrigated family farms dotting the once-barren landscape.[10]

A more recent philosophy valued rivers as scenic, recreational, and spiritual resources. In this vision, a river ran unfettered from its source to the sea, free of engineering. Successful wild and scenic river management bridged this distinction. Encompassing a management unit sometimes less than a mile wide, the wild and scenic river provided an illusion of wildness in the context of a traditional working landscape — a construct sufficiently flexible to incorporate both the quest for pure nature and the Jeffersonian tradition of freehold farming and petty resource use. River preservation is thus a story of accommodation between the urban quest for purity, freedom, and permanence and the pastoral perception of the river as useful, familiar, and engineered.

As early candidates for wild and scenic designation, southern Oregon's Rogue and northern Maine's Allagash highlight the way these polarities — urban and rural, wild and pastoral — coalesced into a system for protecting selected portions of America's free-flowing rivers. Considered the premier wild rivers of Oregon and Maine, the Rogue and the Allagash brokered these conflicting ideals by fusing seemingly exclusive uses and perceptions into a unified vision and a richer environmental imagination. As an example of place-centered grassroots activism, river preservation brought some notable environmental victories, but it required a difficult mediation between these conflicting landscape traditions, and thus it also reveals the developing tensions in people's perception of their proper relation to nature and place.

DESIGNING A WILD RIVER: THE ROGUE

The story of the Rogue River represents the environmental imagination in its broadest scope, since the river has witnessed a full spectrum of human interaction with nature: unbridled exploitation, utilitarian conservation, redemptive recreation, and ecological and wilderness preservation. The Rogue emerges from springs and melting snowfields in the volcanic slopes of the high Cascades near Crater Lake in southern Oregon. Carving a deep canyon through the thick Mt. Mazama pumice in its upper arc, it meanders southward across a forested basin, finally bending west to slice through the deep rock gorges of the Coast Range on the final leg of its 210-mile journey to the sea. Here it becomes stunningly wild, dropping some 950 feet in 126 winding miles from Grants Pass to the coast.

Below Grave Creek the waters of the Rogue enter the "Grand Canyon of the Rogue," bounded by heavily timbered ridges rising up to 4,000 feet.

This steep canyon is inaccessible except by foot trail or water, and trapping, farming, logging, and hunting were circumscribed by an overarching fact of the river's history: "you can't get in or out ... with much more than you can carry on your back."[11] Thus the area retained an image as "one of the most primitive areas left in the United States."[12]

Although most recreationists saw rivers like this as timeless, the geomorphology of the Rogue River is anything but permanent. River levels fluctuate wildly, and the shifting bars and boulders, the changing water levels, and the new channels carved by twisting currents create a constantly changing riverscape. As the river wends its way to the sea, it moves through vastly different worlds: slopes of tall fir and pine on the western Cascades; dry madrona, oak, manzanita, and buckbrush in the central basin; and steep, lush canyons of conifer, oak, maple, cascara, and myrtle in the Coastal Mountains. The river knits together a land of varied topography, precipitation, temperature, biology, and geology, its mood changing abruptly in space and time.[13]

The Rogue's changing face is also a product of long human use that both fostered and compromised its primitive character. Euro–American lore dates from the explorations of Jedediah Smith, whose encounters with Indian raiding parties inspired the river's Anglo name. The 1850s brought a small-scale gold rush along the river and the first substantial watershed changes, as miners washed tons of silt into the river and burned the forests to expose mineral soil. Beginning in the 1850s, logging, stock raising, farming, and irrigation diversions further altered the river, and by 1890 the prosperity of the small farms and ranches in the interior relied on the irrigation water available for crops and the assurance of receiving it on demand.[14]

The river's reputation as wild began to congeal as early as 1906–1908, when three national forests were created in the basin, attracting recreationists to the area. Upriver recreational boating began in 1907 with regular mail-boat runs from Gold Beach to Agness, and downriver drift-boating was initiated on August 10, 1915, when Glen Wooldridge and Cal Allen floated a homemade boat from Grants Pass to the coast. By the 1930s several drift-boat guides were operating on the river. Western writer Zane Grey's stories about Rogue River fishing, many written at his cabin at Winkle Bar, boosted the river's image as an angler's paradise. Recognizing the importance of recreation as a source of income for the valley, in 1921 the state legislature prohibited dams on the main river below Butte Creek. In 1935 a statewide referendum banned commercial salmon fishing as well, beginning, in effect, the process of wild-river management.[15] Once the concern of farmers, miners, and salmon fishers, the Rogue had seemingly become the common property of all recreation-minded Oregonians.

Rapid urban growth in southern Oregon accentuated the river's appeal. During the 1950s the Rogue Valley underwent a dizzying series of changes as its 150,000 "contented souls" were connected to California by a new four-lane interstate highway. Balmy climate, relatively low real estate values, and opportunities for outdoor recreation precipitated a land boom that brought concerns about "creeping Californianism." There was, the *Medford Mail-Tribune* announced, "danger in Paradise."[16] Along with smog, sprawl, rising property taxes, and other problems associated with rapid population growth, valley residents noticed signs of river degradation: pollution, temperature elevation, and siltation. Despite these changes, the Rogue, among the state's last free-flowing rivers, remained wild in the minds of most westerners.

Some local people, however, valued the primitive ambiance far less than they valued the utilitarian benefits of the river. During the 1930s local farmers had learned to articulate their utilitarian views of the river when the state took up the question of developing hydroelectric power in the Columbia and nearby rivers. The question at the time was not dams or recreation, but rather how to allocate the potential hydroelectric power from the dams — to farms or to industry. "The rate system put into effect at Bonneville Dam will be the decisive factor in determining whether the Columbia River basin will become the world's greatest manufacturing region or an area of completely electrified farms and small towns," Richard Neuberger explained. "What would unsightly factories and sprawling industrial centers do to the beauty of the hinterlands? Who want[s] ... rolling hills and sylvan lowlands blanketed with manufacturing plants?" This political battle linked the Jeffersonian pastoral vision to the idea of river development, and all across the West, as Ian Tyrrell notes, the idea of "transforming nature into a garden [through irrigation] quickly attained the dimensions of a popular crusade to create a middle-class utopia."[17]

As proponents of rural electrification, Oregon farmers contrasted the slums, crime, and poverty that came with industrial cities with their own traditional images of pastoral Oregon. Properly deployed, electricity from the dams would ensure pastoral stability in the Pacific Northwest, just as irrigation would permit the spread of agriculture, validating the farmers' sense of worth and their notions of worthy land use. Fear of marginalization gave poignancy to this traditional pastoral vision; with western agriculture shifting in the direction of large-scale commercial farming, and with the countryside becoming more and more an adjunct of the city, modernity seemed as much a curse as a blessing to small farmers and ordinary rural citizens. Rural people benefited from the cultural, educational, and technological advances that made them more like their urban counterparts, but they remained ambivalent about the dramatic changes in their familiar landscapes. Thus they clung to the vision of Arcadia as a buttress against urban culture and corporate capitalism.[18]

The politics of Rogue River preservation were shaped by this complex and contradictory relation between wilderness purity and nostalgic, Jeffersonian pastoralism. While urban devotees celebrated the lower river's aura of primitiveness, the federal Bureau of Reclamation laid plans for dams upriver to augment the pastoral aspirations of farmers and ranchers in the central basin. From a utilitarian standpoint, the Rogue's problem was its erratic seasonal flow, which brought destructive floods in winter and too little water in summer. Responding to a postwar rise in water-intensive farm and orchard crops, the bureau sought remedies. Its plans, however, threatened the primitive qualities of the lower river. This contrast in values set the scene for "one of the most bitterly controversial discussions ever held in Oregon over development or preservation of natural resources."[19] The struggle against dams on the Rogue — and elsewhere — expanded the environmental imagination by melding conflicting meanings of nature into a single vision of useful flowing waters, while a careful balance between pastoral and wilderness values consummated American's love affair with "pristine" nature.

WILDERNESS OR PASTORAL LANDSCAPE?

In 1948 the U.S. Bureau of Reclamation held hearings in Medford about plans it had developed for a high dam on the upper Rogue and smaller dams on several tributary streams. Intended to gather public opinion about three dam options, the hearings became a forum for the emerging wild river movement, drawing commentary from western fish and wildlife organizations and interested citizens from as far south as Los Angeles.

The Medford hearings revealed a complex set of meanings for the Rogue River. Resident farmers did not share the view of wilderness advocates that the river should remain undammed simply to become a "permanent playground for the metropolitan sportsmen," yet both sides viewed water as a cornerstone of the valley's livability and stability.[20] At a time when "refugee Californians," as all newcomers were called, brought a variety of unsettling changes, valley people on both sides of the river debate were concerned about sustaining a sense of regional identity and both sides saw the river as a key to protecting the landscape against the encroaching urban-industrial society.[21] Mountain ranges isolated the basin from urbanized areas to the north and south in Oregon and California, and valley residents in this borderlands region periodically contemplated, only partly in jest, organizing a new State of Jefferson separate from both states.[22]

Still, preservationists and irrigationists took their anxiety about growth and regional identity in opposite directions. River defenders saw the dam proposals as one more threat to the area's idyllic insularity; "shackling" the Rogue would simply encourage the spread of urban blight. Anglers, recent arrivals, and retirees had left California or the East to escape the "phony slogan of 'more production,'" as one put it, and they rallied to prevent the river from becoming a similar "industrial nightmare." These defenders articulated a vision of the river as a recreational asset that affirmed natural values over crass commercialism. "We have, God knows, more grain, fruit and agricultural products than we can ever hope to consume now or in the foreseeable future," one said. Resort proprietors also depended on this recreational construct. Towns like Gold Beach, at the river's mouth, placed their faith in tourism after the 1935 commercial fishing ban. If the dams ruined the sport fishery, they protested, "our livelihood will be taken from us again."[23]

From the farmers' point of view, salmon, steelhead, and primitive ambiance hardly compared in social value with irrigated lands yielding a bounty of field, pasture, and orchard products. In a dry country, they believed, nature unassisted was barren and unsightly; irrigation transformed a land of "brush and dry fields" into a productive pastoral world that mirrored the harmony of human and natural purpose.[24] Irrigationists felt that using the river's waters more intensively would stabilize the valley's pastoral landscape in the face of mounting demographic pressure. Suburban growth and hundreds of new subdivided amenity farms consumed precious water supplies, and only more irrigation diversions could secure the existing farms, ranches, and orchards against this use.

The Rogue River controversy was one of several debates in Oregon over rivers as recreational resources. Troubled by the diminishing number of trout and salmon streams, the state's primary conservation organizations, the NWF and the IWL, insisted that citizens look to their rivers as symbols of Oregon's distinctive way of life.[25] But the issue extended well beyond Oregon. The president of California's IWL told officials at the 1948 hearings that Californians "own a large share of the lands along the Rogue, pay taxes to the State of Oregon, and spend hundreds of thousands of dollars yearly there [and] ... look to the Rogue River with some feeling akin to that with which they regard the Yosemite Valley, and other unspoiled areas." The river belonged to "all the people regardless of where they come from."[26]

Besides nationalizing the Rogue as common property, dam opponents helped popularize the idea that rivers should be valued for their wildness. The Rogue's reputation as a "sporty stream [that] breeds sporty fish" depended on a vision of the river "bursting from the flanks of ancient Mt. Mazama" and running free to the sea. A "lazy fish," one biologist pointed out, simply "couldn't live in the ... Rogue."[27]

The widely reported 1948 debate over damming the river helped redraw the boundaries of river conservation in postwar Oregon and beyond. However, the outcome was inconclusive. Congress refused to fund the project initially, and a 1949 study by the U.S. Fish and Wildlife Service concluded that dams would be disastrous to the Rogue's sport fishery. In 1950 the Bureau of Reclamation dropped its plans, but early in the next decade local irrigationists persuaded the Army Corps of Engineers to design a similar multiple-use project for the river focusing on flood control. The fight for the "roaring Rogue" was far from over.[28]

In 1954 the National Park Service surveyed the controversial river's recreational potential, and shortly thereafter, the U.S. Forest Service (USFS) and Bureau of Land Management (BLM) established a 200-foot

"recreation zone" in federal lands on each side of the lower Rogue, with-
drawing the corridor from mining, homesteading, and logging claims. To
preserve the river's "special values," state officials also asked communities
and mill owners in the lower basin to keep refuse and sewage from its
feeder streams.[29] These directives began the process of segmenting the
river — dividing its waters between wilderness recreation and classic multiple uses. Responding to similar grassroots pressures, federal agencies
designed parallel policies for other western rivers running through public
lands. Here, along these carefully demarcated streams, development
would be deferred until water power became "of such vital importance to
the national welfare that scenic, sentimental, and recreational values have
to be sacrificed to grim necessity."[30]

While the need to protect rivers had been heralded for many years, the
idea of a wild-river system crystallized in the early 1960s in a series of
reports by the Outdoor Recreation Resources Review Commission, headed by Laurance Rockefeller, and by the Senate Select Committee on
National Water Resources. Following the recommendations in these
reports, the Bureau of Outdoor Recreation (BOR) made a two-year study
of more than 600 rivers and found 73 still "worth saving." The list was
pruned, and the secretary of the interior passed the recommendations to
Congress.

In 1964 the National Park Service protected the Current River in the
Ozarks as a "new kind of national park area," providing a prototype for a
larger system. Idaho Senator Frank Church presented a National Wild Rivers
bill in the waning days of the 1965 session, and in 1967 the House
Committee on Interior and Insular Affairs debated four wild river bills, the
most promising introduced by Pennsylvania's John P. Saylor, ranking minority member of the committee.[31] Although local opposition reduced the initial list of potential rivers to eight, testimony on the Rogue River was overwhelmingly positive. Governor McCall considered federal designation to be
"in the national interest," and Charles S. Collins of the California–Oregon
Recreational Development Association advised that federal designation
would give the Rogue "a unified, purposeful management which [would] ...
provide the best possible use of this unique section of river."[32]

In October 1968, Saylor's Wild and Scenic Rivers bill became law, and a
legacy of grassroots activism positioned the Rogue, as feature writer John
Hart Clark put it, as one of the nation's eight "lucky rivers" — Oregon's only

selection.[33] According to the act, designated river sections were to be at least 25 miles long, of high aesthetic or recreational quality, and free-flowing or capable of being restored to that condition. The rivers were to be managed to protect their "natural, wild, and primitive condition essentially unaltered by the effects of man." Yet despite these provisions, the bill actually embraced a complex set of local and national aspirations, leaving critical questions unanswered. Were rivers to be conserved or preserved? Made pristine or simply cleaned up? Kept truly wild or simply eye-catching?[34] "There persists among even the most avid outdoorsmen and conservationists," Saylor observed, "many points of difference on the program."

Some testified that the rivers should be extensions of the recently created Wilderness System. Conversely, one West Virginian sought federal designation for a "wilderness" pond 500 feet from a busy highway; it "would have to be cleaned out" a bit, he admitted. Others saw wild rivers as riparian national parks, and thus a means to attract more tourist dollars.[35] The Suwanee River, by one interpretation, needed no wilderness protection because it already offered, in addition to three state parks and dozens of free boat ramps, a huge commercial campground and "adequate numbers of motels, restaurants, fishing camps ... and other commercial enterprises."[36]

In the mind of the urban recreationist and nature lover, federal designation conferred upon these rivers all the deep symbolic meanings associated with America's burgeoning love affair with wild nature. The chosen rivers came to represent freedom, timelessness, purity, and the authentic life their forbears lived in a forgotten Arcadian world. But to ease the incorporation of these new values into federal thinking about river management, sponsors were careful to build in a certain amount of flexibility. Unlike the Wilderness Act, wild and scenic river designation addressed both the spiritual needs of the urban recreationist and the traditional pastoral perspectives of the bordering rural communities. The resulting compromise was, according to Senator Church, "an *activity* bill rather than a preservation bill."[37]

The system accommodated conflicted meanings by tailoring each river's administrative rules to local established uses, and then buffering the river as much as possible from these local uses through careful landscape screening. As BOR Director Edward C. Crafts explained to a congressional committee, the system would protect a "narrow strip along the shore so that as you travel the river, you *appear* to be in a natural environment." The Suwanee could have its motels, restaurants, and fishing camps, if they could be screened out of the river experience. Wilderness, in short, would be a carefully constructed illusion, using the vague congressional mandate to blend metropolitan concepts of wildness with local pastoral perceptions of useful, familiar flowing waters.[38]

Under federal designation, the lower Rogue, from its confluence with the Applegate River downstream to Lobster Creek Bridge, belonged, as Agriculture Secretary Orville Freeman put it, "to all the people of the United States." The upper 47 miles of this section lay primarily in the BLM's jurisdiction, while the lower 37 miles flowed through the Siskiyou National Forest. The least accessible sections, mostly in the BLM lands, were classified wild, while others were designated scenic or recreational.[39]

To an urban generation fascinated with the primeval, this stretch of river was timeless. Yet the Rogue is, geologically speaking, a youthful system. Its huge charges of winter storm water carried an enormous burden of surface sediments through a series of alluvial valleys, eroding soils in one place and depositing them elsewhere. Changes in one section reverberated throughout the watershed in the river's ceaseless effort to achieve a more uniform gradient. In this restless riverscape, state and federal managers set out to create a romantic illusion of enduring wildness. This stasis, moreover, was to be maintained while other engineers were at work transforming the upper river into an efficient water and power delivery system for the valley's cities, farms, and orchards.[40]

In the river corridor itself, state fish and wildlife officials carefully engineered a wild experience by banning or removing evidence of human intrusion, such as road construction, logging, farming, gravel and gold mining, and revetment work. To encourage wildlife, they created a complement of meadow, brushfield, and burn sites, leaving behind cavity trees and decaying logs. Where snags were scarce, they posted nest boxes for birds and mammals, carefully hidden from public view. Osprey and eagle nesting trees were marked and recorded on maps. State officials mandated small clearcuts on the lower slopes and altered streamside habitat to draw deer down from the ridgetops, making them "more available for observation." They released Roosevelt elk, wild turkey, pheasant, quail, and grouse into the corridor. To maintain wildlife populations near "viewing spots," they irrigated, fertilized, clipped, and reseeded pastures with various grains and with fruit, berry, and browse-producing shrubs, adding nesting boxes and salt and mineral licks.

Such efforts were not out of character with public land management for natural areas. Historian James A. Pritchard notes that even Yellowstone National Park's natural landscapes and wildlife populations were carefully cultivated; they were "not quite the essence of wild nature [visitors might] think." Yet the Rogue's subsequent history illustrates the importance of these carefully crafted illusions in reconciling traditional pastoralism and new wilderness and recreational values. What might have seemed a rather straightforward preservation policy — in Yellowstone and on the Rogue — was thus in fact a detailed management plan designed to enhance the illusion of free, timeless, and authentic nature.[41]

Edging the concept of wildness beyond the intentions of the federal act, in 1969 environmentalists rallied in Eugene and Medford to demand more explicit prohibitions on logging and road building in the canyon corridor. Wild and scenic designation, environmental activist Lawrence Williams felt, should "make it possible ... to know and enjoy a river much as it was prior to man's introduction." The entire lower canyon, according to State Treasurer Robert Straub, should be left in its "pristine condition."[42] At the same time, however, others expressed concern that the Rogue's "human history" — Indians, explorers, trappers, miners, boat pilots — would be lost in the scramble to make the river "wild." This history, the *Medford Mail-Tribune* editorialized, was "part of the river's mystique, and should be preserved for retelling before they are forgotten."[43] The compromises between these perspectives — wild and pastoral — were clear in the coming debate over dams on the upper river as well as throughout the country.

DAM BUILDING IN OREGON AND THE NATION

In the mid-1960s the Corps of Engineers placed its basin plan for the upper Rogue before Congress, reopening public controversy. The proposals came at the apogee of America's conflict over rivers. During this decade — the "hydraulic age," as historian Donald J. Pisani puts it — Congress debated some of the most ambitious water projects ever conceived. These included the Pacific Southwest Water Plan for dams at both ends of the Grand Canyon, a diversion of far northern California's rivers to the corporate farms in the San Joaquin Valley, a plan to draw off waters of the Columbia to quench the thirst of southern California, and the Central Arizona Project, linking the Colorado River with Tucson and Phoenix to irrigate some of the driest land in the country. The era saw pork-barrel politics on a grand scale.[44]

The boom years of western water development came to an end a decade later. If Echo Park marks the beginning of the crusade to save America's rivers, the collapse of the Teton Dam in Idaho, the first earthen dam to fail in the Bureau of Reclamation's history, effectively brought this epic struggle to a close. When the Teton Dam was proposed in 1971, environmentalists criticized the site as geologically unstable and challenged the bureau in court. Despite extra precautions to seal the foundation, on June 5, 1975, the dam breached, draining its 17-mile reservoir and causing 11 deaths and leaving 15,000 homeless along the Teton and Snake river valleys. Predictably, residents below other proposed dams began worrying about their own safety, especially as, with the nation's best sites already developed, the proposals became more risky.[45]

There were other signs that the era was coming to an end. In 1974 Colorado Governor John Vanderhoof spoke out against a dam on the South Platte because it "might lead to too much growth in the Denver metropolitan area." Landowners along the Ohio River expressed dissatisfaction when a series of Corps of Engineers projects flooded highways, homes, farmlands, and whole islands, some with historic importance. When opponents defeated the Grand Canyon dam proposals, even conservative Arizona Senator Barry Goldwater, a proponent of projects like these, doubted that he would "ever ... see this country build another big dam."[46]

Resistance to dam building reflected the rising importance of ecological and recreational interest in America's rivers. But as historian Mark Fiege points out, irrigation projects were also burdened by an underlying contradiction: they promised a Jeffersonian agrarian republic and delivered a corporate-dominated and mechanized landscape. Once viewed as the salvation of the family farm, by the 1970s dams were seen as a bulwark of corporate agriculture.[47] The inflationary pressures of the Vietnam War and the 1973 Arab oil embargo also drove up the cost of dam building, and in a nation plagued by agricultural surpluses, reclamation seemed an unnecessary boondoggle. And with the most obvious dam sites already taken, several remaining projects threatened to flood as much land as they irrigated.

In the public eye, the disadvantages of massive water projects — pollution, temperature elevation, siltation, alkali buildup, destruction of wildfowl and fish habitat — loomed larger than the benefits.

At the same time, environmental organizations became increasingly effective in pointing out the financial as well as the ecological drawbacks of each new project. With pressures from environmental groups mounting and President Jimmy Carter's battle against pork-barrel politics in full swing, the Army Corps of Engineers and the Bureau of Reclamation entered an era of caretaking. The battle over America's rivers continued, but Echo Park and the Teton Dam bracket a period during which free-flowing rivers played a significant role in environmental politics.[48]

The three multipurpose dams proposed for the Rogue River, although smaller than some in the federal backlog, indicate the changing values that helped bring the era of big-dam building to a close. First to come under discussion in 1967 was a high dam on the main Rogue at Lost Creek, 26 miles north of Medford. Responding to anglers' concerns, the Corps agreed to build a massive salmon and steelhead hatchery just below the dam. The hatchery fish, plus the uniform flow during summer months and the cooling effect of spill waters drawn from the depths of the reservoir, would realize the anglers' dream of a restored summer Chinook run. The steady flow of reservoir waters would also dilute downstream pollution.[49] Thus while the project promised to reinforce certain pastoral themes in the central valley — greater irrigation potential — it would also enhance the wild and scenic ambiance of the lower river, providing, as one dam advocate put it, a "clean, wholesome, livable environment, free of pollution and the debris of a wasteful and sometimes wanton society." Irrigationists and federal administrators were learning to balance traditional conservation formulas with newer recreational and ecological demands. The IWL, for instance, endorsed the Lost Creek dam, thus straddling the gap between Oregon's new wilderness values and its older pastoralist impulses. In July 1972 Congressman John Dellenback dug the first spadeful of earth at the damsite, marking the end of the Rogue's history as a free-flowing river.[50]

The Lost Creek compromise was not entirely effective. Environmentalists objected that the new stream-flow characteristics altered riparian ecology all along the river, and recreationists complained that the hatchery changed the nature of fishing on the Rogue. Hatchery fish, as outdoor writer Don Holm pointed out a few years later, were mass-produced to meet the challenge of serving ever more anglers on ever fewer rivers, providing "the most trout for the most people." Yet they were poor substitutes for the stream-smart, tough, and wary salmon and steelhead that gave the Rogue its national reputation in the 1940s.[51] Still, the dam preserved the illusion of wilderness in the lower section, and the compromise accentuated the importance of recreation in the new pastoral riverscape.

With the Lost Creek dam nearing completion, the Elk Creek dam came under scrutiny. This project, on an upper tributary of the Rogue, threatened existing spawning grounds, but as with Lost Creek, some angling groups, including the Oregon IWL, supported the reservoir as a way to enhance summer stream flow and water quality in the downstream wild and scenic section. But in 1975 state foresters discovered that silt from the clay soils above the dam site could remain suspended in reservoir waters for years, and logging companies suddenly realized that they would be forced to use expensive precautions to prevent erosion in the upper watershed if the dam were built. A formidable combination of federal and state conservation officials, the Oregon Environmental Council, and the timber industry lobbied against the dam on ecological and economic terms. After a decade of reversals, a court decision brought the project to a halt after the dam was two-thirds completed.[52]

A third and even more controversial project involved a small dam on the upper Applegate River, a tributary that entered the Rogue just above the wild and scenic section. Although in many ways the Applegate project was the least intrusive of the three proposals, its timing was such that it met the greatest resistance. In the fluid political climate just before the Teton disaster, the balance between traditional pastoral values and newer recreational and ecological sensibilities was tipping. The Oregon IWL again endorsed the project, given its potential for enhancing fish migrations, but the Sierra Club waged a vigorous campaign against it. Senator Robert Packwood, after reversing his decision twice, endorsed the dam, but in 1974 Jim Weaver was elected to the House of Representatives running on a promise to kill it. After heavy Corps of Engineers campaigning, a local referendum supported the project. However, in Jackson County, where the dam was to be located, support was thinner than in Josephine County, below the dam site.

The controversy epitomized growing ambivalence about the nation's rivers.[53] In 1977 the Carter administration condemned 22 dam proposals nationwide, including the Applegate, as environmentally and fiscally unsound. Heavy protest from western congressional delegations restored much of the funding, but the attempt to hold back the flood of new projects gave defenders of the Applegate one more hearing.[54]

As in 1948, the 1977 hearing provides clear evidence that river values were changing from classic multiple-use approaches — mixing recreational and utilitarian purposes — to aesthetic and to some degree ecological designs. Opinion in the valley was split. An older generation of farmers and other dam supporters, holding to a traditional pastoral image of gardened landscapes, castigated preservationists as "late-comers" who failed to appreciate the need for irrigation and flood control. They understood that after decades of clearing, mining, and logging, the river no longer worked in harmony with their fields, pastures, and meadows as it had when their forebears arrived in the valley. Late-summer flows dropped to a trickle, reducing the stream to a series of algae-laden pools, while winter floods tore through the riparian woodlands depositing gravel and sand in the streambed, widening the floodplain and carrying away topsoil. The dam would stabilize the valley's agrarian landscape, controlling the errant river, improving water quality, and making the Applegate, and subsequently the Rogue, "the river it was back in the 1920's — a real beautiful wild river, in its best sense." Killing the dam, Myrtle Krouse advised, would "write *finis* to a way of life." The valley needed "*less* ecology," if indeed that meant more floods and more siltation.[55]

A newer generation of exurbanites, in contrast, defended a rural atmosphere anchored by timeless and untrammeled nature against "greedy chamber of commerce types" and their "red-neck" allies.[56] These forces opposed the dam because they thought the project would encourage construction on the floodplain and hasten recreational development, ushering in "another California–style smog basin." Some opponents urged ecologically sensitive alternatives, including better watershed management, reforestation, floodplain zoning, and channel clearing, and asserted the rights of the river as a "natural ecosystem."[57] A spokesperson for Friends of Oregon Rivers pointed out that damming a major tributary of the Rogue would upset a delicately balanced ecosystem all along the river.

In May 1980 Senator Mark Hatfield, standing on the shores of the new Applegate "Lake," dedicated the dam project by reiterating a phrase long familiar to his constituents: "it was not the six-gun, it was water impoundment that won the West."[58] But Hatfield's appeal to utilitarian multiuse conservation was an anachronism in the post–Teton Dam era. Advocates got their dam, but they did so in part by adopting the premises used by opponents of dams all across the nation: rivers should serve the recreational needs of urban America. Officials from the flood-control district and the IWL agreed with ecologists that "what happens on the Applegate affects the Wild and Scenic Rogue," but also pointed out that reservoir waters meant more fish and better rafting when the Rogue watershed lay parched under a warm summer sun.[59] Like the Lost Creek dam, the Applegate project enhanced the illusion of wilderness on the lower Rogue, even while it fed the illusion of traditional pastoralism in the middle basin.[60]

The trouble with this carefully crafted illusion, as historian William Cronon points out, is that viewing wilderness as a discrete realm of nature obscures our responsibility to the rest of our environment.[61] Imagining the Rogue as wild helped southern Oregonians keep fears of creeping Californianism at bay even while they carpeted the central valley with subdivisions and orchards.

Yet even this abstracted version of wilderness was instructive. The recreational experiences rivers provided were immensely useful in educating Americans about the importance of nature, as the changing rhetoric of the river engineers themselves suggests. Through redemptive play on wild rivers, Americans came to know nature better, and they developed a powerful claim to the nation's rivers as recreational treasures. These recreational and ecological values enriched the environmental imagination, even though dam proponents and developers prevailed one more time on the Rogue.

On the Allagash and St. John in Maine, pastoral and wilderness ideals similarly blended into a new imaginative and cultural understanding of rivers. But there these ideals intertwined to yield a different outcome, as builders and developers lost their bid to engineer the watershed.

THE ALLAGASH: A LAST CHANCE FOR WILDERNESS

On a misty fall morning in 1960 Supreme Court Justice William O. Douglas and his guide, Willard Jalbert, pushed their canoe into the still waters of Round Pond, a small still-water stretch of northern Maine's renowned Allagash River, and set out for Rankin Rapids, site of a proposed dam just below the river's confluence with the St. John. Slipping through the quiet waters, enveloped in fog, Douglas could feel the pulse of the brooding spruce–fir forest along the riverbanks: "objects come into focus and then quickly fade away, as in a land of fantasy." As they passed downriver, elements of a grander scheme — banks of sweet grass, gravel bars, water willows, loons, ducks, moose, deer — emerged and disappeared in the morning fog.

When the skies brightened and the land became more distinct, Douglas's thoughts turned to the proposed Rankin Rapids dam and the huge reservoir that would flood the meadows, the islands, the St. John, the Allagash itself — all "wiped out in the name of hydroelectric power." Douglas and his party resolved to do what they could to preserve the river, knowing that their struggle would be a "chance to redeem in the Allagash some of the values" they had lost to civilization elsewhere.[62]

As with the Rogue, the idea of redemption on the Allagash expressed a growing devotion to remote rural rivers emerging all across America. In Maine as elsewhere, Americans mixed new wilderness values with an older set of pastoral traditions as they took a hand in preserving these rivers. The result was an expanded vision of rivers as recreational assets, a part of the national commons.

In this eastern context, pastoralism involved a different set of traditions. Maine's north woods was not a farmers' world like the Rogue River basin; rather, it belonged to a relatively small group of corporate timberland owners, most of them involved in pulp and paper production. Here the pastoral idea of a mythic working landscape was carried forth by an army of woods contractors, jobbers, and loggers who followed a tradition of rural work dating from colonial times. This work was labor-intensive, dangerous, and Homeric, and it linked these rural producers — many of them also part-time farmers — firmly to the tradition of folk and nature.

Unlike the Rogue Valley farmscape, Maine pastoralism also embraced recreational use. The privately owned Maine forest supported a vibrant tradition of trapping, hunting, fishing, guiding, and camping that was integrated into the working woods. For generations, Maine people had used the north woods without much regard for formal ownership or legal regulation, and these traditions of unrestricted recreational use defined, for Mainers, a proper ethical and aesthetic relation to woods, waters, and wildlife. In this context, recreation offered authenticity at least in part because of its association with the logging lore of the north woods, and a sense of freedom because it was noncommercial and completely unrestricted. As a state legislator explained in 1966, "people up here in our country when they want to go fishing ... can go in there ... any time ... and stay as long as they want to, camp wherever they want to and come home whenever they want to; nobody bothers them."[63]

Modern wilderness advocacy departed from Maine's traditional pastoral sensibilities, particularly from Maine's emphasis on the folk and productive engagement with nature. A century of logging activity had littered the north woods with abandoned camps, decayed log-driving dams, and rusting logging equipment, and these ruins achieved iconic status in local perspective. By contrast, Justice Douglas saw this debris in negative terms: "wooden dams, now in disuse and disrepair," marred the landscape; rusty spikes menaced canoes in the old sluiceways, and "ugly remnants of the old structures" defaced the lakeshores. Like the fog-shrouded riverscape he drifted through on that fall morning in 1960, Douglas's vision of the Maine woods was clouded by romantic anticipations. As activists from both traditions came head-to-head with dam-builders in this "last frontier" of New England, they rethought the wilderness values espoused by Douglas and the traditional pastoral vision that left room for a working woods.[64] Here again was a tale of accommodation.

As a milestone in the emerging river preservation movement, the Allagash debate pioneered three wilderness ideas appropriate to the East, where the boundaries between nature and culture, common and private, trammeled and pure, were less distinct than in the West. First, unlike the Rogue, the Allagash flowed through private commercial forest. Legal traditions dating from the early colonial period secured public access to these extensive timberland holdings, and what one federal report called a "kind of cold war [policy] in which the landowners make concessions so that the public will not confiscate a much larger area under the right of eminent domain" sustained these rights. Still, the suggestion of *de jure* public control over a portion of this huge private landholding — several million acres held by some two dozen firms — was an untested proposition in wilderness legislation.[65]

Second, Allagash preservationists advanced the concept of a "recovering wilderness" as a preservationist ideal. Wilderness policy had evolved as a management option for roadless federal lands in the West, where vast ecosystems had been spared almost all direct human impacts. The East contained no such pristine lands. Nor did ecological succession fit the western wilderness ideal, where altitude, severe climate, and competition for soil moisture created open, parklike forests of relatively stable — and seemingly timeless — composition. Eastern forests were more dynamic and changeable, and in some cases recovering wilderness became virtually impassible — at least in the short term — owing to explosive growth of pioneer species. Eastern forests lacked the timeless qualities we associate with wilderness. Yet uninhabited northeastern areas were wild in their own right. A century of intensive logging had altered the forest in terms of character and composition, but no important species had been lost, and the woods rebounded vigorously. Thus despite a century of hard use, the Allagash still looked primeval to travelers like Justice Douglas, and it was capable of eliciting the fierce defense of wildness that shaped management policy on the Rogue.[66]

The wildness of the Allagash country was an accident of history and geography. In the nineteenth century, remoteness and logistical disadvantages dictated a conservative harvest of the dominant spruce and fir; taking only the most valuable trees, loggers left behind an almost continuous forest canopy. Late in the century, resort owners and railroad passenger carriers began publicizing the waters of the north woods as America's classic wilderness canoe trip. In 1936 Linwood L. Dwelley of Auburn, Maine, formed the St. Croix Voyageurs and began guiding teenage boys down the Allagash. These "adventures in living" enshrined the river in the hearts of thousands of youngsters over the next three decades. By the 1960s the Allagash had become an icon to the northeastern sporting set, combining the literary imagery of Henry David Thoreau's Maine Woods (1864), the lure of the region's lumbering legacy, and the ageless mystery of the deep north woods.[67]

The first in a long series of attempts to federalize this northern wilderness came with passage of the Weeks Law in 1911, which sanctioned federal purchase of eastern timberlands for national forests. Imbued with a powerful sense of place and an exaggerated faith in Maine institutions as the "last citadels of democracy," legislators remained deeply suspicious of federal proposals to buy parcels of northern Maine forest. "The people of Maine are ... honest people," a legislator proclaimed, and "the waters of the

State are ours."[68] Although New Hampshire's White Mountain National Forest included a small portion of Maine's western mountains, calls for similar federal purchases in the northern and eastern woods fell on deaf ears. In 1931 Congress proposed federal acquisition of tax-delinquent timberlands for a national forest in Maine, as was occurring throughout the eastern United States. The proposal was so unpopular that no state legislator would sponsor an enabling bill.

In 1933 Congress returned with a proposal for purchases on a willing-seller basis. Paper company officials, aware that federal purchases would bid up land and labor prices, opposed the measure. The enabling bill passed the state legislature, but in a form so amended by company lobbying that Congress refused to act under its provisions. Similar plans succumbed to antifederalism in 1948, 1957, and 1959.[69]

While some federal agencies offered to conserve the north Maine woods, others developed plans for a dam that would obliterate sections of the forest altogether. Since the 1920s state and federal officials had dreamed of harnessing the 20-foot tides in eastern Maine's Passamaquoddy Bay to create a TVA–type development program for the economically depressed region. The project spurred a search for suitable inland reservoir sites that could be dammed, and the resulting hydropower used to compensate for slack-tide periods in the Passamaquoddy dams. In 1955 the Corps of Engineers identified a potential site at Rankin Rapids, on the St. John River near its confluence with the Allagash. The dam, more than a mile long, would inundate much of the St. John and Allagash river valleys.[70]

The idea of flooding the Allagash galvanized Maine outdoor recreationists. In 1956 officers of the Maine Fish and Game Association met with other conservation leaders and proposed a "wilderness" corridor running the length of the river — a scenic screening device similar to the management plan the Forest Service and BLM implemented on the Rogue that same year. In 1959 a similar coalition formed the Natural Resources Council of Maine (NRCM), Maine's chapter of the National Wildlife Federation, with a three-year grant from David Rockefeller. The NRCM and its affiliated outdoor, civic, and garden clubs challenged the prevailing political viewpoint that the woods "existed for the pulp and paper industry, and the lakes and streams for developers," and took on the 1956 Allagash plan as their primary concern.[71]

Maine outdoor enthusiast James Carr urged the National Park Service to intercede in the dam proposal, but he stopped short of endorsing the area as a part of the national park system, with its inevitable crowds and commercial development. Acting on a tradition of low-intensity outdoor recreation in a working-woods context, Carr envisioned a "primitive" fishing and canoeing park free of the commercial trappings of urban civilization, with "small country roads leading into wilderness areas" and "no feather beds." He took inspiration from nearby Baxter State Park, a 200,000-acre protected wilderness surrounding Mount Katahdin, created through a series of timberlands purchases beginning in 1930 by wealthy Portland conservationist and former governor Percival P. Baxter.[72]

The Park Service responded by sending in field parties with an eye to "sheathing" the Allagash in a wilderness "reservation." Like Carr's plan, it would be accessible, as a local paper put it, only to "rough-its." Yet this was not exactly a wilderness proposal: officials projected that national park designation, coupled with a $2.8 million investment in visitor centers, camping facilities, trails, interpretative displays, and access facilities, would draw some 90,000 canoeists to the region during the 90-day canoeing season — a figure raised a few years later to a million visitors. At the time, fewer than a thousand canoed the river each year.[73]

These conflicting proposals for dams, parks, and reservations confused Maine people, and the lack of unanimity created a sustained debate in the media that popularized the north woods wilderness. Journalist Gene Letourneau noted that only a few canoeists used the Allagash each year, but those who did "would rather have it that way." Make it easy, he cautioned, "and you'll destroy it."[74]

Only Maine's paper company officials were completely clear on the issue of federalizing the river. Unlike the Rogue Valley farmers, Maine's dominant landowners rejected the philosophy of multipurpose dams and their developmental assumptions. Lack of year-round residents and public infrastructure in the northwestern townships gave corporate landowners one of the most favorable timberland tax assessments in the country. Federal projects would bring development, increasing the tax burden and encouraging mass recreation, with unpredictable results.[75] Moreover, the industry would face a labor market in which "people used to minimum-wage jobs in the forest [would] ... have the chance to make three times that" on federal dam and park projects. There were, as state Forest

Commissioner Austin Wilkins gently put it, "reports of growing sentiment to keep the Allagash 'in the Maine family.'"[76]

Shortly after the federal proposals surfaced, the principal landowners in the Allagash formed the Association for Multiple-Use of Maine Timberlands (AMUMT) and proposed a "working wilderness" concept for the Allagash corridor, secured through the "principles of scientific forest management." The oxymoronic term, made even more ironic by an ambitious woods-road construction program then under way, resonated with traditional recreational values in Maine. AMUMT offered to guarantee a forested border along the waterway and promised several other measures to maintain the natural appearance of the river, in return for state — not federal — regulation of recreational activity on the river itself. In Maine's dense spruce–fir forest, American Forestry Association official Henry Clepper argued, "an army of loggers might be felling pulpwood a mile inland, and unless the canoeist heard a power saw he would never be aware of the logging or see evidence of it." A sympathetic news writer captioned an aerial photo of two Allagash lakes in the heart of the lumbering district: "what could be lovelier? ... Virginal, placid, and unspoiled by man."[77]

The working wilderness idea served the needs of the paper industry admirably. By the 1960s, problems of campfire control, public traffic on private logging roads, lost hunters, litter, and general safety were beyond the scope of traditional company management. The state, landowner Bradford Wellman concluded, should "assume the responsibility of taking care of these users — and incidentally picking up after them." At a time when canoe parties were complaining about the "Coney Island" atmosphere along the waterway, company foresters insisted that the "Allagash has not been, is not, and never should be a 'mass recreation' area."[78] Tacking the working wilderness concept onto industry's traditional multiple-use slogans, company officials found common ground with recreationists and rural river preservationists who feared that federal management would bring either dams or a complex of motels, restaurants, curio shops, and other forms of commercial clutter.[79]

In 1961 the legislature debated the working wilderness concept as a means of preempting federal plans and asserting state sovereignty over the Allagash. The NRCM endorsed a similar, somewhat more protective plan for a mile-wide wilderness corridor along the river, and paper companies launched a "full-scale battle to keep the Allagash in ... private ownership."[80] Preservationists and industry officials, united on the issue of state sovereignty, began moving toward a distinctive wilderness ideal for New England's last frontier.

In addition to the overlapping state, federal, and private wilderness proposals, plans for two more dams emerged in 1963. The first, eventually known as Dickey–Lincoln, involved an earth-fill dam on the St. John River at the township of Dickey, above its confluence with the Allagash. More than two miles long and higher than either the Hoover or Aswan dams, Dickey–Lincoln would inundate most of the St. John but leave the Allagash largely intact. The second, called the Cross Rock dam, was even more destructive. Put forth by a consortium of northern Maine business leaders, the proposal called for a mile-long dam downstream from the confluence of the Allagash and St. John. The Cross Rock proposal offered a favorable cost–benefit ratio and directed revenue to the state rather than the federal government, but unlike Dickey–Lincoln, it would, opponents argued, turn the Allagash into a "vast deadwater reservoir with stinking mud flats and barren gravel bars."[81]

The 1963 legislature faced a complicated decision involving three dam proposals, a federal recreation area, the NRCM's state management plan, and the private working wilderness concept. Most legislators agreed on keeping the Allagash out of federal hands, where it would be "as subject to influence by a California Congressman under pressure from the Sierra Club as by [Maine's] own elected or appointed officials." But Maine had not yet reached a consensus on the alternatives. State Senator Edward Cyr, a Cross Rock proponent, pointed out that between 1959 and 1963 state game wardens counted only 808 canoeists on the lower Allagash; of these, 250 were state officials, 255 were Linwood Dwelley's boy–adventurers, and 150 were local people going to and from their summer camps. Given the river's limited recreational use, it was difficult to dispute Cyr's contention that wilderness was simply a tool landowners embraced to save their forests from federal intervention.[82]

When the Allagash bill finally reached the floor of the legislature, it had the unmistakable imprint of industry lobbying. It offered weak state enforcement powers for selective cutting in a thin buffer along the river, in return for substantial tax concessions in a wider multiple-use zone.[83] To bring dam supporters on board, legislators accepted an amendment declaring that wilderness designation would not "prohibit the cutting and harvesting of timber or removal of minerals, and shall not be applied to restrict the exercise of those rights commonly known as flowage." Some worried that this ridiculously broad definition of wilderness would fail in its primary purpose: "what you are doing by passing a bill such as this is ... just asking the federal government to come in. They are not stupid down there."[84] Indeed they weren't: BOR Director Edward Crafts declared the bill unacceptable, and Interior Secretary Stewart Udall threatened federal intervention if Maine "didn't come up with either a federal park or a state park proposal pretty soon." Paralyzed by these mutually exclusive options, the legislature turned the Allagash back to an interim committee with instructions to hold public hearings.[85]

By this time, forest landowners across the nation were following the matter closely. The federal wild and scenic proposal listed more than 60 other waterways, and the Allagash woods, considered "one of the crown jewels in the private forestry firmament," was industry's best example of the private landowner's ability to meet the challenge of surging public use on the nation's commercial woodlands. On the other hand, the idea of an Allagash wilderness had taken hold nationally. Editorials favoring its preservation appeared in *The New York Times*, *Readers Digest*, and sporting magazines across the country. To urbanites living in the Northeast, it was important to know that Maine still harbored a "vast forest world ... nearly the same today as when Thoreau saw it." As much as Maine people would have liked to keep the river for themselves, the Allagash, like the Rogue, had become part of the national commons.[86]

Debate over the Cross Rock dam helped crystalize Maine's choices. Hoping to broaden its appeal, Cross Rock proponents laid out a vision of a 20,000-acre woodland park north of the dam site to serve as a recreational gateway to what promoters were now calling Grand Allagash Lake, a giant reservoir lying over most of the Allagash–St. John drainage. At the dam site, the mass-recreation park would offer marinas and boat basins, campsites, trailer parks, cottages, lodges, nature trails, a viewing tower, and a fish hatchery and biological station. As the largest artificial water body in the East, the reservoir would encourage a "steady flow of tourists," which would in turn foster motorboat dealerships, vacation-home construction, and numerous service activities.

To replace the drowned Allagash, the authority proposed damming the St. John headwaters to provide even, all-summer flow for a new Allagash

Canoe Way running 67 miles from the upper St. John ponds, down the unflooded portion of the St. John, to Grand Allagash Lake.[87] The differences between Grand Allagash Lake and the Allagash wilderness waterway could not have been more stark. As a multipurpose water conservation project, the dam would help industrialize New England's "last frontier" and open a vast recreational area to millions of tourists "who might never entertain the idea of visiting the wilderness region otherwise."[88] The wilderness waterway, by contrast, would maintain thousands of acres of riparian forest in its natural state, available to a few hundred intrepid outdoor enthusiasts during a short three-month canoeing season. The choice galvanized old-style recreationists and wilderness advocates, both warning against designs that would create a "honky-tonk, candy wrapper paradise up there in the Allagash." At the same time, officials from Maine's three largest private power companies, who had much to lose from the public power proposals, joined paper industry officials in the campaign to keep the Allagash wild.[89]

In spring and summer 1965, as another legislative study group traveled around the state collecting public opinion, Maine people moved toward consensus, and the Allagash Authority fleshed out plans for its "wilderness showpiece" in northern Maine. By this time debate centered on the plan's details. Some argued that a 300-foot buffer would sustain the illusion of wilderness in the thick north woods, but guides complained that they could already "see daylight" through the trees on the banks where pulp operators were cutting. The NRCM hoped that logging beyond the buffer would be done "in the old fashioned way," without "bulldozers [scoring] ... the earth down to bedrock, and the mud run[ning] into the river."[90] In October 1965 Congress gave tentative approval to funding the Dickey–Lincoln project on the St. John, and the Cross Rock proposal was set aside. With the Allagash suddenly free from the threat of inundation, legislators quickly pieced together a wilderness proposal that would shield the river from the "people down in Washington [who were] just itching to get their hands on this area."[91]

In January 1966 Allagash Authority Chair Elmer Violette presented a plan calling for a $1.5 million state bond issue to purchase, with matching federal funds, a corridor ranging from 400 to 800 feet on each side of the waterway. In a half-mile zone beyond the strip on each side, cutting would be regulated by state park officials. On November 8, 1966, Maine voters decisively endorsed the bond issue, making Maine the first state in the nation to specify rules protecting the natural character of a river. After a lengthy delay to survey and appraise the lands and negotiate purchases, the Wilderness Waterway was dedicated in July 1970. Ironically, the ceremony took place on the site of a newly rebuilt dam near the head of the Allagash. The Churchill (or Huron) Lake dam was designed strictly for recreational purposes — to provide even flow all summer for canoeists.[92] Although planners were sensitive to natural processes in the riparian corridor, the dam suggests that ecology was still a minor factor in the compromise between wilderness, recreation, and the old pastoral ideal of a working woods.

With the Allagash secured in 1966, the politics of preservation shifted to the Dickey–Lincoln dam; the subsequent fight to protect the St. John River brought ecology to the fore. Despite congressional authorization, a coalition of private utilities lobbyists, fiscally conservative Republicans, southern Democrats, legislators from coal and oil producing states, and 17 of the region's 25 members of Congress blocked full funding for the dam. The proposal was revived in the mid-1970s by a series of destructive floods on the upper St. John River and by the 1973–1974 oil crisis, which convinced utility companies to reverse their stand on public power.

With the Allagash protected and a long battle against the Alaska oil pipeline fading into the background, national environmental organizations took up the St. John as a focal issue. As with the Allagash, preserving the St. John drew together a complex alliance of outdoor recreation advocates, antifederalists, landowners, private power officials, women's club leaders, and tourist interests, united around a wilderness ideal that in this case depended heavily on a concept of ecological integrity rather than recreational aesthetics.[93]

In 1976 crews preparing an environmental impact statement (EIS) for the Dickey–Lincoln dam proposal found a clump of rare snapdragons in the reservoir area, and environmentalists raced to include the Furbish lousewort on the endangered-species list, providing an ecological rationale for opposition to the dam. As it turned out, the plant was not confined to the St. John habitat, but the prospect of endangered-species listing continued to taint the dam project in the public mind. A year later Dickey–Lincoln came under President Jimmy Carter's moratorium on new dam construction. Heavy pressure from Senators Edmund S. Muskie and William Hathaway restored some funding, but only enough to complete the EIS.[94]

In this struggle of larger forces over a national river commons, the people of the remote towns in the St. John Valley below the dam site seemed to play a minor role. Living on homesteads their Acadian, French– Canadian, and Scots–Irish ancestors had settled nearly two centuries earlier, local residents seemed perpetually isolated from Augusta and Washington and somewhat clannish by the reckoning of other Mainers. Nonetheless, their opinions flowed with the tide of public discourse on Dickey–Lincoln. A stagnant northern Maine economy dictated initial local support, yet valley residents feared the influx of construction workers and urban recreationists into their pastoral world. "The way things are now," one mused, "we don't even knock when we call on our neighbors. We just walk in and say hello."[95] Still, the 1973–1974 floods tipped local opinion in favor of the dam, and the widespread feeling that flooding resulted from poor logging practices upriver left valley residents unsympathetic to the companies' appeal to save the timberlands slated for flooding.[96]

Valley opinion on the dam was in fact inseparable from resentment toward the timberland owners who employed Canadian rather than Maine woods workers, then sent the sawtimber across the border to Canadian mills, to be reimported as lumber competing with the products of Maine's own mills. Nor were valley residents receptive to the "armchair environmentalists" from Boston or southern Maine.[97] The St. John country had been logged for more than a century, local attorney Robert Jalbert pointed out; urban wilderness advocates were fighting to preserve "something that's [already] nonexistent."[98]

Traveling a path apart from urban wilderness advocates and corporate proponents of the working wilderness, valley residents eventually arrived at similar conclusions: the St. John, offering a canoe trip of seven days without a sign of civilization, was the East's last chance for wilderness. Greg Jalbert, grandson of William O. Douglas's guide, told officials at a public hearing that, for him, the value of wilderness could be measured only by "being on the river" and listening at night to the cry of the coyote and loon. Jalbert's speech, which "struck a resonant chord" with the valley audience, dwelled on the direct, unmediated, and unregulated experience of place: the valley residents' unique claim to the wilderness. While Jalbert distanced himself from the logging companies' idea of working wilderness and from urban preservationists who had never actually seen the river, his opinions expressed an important dimension of the Maine wilderness idea: a primitive, noncommercial relation to nature — the north woods as a kind of wilderness backyard.[99] "They already took the

Allagash and gave it to the people from New York and Connecticut," another Ft. Kent resident claimed; the St. John belonged to the people of northern Maine. And indeed, the river's brief canoeing season, its swift water, its legendary black flies, and its enormous logistical disadvantages as a recreational river left it largely in the local common domain.[100]

Throughout Maine, opposition to Dickey–Lincoln solidified around the issue of wilderness. Paper companies, the state Chamber of Commerce, the environmental establishment, the Maine League of Women Voters, the state Grange, L.L. Bean Co., and the Sportsman's Alliance of Maine joined the opposition. With a nuclear generator under construction on the coast at Wiscasset, promising an output 13 times that of the dam, support for Dickey–Lincoln evaporated. Edmund Muskie, Dickey–Lincoln's most conspicuous supporter, left the Senate in 1981, and in 1983 Congress quietly deauthorized the dam.[101]

The Allagash and St. John controversies helped refine the wilderness concept in ways that made sense for a well-trammeled eastern forest. The example was useful, for instance, when environmentalists began calling for wilderness designation in eastern national forests in the late 1960s as a way of meeting demand for outdoor recreation convenient to large metropolitan areas. Preservationists demanded full protection for these recovering forests, and in 1972 Senators John Saylor and James Haley proposed an Eastern Wilderness Areas Act, designating 28 new units from eastern, midwestern, and southern national forests. When the bill passed in 1974, eastern wilderness management differed little from policies outlined for western lands in the 1964 act.[102]

The quest for authenticity, permanence, and freedom through redemptive play on America's wild rivers helped modernize the environmental imagination. On the Rogue, preservationists balanced the illusion of wildness with an older pastoral vision of a working river, reconciling a western tradition of aggressive water management with the spiritual needs of those who saw the river's tumbling waters as a means to wilderness renewal. On the Allagash and St. John, careful coalition-building created an eastern wilderness concept that embraced the older idea of primitive recreation in a working rural landscape and the newer concept of redemption through natural purity. Goaded to action by a broad, inclusive wilderness vision, activists in Maine and Oregon helped expand the environmental imagination to embrace a far more complex world beyond the waters of the urban river.

Cannon Beach, Oregon

"*To those of us who live in the Northwest, the ocean shore is a retreat, a place to visit in search of fish and clams or beauty and isolation.*"

—*Murray Morgan, writer*

CHAPTER 4

SAVING NATURE'S ICONS: THE MAINE COAST AND OREGON BEACHES

O n February 3, just months before Earth Day 1970, Rep. Marion Fuller Brown stood before the Maine legislature and lashed out at the "exploiters and abusers of the public air, water and land" who now laid siege to Maine's premier nature icon, its rockbound coast. In language soon to become familiar all across America, she insisted that an unspoiled environment was "one of our inherent freedoms," a right she would defend in the seventies as others had fought for civil rights in the sixties. On the opposite coast Oregon Governor Tom McCall voiced a similar concern that developers from California, a state already "filled up" with people, smog, and concrete, were headed north like a "ominous band of human locusts ... even now zooming in on what's left of ... our remarkable farmlands, mountains, forests, and coasts."

As rhetoric like this suggests, the environmental imagination had become a mainstream political idea by 1970. The Audubon Society's Charles Callison wrote in 1969 that *the environment* was the new "in" phrase, appearing in political pronouncements "almost as frequently as such old standbys as *peace, prosperity,* and *national security.*" Another

Audubon writer noted that "we hear about the environment wherever we turn, from a speech by the President to network television specials, from the comic strips to full-color advertising spreads by big corporations anxious to cleanse their image."[1] A variety of political events helped move environmentalism into the mainstream, but certainly among the most important were campaigns to protect the landscape features that defined regional identity for those living around them. These scenic icons, charged with meaning derived from both pastoral and wilderness traditions, were central to the process of popularizing the environmental imagination.

In Delaware, for example, Russell Peterson, DuPont executive for 26 years and now governor, shocked the nation's business leaders by blocking a $360 million, 5,800-acre oil–coal–steel complex planned for a relatively undeveloped section of Delaware Bay, citing his state's "obligation to provide recreation facilities for the whole East Coast region." Under Peterson's urging, a state task force initiated the Delaware Coastal Zone Act in 1971, barring all heavy industry from the state's 115 miles of shoreline. People, Peterson remarked, were "drawing lines around choice pieces of real estate and say[ing], 'Look, this is off limits for certain kinds of operations.'" Defending icons like the Delaware coast made sense to Americans who questioned growth politics in a society already "bursting at its seams" with industrial development.[2]

Like the clean-waters and wild-rivers campaigns, the defense of iconic landscapes fed on the contrast between redemptive nature and the doomed city. After a half-decade of destructive riots and fiscal crises, city dwellers harbored a dark assessment of their urban environment. Typical was a planning consultant's forecast for western Oregon: "new factories would pollute the Willamette River and destroy its salmon runs; "clouds of smog from busy Portland" would drift down the valley; roads would multiply and "clog ... up with traffic"; urban centers would die and decay; and suburban sprawl would "march steadily across the landscape." Others predicted that future city dwellers would depend on "artificial oxygen at night to keep them breathing."[3] To landscape architect Ian McHarg, the city at its worst was a "smear of hot dog stands, gas stations, diners, and subdivision ranchers spreading a blight over a beautiful native landscape." By contrast to the chaos of modern urban life, an old-growth forest, a national park, and an idyllic ocean and lakeside vista seemed immutable, tranquil, and lucid. Social unrest encouraged urbanites to incorporate prominent aspects of their natural hinterland into their sense of regional identity, and to insulate them from the progrowth themes in political life. If these "God-given retreats" were degraded, urban dwellers understood, the city would also lose its "desirability as a place to live."[4]

Samuel P. Hays, who surveyed the environmental movement region by region in his *Beauty, Health, and Permanence: Environmental Politics in the United States, 1955–1985,* found a "close connection between the positive expression of environmental values and the proximity of natural landscape features that served as reference points for [environmental] action." Identification with these natural features grew in the late 1960s as roads and highways improved and as second homes, resort accommodations, and recreational equipment became more affordable. These same circumstances, of course, brought commercial and industrial threats to these iconic features. Popular campaigns to protect them became, then, in Hay's words, "baselines from which environmentalists sought to spread environmental values to the wider urban society."[5]

Campaigns to defend these iconic landscape features politicized the environmental imagination, and a multiform challenge to the politics of growth brought together civic, scientific, and conservation groups and spread the environmental ethic to new arenas. In 1964, New Jersey's Great Swamp, one of the largest unspoiled wetlands on the East Coast, gained protection when citizens purchased land and turned it over to the federal government as a wildlife refuge. From 1968 to 1969, Florida activists waged one of the most intense conservation campaigns of the decade to halt plans for a huge jetport adjacent to Everglades National Park. In the Midwest volunteer groups campaigned for the Indiana Dunes National Lakeshore, and in California they worked with the Sierra Club to block Walt Disney's plans for a $35 million resort complex in the Mineral King basin, near Sequoia National Park.

Perhaps the nation's boldest statewide preservation effort occurred in California, where a 1972 citizen referendum imposed sanctions on development along the state's 1,000-mile coastline. In less publicized campaigns all across the nation, local groups blocked shorefront construction, battled air, water, and scenic degradation, protested ocean dumping, tightened rural zoning laws, and placed moratoria on housing projects. According to a 1971 *Forbes* magazine article, it was "nearly impossible to build an electric power plant, a jet airport, an open-pit mine, or a resort complex without strong protest from keep-out forces."

Taken aback by this onslaught against the politics of growth, an Oregon Pacific Power and Light official articulated the rationale for industrial expansion in guarded tones: "I know that ... development ... conjures up the idea of smokestacks, lunch buckets, etc. and this is in conflict with the uncluttered, tranquil landscape that all of us would like to have. But I think you will all concede ... that what you, I think, expect as to the future is that we will have lots of people here. We must plan for them, and they have to have jobs." The *Forbes* editor mused nostalgically: "remember when the high school band — not pickets — met those visiting industrial delegations?"[6]

For Maine and Oregon, the campaigns that popularized environmentalism involved coastal areas, although in different ways. As Robert P. Tristram Coffin had suggested two decades earlier, the Maine coast, a pastoral world of fishing villages and seascape vistas synonymous with the state's flinty individualism and its traditional way of life, served as a reservoir of authenticity, freedom, and permanence for the heavily urbanized and industrialized Northeast. While the idea of the folk and their pastoral heritage enriched Maine's coastal iconography, the Oregon coast was linked to the wilderness tradition. It epitomized Rachel Carson's description of the authenticating natural experience: the shore, she wrote in *Edge of the Sea*, "fascinates us who return to it, the place of our dim ancestral beginnings. The edge of the sea is a strange and beautiful place ... always it remains an elusive and indefinable boundary."[7] The eternal confrontation of land and water, the wildness of the subsurface world just offshore, the screech of seabirds overhead, the pounding surf, and the solitude of the long, windswept beaches offered an experience of raw nature that pulsed with feelings of freedom and authenticity.

Just as they drew inspiration from different aspects of the environmental imagination, Maine and Oregon perceived coastal environmental issues differently. Beginning in 1968, Maine became embroiled in a series of oil-refinery and tanker-port proposals that threatened to despoil a scenic and folk ideal at the very core of Maine identity. What kinds of work and industry, citizens were asked, were appropriate to this working coastal landscape? The issue, put bluntly, was not the possibility of "spill[ing] a little oil," but rather preserving a pastoral icon: "a water view ... is a refreshment to the heart. ... We do not want to stare at oil tanks, nor have them stare back."[8] Oregon coastal preservation, in contrast, was shaped by a pressing need to distinguish Oregon's character from that of its dynamic southern neighbor California. In the mythical founding of the Far West, emigrant trails led to an agrarian utopia — Oregon's fertile Willamette Valley — or to the quick fortunes hidden in the Sierra gold fields, precursor to modern California's volatile and hedonistic economy. Oregon cultural identity stressed this distinction, and when a burgeoning mid-1960s economy threatened to "Californicate" its coast, citizens mobilized.

Still, for all their differences, the coastal campaigns in Maine and Oregon followed remarkably similar life cycles. In both states threats from an easily identifiable "outside" despoiler invoked a fierce defense of these regional icons; in both states this controversy brought environmental rhetoric and organizing to its apogee; and in both states it forced officials to respond to popular activism with planning proposals that channeled this energy in safer directions.[9]

Concern for the coast was not unique to Maine and Oregon, of course. The early 1970s brought a rush of protective laws responding to the fact that these fragile areas were "under attack" from polluting factories, oil industry developers, and hotel, apartment, and second-home builders interested in filling in bays, harbors, and estuaries.[10] The battle to save the coasts, pursued by grassroots groups and conservation organizations and followed closely by journalists and state politicians, extended the reach of the environmental imagination from common riverine areas to broader natural landscapes.

MAINE'S WORKING COAST

In Maine, the dilemma of coastal development emerged during the administration of Democratic Governor Kenneth M. Curtis, who held office between 1967 and 1973. Like most Maine Democrats before environmentalism became polarized along party lines, Curtis was an enigma: "no governor since the late Percival P. Baxter has spoken out more forcefully ... on conservation matters," environmentalists acknowledged; yet like all postwar Democrats, Curtis was a strong advocate of state-sponsored economic programs in the New Deal tradition. Maine's poor were still unable to find decent jobs, and the exodus of rural youth to growth centers outside the state undermined Maine's future. Environmentalists, having cast the paper industry as an arch-villain in the fight to clean and save the rivers, now took aim at other corporations, challenging this liberal growth standard. Early in his administration Curtis was inclined to characterize environmentalists as preservationist-minded "summer people," who, he felt, made their money off industrial growth elsewhere, then helped block similar developments in Maine. "Their fall, winter and spring habitats have become unbearably foul and now they're worried about the[ir] summer nest."[11]

Rhetoric like this polarized Curtis's Democratic party. While labor, the core of the party's constituency, remained lukewarm on environmental issues, other Democrats saw their party as heir to the environmental vote. Curtis straddled this gulf by turning to statewide multiple-use land planning, beginning with a survey to identify those natural sites "that make Maine different from other states." Planning could ensure that these features would not be "trampled to death in the state's rush for development profits," and with selected nature icons encased in protective law, Maine could move ahead to "broaden its economic base without detracting from its environmental integrity."[12]

Coastal preservation provided a framework for testing this strategy. The controversy began in June 1968 when company officials announced a proposal to build a tanker port and refinery in the small town of Machiasport, near the eastern border of Maine. Later that year, additional port and refinery proposals surfaced for the state's heavily populated but still scenic Casco Bay, and in 1970 for Penobscot Bay, the state's most picturesque coast. These projects were motivated by new oil tanker technology in place by the late 1960s. East Coast ports could handle ships up to 60,000 deadweight tons, but this was only one-fifth the size of the new supertankers appearing on the sea lanes. Maine's natural deepwater harbors offered the only obvious alternative.[13]

Among the earliest advocates of the Machiasport project was the New England Council, a Boston-based regional chamber of commerce organized in 1925 by the six New England governors initially to deal with the textile depression. Speaking for the council in Portland in October 1968, A. Thomas Easley argued that Maine, by accepting an oil refinery on its coast, would begin to contribute its fair share to New England's overall economic development. An oil-poor, energy-intensive region, New England needed the port to attract independent oil producers, Easley asserted, and to break the pricing monopoly of the major oil companies. Thinking in regional terms, Easley dismissed the environmental hazards of a super-tanker oil port. Machiasport's remoteness recommended the proposal, he asserted; it could "handle the super-tankers ... with a minimum hazard to densely populated areas along the New England seacoast."

An obvious allusion to the recent Torrey Canyon supertanker disaster off the coast of Cornwall, England, Easley's statement was equally obvious in his willingness to sacrifice Maine's sparsely populated and politically impotent eastern coast to benefit the rest of New England. Easley closed by returning to the regional theme: "this project is the only real hope for a deep water port and refinery on the East Coast that will solve the high energy costs of our forgotten region."[14] Laying aside a bleak picture of regional victimization, Easley painted a vibrant future based on competitive energy prices and regional economic growth, all resulting from the Machiasport proposal.

Maine's metropolitan neighbors generated initial opposition to the Machiasport refinery. A pamphlet prepared by the Boston-based New England chapter of the Sierra Club identified the eastern Maine coast as a New England icon: a landscape of small farms and fishing villages that remained "as it must have been a hundred years ago." A Sierra Club member who rented "snuggle home cottage" near the proposed refinery site extolled the pastoral world of simple, weather-worn houses, rough and twisting roads, and wild, lonely beaches — a regional repository of freedom, authenticity, and permanence defined in recreational and aesthetic terms. After a joyous month exploring historical sites, beach walking in sun and fog, and watching lobster buoys bobbing in the waves, she concluded that Machiasport was "one of the finest unspoiled areas on the shrinking Maine Coast." This summer reverie provided dramatic backdrop for enumerating the "all too familiar ... problems" the refinery would bring. Shipping would be subject to enormous 20-foot tides and to storms, fog, and rocky shoals, ensuring a fate equal to or exceeding the Torrey Canyon disaster.[15]

Because Maine's coast was a working rural landscape, the Sierra Club recognized a need to provide alternatives to the economic stimulus refinery promoters pledged. Officials suggested that low-impact tourism would take up the slack, and for readers bothered by the irony of expanding the number of tourists by promoting the idea of solitude, the Sierra Club offered another proposal: federal subsidies to offset lost income, if the area accepted zoning and preservation controls. Towns would be compensated "for surrendering their right to develop." Those individuals predisposed not to stay and "work hard for relatively meager incomes" could simply "move away."[16]

With the oil issue framed in terms that assumed Maine's continuing colonial status, state officials entered the fray. Governor Curtis approved the Machiasport site for reasons similar to those offered by the New England Council: on the remote and economically depressed eastern coast, local political resistance would be minimal. But locating an oil port anywhere on the Maine coast that was not within the seascape view of at least a few summer residents would have been practically impossible, and Curtis's choice of Starboard Island in outer Machias Bay galvanized those residents.

Gardner Means, a consulting economist based in Washington, DC, was undoubtedly shocked by the announcement that his summer view might now include supertankers, pipelines, oil refinery stacks, tank farms, and related installations, but as a nonvoting outsider he realized that his protest would lack the ring of authenticity. Means initiated Maine's first local response to the oil proposal as an appointed member of Governor Curtis's own Council of Economic Advisors. In September 1968 the council sent the governor a list of 10 questions, drafted principally by Means. How much, the council asked, would development costs rise if the refinery were located inland, to minimize "the deleterious effects upon a particularly beautiful portion of the coastline?" Were the opportunities for ancillary industries — petrochemicals, shipbuilding, a container port — realistic? Had the effects of air and water pollution on clams, lobsters, fish, blueberries, tourism, and recreation been investigated adequately? The questions were not to be taken as a negative view, but rather as an opportunity for "thorough consideration" of a wider range of issues. Curtis responded by appointing a carefully selected ad hoc Conservation and Planning Committee on the Machiasport project, which determined that Mainers favored the oil port as long as no pollution would result.[17] Like Means, the governor voiced his position on the Maine coast with a great deal of circumspection, but even at this early stage the battle lines were clear.

With the Machiasport proposal stalled in the federal permitting process, a second supertanker port proposal, sponsored by King Resources Co., targeted southern Maine's densely populated Casco Bay region. Portland harbor was already the second-largest importer of crude oil on the East Coast, servicing, among other users, a major pipeline to Montreal's refinery complex. The additional facilities, to be located at an abandoned navy oil terminal on Casco Bay's Long Island, would make Portland the largest oil importer in the world. This prospect energized local opponents, who gathered over 2,000 signatures and organized Maine's first grassroots anti-oil organization under the name Citizens Who Care (CWC).[18]

Coming on the heels of the Machiasport controversy and a disastrous oil spill in California's Santa Barbara channel — a spill that "brought environmentalism into its own" in that state — the Portland oil terminal proposal quickly emerged as the most volatile issue in Maine politics. In 1969 the Natural Resources Council of Maine (NRCM) and the Maine Audubon Society (MAS) formed the Coastal Resources Action Committee (CRAC), a full-time lobbying organization led by Republican lawyer–lobbyist Horace Hildreth, son of the former governor, and his Democratic counterpart, Harold Pachios.[19] CRAC's founding underscored the professionalization of the environmental movement in Maine. Having operated in a world of powerful industrial and labor interests, Hildreth and Pachios were familiar with the process of drafting legislation, negotiating compromises, and working within the system of legislative politics. In the ensuing controversy, Maine's environmental organizations gained a remarkably professional veneer, yet unlike the outside organizations that initially defined the controversy, they remained sensitive to the perspectives of those living and working on the coast. Playing on these local concerns, the NRCM quoted oil promoter Robert Monks, who dismissed the threat to coastal fisheries offhandedly: if local fishermen could "run lobster boats, they can run [oil-spill] clean-up boats" as well. Highlighting Monks's apparent disdain for the local fishing community, the NRCM countered with the pastoral wisdom of lobster fisher Jasper Cates, who talked in meaningful terms about the impact of the refinery on "our livelihoods, our environment, and our way of life."[20]

Sensitive to the changing mood along the coast, the NRCM experiment- ed with a new theme gaining popularity in Oregon but only rarely articu- lated in Maine: a local aversion to in-migration. The predicted spinoff industries — petrochemicals, pulp and paper, aluminum, metal products, perhaps shipbuilding — would create more jobs than Maine's "current pop- ulation" could fill, the NRCM warned, meaning a "rapid migration into the area ... overwhelming both the natural environment and the way of life of the present inhabitants." Applying the "limits to growth" debate popular in national and international environmental literature, and unlike Boston promoters and preservationists, NRCM and CRAC learned to articulate their point of view in consonance with the fears, concerns, and ambitions of Maine's coastal residents. [21]

The message took hold. Oil promoters had argued time and again that tanker ports and refineries would allow Maine a "share in the nation's riches." Heavy industry meant a broader tax base for sewage treatment plants, schools, low-cost housing, and streets; it meant better connec- tions, more businesses, and above all more jobs. And yet along the east- ern coast, insularity and sense of place bred a certain apprehension about such changes, especially when initiated by outside promoters.[22] One downeaster worried that economic progress would bring "whorehouses and gambling casinos ... as they have in New York, New Jersey and other places where oil refineries are located." The intangibles of the develop- ment question — noise, traffic, pollution, crime, big government, corpo- rate domination, loss of community control — played on the minds of locals.[23]

Still, NRCM officials realized that they could not "sit back and relish the prospect of a region preserved like some curiosity in amber." Although they valued permanence as a landscape ideal, local environ- mentalists accepted a more dynamic view of the working coast. They offered to cooperate with state planners to develop economic alternatives more appropriate to maintaining the pastoral landscape. Promoting small businesses, perhaps on the model of rural Scotland's electronics industry or Canada's aquaculture experiments, they would transform Maine's fixation on growth by appending the word "clean" to industrial development.[24]

With Maine people closely tuned to the acrimonious controversy over two pending oil proposals, the issue moved inexorably to the state legislature. In 1969 Republicans and Democrats reached a consensus that pro-jected their sensitivity to both economic needs and pastoral values. Oil could come to Maine (the prodevelopment stance) as long as it did no harm to the environment (the anti-oil stance). To effect this balance, in 1969 the House of Representatives created an interim committee, chaired by liberal Republican Harrison Richardson, to study ways of ensuring the safe transportation of oil along the Maine coast. By unanimous vote, the bipartisan committee recommended a measure considered by *Newsweek* one of the "nation's strongest antipollution bills ever": a spill-abatement program lodged in the Environmental Improvement Commission (EIC) and funded by a tax on oil imported at Maine coastal terminals. "Outraged industrial lobbyists sputtered in disbelief," according to *Newsweek*, "especially now that Maine's deep-water ports ... have put the state on the verge of an oil bonanza."[25]

The coastal debate represented an important step in Maine's evolution as a postindustrial natural state. On the House floor, the debate generat-ed a profusion of bipartisan rhetoric proclaiming fidelity to the Maine coast. Intended largely as campaign fodder — approval was a foregone conclusion — the debate demonstrated the growing power of Maine's pastoral images over the legislative process. Richardson opened by point-ing out the national significance of Maine's coast. Economically, Maine was viewed as a "sort of ... weak sister of the continental United States," he conceded, but in this instance its underdevelopment offered an advan-tage: Maine could avoid the mistakes already made by its more industri-alized counterparts.[26] Democrat John Jalbert recalled the days when industry was "in complete control" of Maine's natural landscape — when fumes from the polluted rivers peeled paint from nearby buildings — and contrasted this with his pastoral vision of the unspoiled coast: "we have the most beautiful coastline in America, and a tourist business and fish-ing and lobstering business that will disappear unless we are able to place meaningful controls on a conveyancing of oil." Like Richardson, Jalbert sensed "the eyes of the nation ... upon us today." His speech was followed by several other tributes to the "rockbound coast of Maine, revered by people throughout the world," and the bill passed by the unusual vote of 134 to 1.[27]

Legislation enacted by a vote of 134 to 1 contains a variety of unstated compromises. For his part, Governor Curtis had gone from labeling the

NRCM a "bunch of conservative Republicans [who] ... like Maine just fine the way it is" to active cooperation with the newly professionalized environmental establishment. In fact, by the time the bill passed, all sides had reached agreement. Curtis published statements supporting the legislation, oil and environmental lobbyists joined the negotiations, and committee hearings on the bill generated widespread public support.[28]

While backroom negotiations probably explain the nearly unanimous vote on the Oil Conveyancing Act, this consensus was remarkable given the emotional tenor of the process. The debate and the vote expressed a longstanding frustration with Maine's status as New England's poor country cousin. Mainers of all kinds were beginning to savor the importance of their coast as a deepwater resource — and as a pastoral sanctuary. Big industry, urban New England, and the nation at large seemed to need Maine. The state's nineteenth-century motto, *Dirigo* ("I Lead"), assumed renewed significance as the legislative special session garnered nationwide media attention. Rep. Jalbert's remark about the "eyes of the nation" on Maine articulated an emotional sensitivity entirely new to the chronically depressed state. According to *The New York Times*, "the magnitude of support for these imaginative conservation measures ... demonstrates not only the good sense of the people of Maine, but the power of a good idea whose time has finally come."[29]

The Oil Conveyancing Act was accompanied by another bill that allowed the EIC to regulate the siting of large industrial developments and residential subdivisions. The commission would approve a project only if the developer could ensure that it would fit "harmoniously into the existing natural environment," and that it would not "adversely affect existing uses, scenic character, natural resources, or property values in the municipality or in adjoining municipalities."[30] Endorsed by a two-to-one margin, the Site Location of Development bill attracted far less attention than its companion legislation, but because it mandated public hearings it would have a greater impact on popular environmental consciousness. In a mostly rural state unfamiliar with land-use planning but with a long tradition of participatory and contentious local town meetings, the law was tailor-made to raising grassroots environmental awareness.

A proposal by King Resources to build a supertanker port for Portland's Long Island first tested the law's political implications. A source of public controversy for over two years, the project had already polarized public opinion. EIC Chair Donaldson Koons predicted that the meetings would be a learning experience for everyone, and his expectations were more than borne out.[31] The hearings provide ample evidence of the extent to which the oil controversy had penetrated Maine consciousness and defined the coast as an essential ingredient in the Northeast's postindustrial culture. [32]

On a hot, muggy afternoon in May 1970, hundreds of Portland-area people gathered in a local high-school gymnasium for the first hearing under the new site location law. Problems with the public address system augured for a generally chaotic meeting: speakers shouted, the audience called for louder voices, and tempers rose — all familiar experiences to Mainers used to often raucous annual town meetings. CWC, by now a veteran grassroots fighter, spearheaded organized resistance, while a new Maine Citizens Oceanology Alliance, claiming 500 members, added support. A steady parade of homemakers, small-business owners, summer residents, fishers, and property owners kept the meeting lively through the long afternoon and evening. King Resources got "only two tentative and timid claps," while Koons had to gavel for order several times to halt applause for the opponents. Although he tried to confine debate to how the King pier would affect the environment, Koons was overwhelmed by the "mass of emotion" focused on the broader implications of the tanker port.

Representatives from King Resources began by observing that Portland Harbor was already a major oil port, with extensive tank farms lining its inner shores. The proposed supertanker pier, located in the outer harbor, would have the latest pollution safeguards. In deference to prevailing pastoral sensibilities, storage tanks and buildings would be painted "soft green and white ... to maintain the motif of an island village." Oil terminals, they asserted, were located at "some of the most prominent and elegant beaches throughout the world."[33] A group of Long Island residents presented a petition supporting the project and predicting that if Mainers rejected ideas like this, Portland's youth would "have to move to Philadelphia to find work in a refinery and come up here in summer."[34]

As the afternoon wore on, erudite witnesses and folk philosophers demonstrated varying abilities to express the feelings of the audience, which were clearly weighted against King Resources. These speakers raised concerns about the project's impact on residential property values, scenic integrity, and the Casco Bay quality of life, earning more or less sustained applause depending on the speaker's rhetorical flair. The final effect of the project, as local legislator Mary Payson put it, was akin to "dropping a somewhat overripe mackerel on the doorsteps of Falmouth [a wealthy Portland suburb]." A representative of KOO — Keep Oil Out — attempted to pass out oil-soaked postcards to the commissioners, who refused the gift.[35]

The hearing fully engaged the complex environmental constituency emerging from the oil-port issue. Town officials from Portland's suburbs focused on the lack of comprehensive planning for Casco Bay; fishers and boatbuilders expressed fears of more pollution damaging their industries and rendering seafood "completely unmarketable"; marine scientists spoke of aquaculture as a preferable economic alternative; and residents and recreationists worried about oil spills on their "white, white beaches." Robert Chute, representing the environmentalist State Biologists Association, drew attention to the rights of "non-human as well as human organisms."[36] Long-time resident Ellis O'Brien summarized the new concern over Maine's coastline when he argued that Maine was at a crossroads: "20 years ago, we would have had to jump like a fish at bait if somebody like King Resources [came to Maine] ... There was no alternative." Maine was still economically weak, but O'Brien and people like him had learned to embrace this lightly developed working landscape of small, traditional activities as a place "where people can get out of the cities and go to live." This pastoral scene, O'Brien concluded, was "inconsistent, to say the least ... with industrial development." Defending the bay from "arrogant industrialists," witnesses politicized the rustic metaphors at the core of Maine's emerging self-identity.[37]

The EIC voted against the Long Island project on grounds that it posed too great a risk to recreational assets in the Casco Bay region. Although the Maine Supreme Court eventually reversed the decision on procedural grounds, in the interim the promoter went bankrupt and the project was abandoned.[38] But although the controversy was at least partially decided by broader developments, it was a crucial moment for Maine's emerging environmental constituency. The tanker-port hearings crystalized the major themes of Maine's pastoral ideal: a sense of pride in its unique working rural landscape; a belief that Maine's gift to the nation was the sense of freedom, permanence, and authenticity this iconic landscape conveyed; a reminder of past injustices at the hands of outside interests; and a vision of change that could be directed away from the mistakes of America's industrial development. Such optimism was clearly new to Maine, and it energized people like Ellis O'Brien, spreading environmental concern well beyond the issue of an oil dock in Portland's backyard. Maine's traditional town-meeting dialogue had transformed into an environmental revival meeting.[39]

While the Machiasport and King Resources oil ports were under review, yet another independent oil promoter initiated a third proposal in 1971. New York Consolidated Edison originally created Fuel Desulphurization, Inc., to help its metropolitan generating facilities meet new federal air-quality standards. When outraged citizens in suburban New York blocked a refinery project on Long Island, they turned to Maine, hoping to convert their federal impact license into a project that would serve Boston-area utilities' needs for low-sulfur fuel. Reincorporated as Maine Clean Fuels, the company first approached South Portland, home of Maine's largest concentration of existing oil storage facilities. But King Resources had already spoiled the ground. When thousands of citizens signed petitions opposing the idea, South Portland officials reversed their initial positive response. The promoters looked for a site as yet unencumbered by Maine's new environmental consciousness.[40]

Politically, the company's second choice was even less astute. The upper Penobscot Bay town of Searsport was physically and economically suitable, with an existing industrial harbor and a rail connection, but the supertankers would have to pass through a section of the coast renowned for its inshore fisheries, its yachting, and its summer homes belonging to some of the nation's wealthiest families.[41] The area's summer people began a campaign that landed over 1,200 letters on the governor's desk. Intending to demonstrate that it, too, had public backing, Maine Clean Fuels staged a meeting in Searsport for local supporters. To enforce a locals-only rule, company officials posted uniformed guards at the entrance to the gymnasium, and state police with riot gear stood in reserve. While only about 200 people were allowed to enter the building, another 800 milled outside with placards and anti-oil literature. When local officials decided to open the gym to everyone, the moderator attempted to maintain decorum by redefining the meeting as an informational session. Although the protesters eventually quieted down, they had more than made their point; the event succeeded only in portraying the company as a political novice with villainous overtones.[42]

By the time the EIC held site-permit hearings in March 1971, 11 of the 15 towns along the lower Penobscot Bay had voted against the project. Determined to avoid the town-meeting–style badgering they had experienced at Portland, the commissioners established strict rules to control the qualifications and the testimony of the witnesses. Audience members were required to submit all questions in writing through the chair, and applause was prohibited. Commission Chair Donaldson Koons opened the hearings by warning the audience that all testimony was to be "factual."[43] Under these restraints, the hearing was reduced to a string of "long, tedious, and repetitious testimony"; initial attendance of over 700 persons dwindled to about 200 per day. While the opening scenes of the drama had been vociferous and demonstrative, the middle scenes were more earnest and determined, dominated by a professional presentation style.[44]

During the hearings, the project's supporters challenged the pastoral images raised by preservationists. A Maine Labor Council official reiterated the state's traditional concern for jobs: "many argue that Maine's scenic coastline will be ruined by the refinery. But what does the presence of substandard housing and tar-paper shacks do to enhance our coastline? How do people who are poorly attired, undernourished, and ravaged by ill health contribute to a scenic view?" Branding preservationists as wealthy elites was a common discourse in the debate. Columnist Donald Hansen dismissed Penobscot Bay as a "private domain" for the privileged: to struggling mill-town residents, "all this pious talk about saving the Maine Coast must seem so much baloney when you can't get to it."[45] There was some substance to the argument that wealthy summer visitors dominated the anti-oil campaign, and that their concerns were more private than public, but in fact opposition cut across Maine's heterogeneous coastal population, which included fishers, farmers, small-business owners, and mill, factory, and service workers. Local organizers gathered 23,315 signatures on a petition opposing the project, easily outnumbering those submitted by oil proponents. Notwithstanding the commissioners' disclaimers about a popular plebiscite, this kind of grassroots opposition weighed heavily in the debate.[46]

The most convincing arguments against the refinery, in fact, came from ordinary citizens whose considerable public claim to authority and authenticity lay in their lifelong working relationship to the coast — the "folk," in traditional pastoral terms. With an air of experience and ownership, Ossie Beal, president of the Maine Lobstermen's Association, focused on the threats to his own livelihood, demonstrating a stern self-interest and local understanding that quickly became a standard for effective argument in statewide press coverage. "We who have spent our lives on the coast of Maine are familiar with its strong tides, its heavy fog, its rocky shoals and severe storms, [and] believe it to be one of the riskiest places to handle oil. For this reason we believe that spills are inevitable."[47] Adding to this down-home testimony were several Maine coast retirees with experience in more industrialized settings, who drew compelling contrasts between the pastoral dream and the industrial nightmare: "it can only be the purest wishful thinking to believe that through some mysterious alchemy a refinery and oil port can exist without the attendant industrial, obnoxious atmosphere that is an integral part [of] ... refinery technology."[48] These opponents convincingly grafted aesthetic and recreational arguments for preservation onto traditional pastoral images of the working rural landscape.

The Searsport hearings closed after eight days. The Environmental Improvement Commission found that the promoter had failed to present adequate evidence on more than 10 key points and denied the application. Most important, the company had not demonstrated that an oil complex could fit "harmoniously" into the environment and the existing occupational and recreational uses of the Penobscot Bay region.[49] Using pastoral ideals to bridge the gap between locals and summer people, preservationists won the day.

The public testimony at Searsport represented the high-water mark of popular environmentalism in Maine. Never again would the issues seem so clear, and never again would they be so simply stated as during those eight days. In the political debates that followed, the hard-won consensus began to break down, particularly along ecological lines.

After Searsport, Governor Curtis shifted the focus of the oil controversy from emotional witnessing to rational planning. As long as the siting process depended on oil promoters and their critics, he felt, Maine was open to an endless round of hearings like those at Portland and Searsport, and in fact, the momentum against oil might become politically irreversible.[50] To counter this, in November 1971 Curtis appointed 22 people to a Task Force on Energy, Heavy Industry, and the Maine Coast. In his instructions, he pointed to conflicting pressures from those who looked to the coast "as a source of jobs in a time of high unemployment, recreation and solitude in a time of urban unrest, marine resources in a time of worrisome food projections, energy during an energy shortage, tax dollars to relieve unfair property taxes, and profit in a declining economy." The task force was to address these conflicts by categorizing potential development sites, sparing the state the "costs and confusions of continual heavy industrial speculation over the whole length of the Maine Coast." Planning would provide the overarching political authority that had eluded the administration during the siting hearings.[51]

As the task force was completing its eight-month review of the coastal situation, the oil tanker *Tamano* ran aground off Portland harbor, creating the most serious oil spill in Maine history. While this accident seemed prescient to oil opponents, the task force brushed it aside and recommended Portland as Maine's single oil port. The compromise was clear: Maine would preserve its coast by sacrificing Portland, already the state's most industrialized harbor.[52]

Although calculated to bring opposite sides together, the report succeeded only in dividing coastal residents; some, stressing broader ecological threats to the Gulf of Maine, opposed all development; others worried that a lack of tanker ports would isolate Maine from the economic mainstream, like a "giant state park" — a prospect that threatened the working-coast concept.[53] The planning approach to resolving the oil controversy also split the environmental movement even as environmentalism was achieving its broadest popular support in Maine. John Cole, the outspoken editor of the *Maine Times*, accepted the premise that oil development — somewhere in Maine — was "inevitable." Maine's working coast, he felt, could incorporate industrial development if it was carefully contained. CRAC cofounder Horace Hildreth disagreed. These divisions became clear in May 1972 when the Army Corps of Engineers held hearings in Portland as part of a general survey of East Coast deep-water ports. CWC, the Maine Audubon Society, and the Sierra Club opposed designating Portland as Maine's oil port; NRCM and CRAC were conspicuously absent. When legislators introduced a bill implementing the task force recommendations in 1973, environmentalists categorically opposed to oil anywhere on the coast united with industrialists categorically opposed to state attempts to limit private oil initiatives, and together they defeated the bill.[54]

With central planning temporarily out of the picture after 1973, new oil initiatives appeared, often with environmentalists on both sides of the issue. With strong backing from Governor Curtis, Harrison Richardson, NRCM, and former CRAC lobbyist Harold Pachios, the independent Gibbs Oil Co. proposed a pipeline to carry crude oil from Portland Harbor to an inland refinery at Sanford. Frustrated by the ecological objections raised by opponents, Pachios stressed the lack of social vision among environmentalists, who were, he felt, typically "well off, and they don't want change." After lengthy siting hearings and years of effort, the company abandoned the project when oil prices declined in the years following the end of the Arab oil embargo.[55]

With the coastal planning concept in shambles, Maine received one last oil proposal. Unlike the others, the Pittston Company plan for a refinery at Eastport, near the Canadian border, seemed doomed from the start. It called for construction of one of nation's largest refineries at the end of a narrow coastal passage along a fog-shrouded coast renowned for its huge tides and swift currents. Head Harbour Passage, moreover, lay in Canadian waters, and the Canadian government steadfastly refused to jeopardize its local fisheries. And Pittston's lack of corporate responsibility had become legendary after a dam at a Pittston coal-cleaning plant in West Virginia ruptured in 1972, burying the town of Buffalo Creek in mud and leaving 125 dead and 4,000 homeless. A $205 million lawsuit hung over the company while it negotiated for the oil refinery in Eastport.[56] The company's federal applications were equally suspect. Its air-quality modeling, for example, lacked local meteorological data; in the event of a refinery spill Pittston simply planned to ignite the oil, an idea those familiar with the region's notoriously heavy weather found appalling. And finally, because the site was near Roosevelt International Park on Campobello Island, the EPA permit came with stringent air-emissions requirements, and the area was home to several endangered marine species and nesting bald eagles.[57]

Yet local resistance played a decisive role in the long Pittston controversy, confirming a shift in values even in the hard-pressed communities of eastern Maine. A location too remote to attract influential "summer people," Eastport's fate seemed to hang on local support or opposition to the project and its promise of jobs and prosperity.[58] Throughout the controversy federal officials orchestrated the public hearings as dull renditions of project history and permit processes, giving locals the impression that the decision was in the hands of "the bureaucrats, not the people." Despite this, and a 20 percent local unemployment rate, opponents continued to turn out for state and federal hearings, becoming even more vocal and strident in their anticorporate overtones as the controversy dragged on. At a 1975 meeting before state officials, nearly a third of the 150 who attended rose to speak, all but a few opposing the project.[59]

Given the unending round of permit hearings and the dogged resistance in Eastport, the issue faded from public view when oil prices stabilized in the late 1970s. With the defeat of King Resources, Maine Clean Fuels, and eventually Pittston, it appeared that the Maine coast was free of impending threat. There was, as journalist Richard Saltonstall pointed out, always the possibility that sooner or later the quandary of underdevelopment would drive home the logic of oil-port development, but, in fact, the victory over the oil promoters was part of a nationwide shift in thinking about the costs and benefits of energy development. Opinion expressed in controversies over offshore oil and gas leases on the Atlantic and Pacific coasts, the TVA's Tellico Dam on the Little Tennessee, the Seabrook nuclear energy plant in New Hampshire, the Kaiparowitz coal-fired generating plant in Utah, and the oil-shale proposals in Colorado correlated with Maine's shift in thinking about growth, energy, and quality of life.[60]

Reflecting on the events of the early 1970s, editor John Cole mused about the appearance of a "populist, top-to-bottom, rich-to-poor, right-to-left, universal constituency" aroused to defend Maine's pastoral landscape: "it's as if this corner of America, parked for two centuries in a geographic and economic backwater, has been banked all these years just so it could be here when Americans decided for the first time since the Industrial Revolution that there may be a better way."[61] The statement was noteworthy, given the long tradition of growth politics in Maine. By popularizing pastoralism as an alternative to industrial growth, the oil controversy served as midwife to an environmental movement. Henceforth, nature would be given due consideration in all manner of development proposals.

Maine had gained much in this passage. By the mid-1970s the state had passed landmark legislation that lodged veto power over heavy industry in the state planning apparatus. Perhaps more important, the battle for the coast drew national attention to the distinctiveness of Maine as a postindustrial sanctuary, and in the process advanced the environmental message in Maine and across the nation. Energized by public hearings and multiple threats to an iconic feature of regional culture, statewide organizations and grassroots groups gained confidence, experience, and an audience well beyond their initial upper-class base. This constituency — vacationers, hunters, anglers, hikers, backpackers, members of women's clubs and civic groups, PTA leaders, journalists, scientists, politicians, and the "many Mainers who live in urban areas but who have camps and cottages on the coast, [or] on the lakes" — remained the heart and soul of Maine environmentalism.[62] As Mainers defined for themselves the type of work appropriate to this mythic working landscape, the environmental imagination expanded to include newer concerns for recreational, aesthetic, and ecological integrity. A blend of newer and older landscape ideals, the environmental imagination emerged from the coastal campaigns as a popular political ideology.

OREGON'S RECREATIONAL BEACHES

Like Maine's granite coast, Oregon's sand beaches were central to civic identity. Pictorial representations of Oregon invariably include a long, lonely, windswept beach stretching between crashing surf and wild headlands. Here, too, threats to the coast brought a full flowering of the popular environmental imagination, and here too the concept of planning — separating the sacred from the profane, the pristine from the mutable — helped channel a burgeoning environmental consciousness in directions considered safer by public administrators. Two coastal controversies with statewide implications, each beginning in 1966, galvanized environmentalists in a manner similar to the threat of oil on Maine's rockbound coast. One focused on a scenic highway proposed for a northern beach; the other concerned appropriation of the dry-sand portions of Oregon's beaches by private developers. The two became intertwined in a larger debate over recreation, commercial development, and ecology that popularized the environmental imagination but at the same time, as in Maine, fractured the environmental movement.

Penobscot Bay, Maine

"[A] private domain ... all this pious talk about saving the Maine Coast must seem so much baloney when you can't get to it."

—Donald Hansen, columnist

Unlike Maine's coast, Oregon's 350 miles of ocean beaches were unambiguously public property, at least from the sea to the dry-sand, high-tide mark. When Oregon was admitted to the Union in 1859, the new state reserved its coastline for common use, and in 1899 Oregon declared most beaches "public highways" and withdrew them from sale. Under Governor Oswald West in 1913, the legislature expanded this principle to include all Oregon beaches, and in 1965 the legislature declared the shores "state recreation areas" managed by the Highway Commission. Oregonians used their Pacific shoreline for a variety of recreational purposes — clamming, fishing, swimming, driving, camping, picnicking, promenading — and viewed these pleasures as essential components of Oregon's livability.[63]

In Maine, the coastal controversy represented a fusion of older pastoral constructs of a working rural landscape and newer recreational values. Oregon's controversy focused on wild, solitary, and clearly nonindustrial scenery, although, as in Maine, the coast included several working ports. In the public mind, the issue in Oregon was not the nature of work in a working rural landscape but the boundary between commercial and natural recreation.

The 1965 state law had defined recreation broadly enough to include scenic highways, and it was just such a highway proposal that triggered Oregon's beach controversy. Business interests had long advocated straightening and widening U.S. Route 101 to provide better access to tourist points along the coast. In February 1965 the Highway Commission held hearings on a proposal to bypass a particularly tortuous section by constructing a four-lane extension across Nestucca spit, a natural area well within recreational reach of Willamette Valley cities. The hearings focused on four alternatives for some seven and a half miles of U.S. 101: two inland routes that cut through expensive development property and were thus bitterly opposed by real-estate brokers, another across the spit itself, and a fourth alternative to simply widen and straighten the existing highway.

The spit route would cut Pacific City in two, and the prospect of more traffic and the loss of a scenic strip of ocean beach appalled some locals, who formed Citizens to Save Our Sands (SOS).[64] Other coastal residents, however, supported the spit location. The publicity campaign quickly spread inland to the Willamette Valley, developing into a debate principally between Pacific City residents who supported highway relocation and a group of Willamette Valley "weekenders" who owned homes on the spit's north end. However, as with Maine's oil controversy, the dispute broadened over the next two years into a statewide debate over Oregon's most iconic natural feature. This contentious process helped transform the localized initiatives typical of mid-1960s environmentalism into a full-fledged statewide movement.[65]

Also like Maine's oil controversy, the Nestucca spit issue was charged with emotion from the start. A letter published in Portland's *Oregonian* formulated the issue in crisis terms, describing the Highway Commission as "a colossus rearing its ugly head to devour this beach ... its black, asphalt, serpent-like tongue ... desecrating ... the homes, the land, the dreams with a thick black crust." Willamette Valley resorters and local antidevelopment forces filed a lawsuit claiming that the land had already been committed to park purposes, while other groups supported highway improvement with their own publicity campaign.[66] In 1966 Democratic State Treasurer Robert Straub adopted the Nestucca spit issue as part of his environmentally charged gubernatorial campaign. On Mother's Day, he led 200 activists on a hike along the proposed highway route. Carrying signs, they marched four miles up the beach, built a roaring campfire, roasted hot dogs, and sang songs. On the return hike the protesters found their route "booby trapped" with signs reading "You are being misled" and a figure representing Straub hung in effigy. As Straub made light of the effigy, a man holding a gun appeared to threaten the crowd.[67] The heavily politicized Mother's Day march set the tone for a contentious debate over coastal preservation.

With SOS stepping up the campaign against the Highway Commission, a second controversy erupted about 50 miles up the coast in Cannon Beach, where the owner of the newly built Surfsand Motel outlined the dry-sand area in front of the property with drift-logs and signs to provide a "private beach" for his customers. This commercial expropriation of a popularly perceived public space immediately drew the attention of Cannon Beach residents and visitors. J. Richard Byrne, a Portland professor of mathematics, wrote a letter to Highway Commission Chair Glenn Jackson demanding clarification of public rights to the beach. Jackson's interpretation was that public rights included only the wet-sand areas below the summer high-tide line, but he launched an inquiry into whether the state could claim ownership of the dry-sand areas on the legal principle of adverse possession. Beaches in California were often fenced off at the high tide line, Jackson noted; to allow this in Oregon would "deprive us of our greatest recreational asset."[68] Beach-goers complained to Governor McCall that such incidents would "lead to the eventual defacing of all Oregon beaches and the take-over by commercial enterprises."[69] Like Nestucca Spit, the Surfsand Motel issue demonstrated Oregonians' emotional attachment to an important natural icon.

The Cannon Beach controversy highlighted other instances of commercial development along the coast, spurring the Highway Commission to complete its study of the beachfront property-line issue in time to submit a bill in the 1967 legislative session. The bill sought to confirm lawful public claim to the dry-sand portion of the beaches extending up to the "green vegetation line." Introducing the bill in the House Highway Committee, State Parks Director Loran Stewart predicted that failure to pass would result in more No Trespassing signs and more fences; he concluded by comparing Oregon's coast, "the finest beach recreation areas in the nation," to the "impossible ... situation in California," where commercial sprawl and developers' greed limited beach access. Lawmakers, the *Daily Astorian* observed, "report they have received more mail on the question than on any other issue — including tax relief."[70]

Since the bill created no new rights and threatened no adjacent property values, Stewart hoped it would pass without controversy. He was mistaken. An explosion of opposition resulted, providing a sobering impression of just how many landowners had development plans for the dry-sand areas. Opponents argued that the bill violated property rights and predicted a rash of lawsuits as the Highway Commission moved to claim the beaches.[71] Some ecologically minded environmentalists also opposed the bill, but for a very different reason: it did not include the "green vegetation" zone that was ecologically part of the living beach–and–dune system.

Despite widespread popular understanding that the dry-sand beaches were public recreational assets, the bill fared poorly in committee. Arguing that it would "make a playground" out of their "front yards," landowners made a surprisingly strong showing. Meanwhile, the owner of the Surfsand Motel planted grass west of his enclosure, "to show how stupid the [green] vegetation line is," and reinforced the log barrier by driving pilings into the beach. The sound of the pile-driver, as the *Daily Astorian* put it, "echo[ed] through the capital in Salem and up and down the coast and Willamette Valley."[72]

A vigorous grassroots letter campaign persuaded committee chair Sidney Bazett to schedule another public hearing on May 2, 1967. This time, the hearing was dominated by Willamette Valley citizens, who brought photos of more barricades and more posted areas. Some stressed ecological values, arguing that the dynamic dune areas above the green vegetation line were essential to the long-term stability of the beaches. As in Maine, the hearing touched off an unprecedented outpouring of public sentiment, articulated through letters, hearings, and media coverage. Bazett received 11,000 communications during the two weeks the bill was active; other legislators described similar volumes of mail.[73] The defense of public rights to the beaches helped transform Oregon environmental politics from a series of local controversies into a statewide dialogue on the Oregon way of life.

As the sandstorm of public opinion advanced across the landscape, Oregon's two leading politicians staked out their positions. Governor Tom McCall submitted a revised bill to establish the boundary of public ownership at a specific elevation above the high-tide line and announced that he would visit the coast with a group of oceanographers to "scientifically determine" the appropriate elevation. Not to be upstaged, State Treasurer Robert Straub took up the ecologists' position by suggesting a state bond issue to compensate landowners for the entire beachscape, including the dune areas above the green vegetation line. These political maneuvers reflected a subtle split in the environmental community, McCall representing the more traditional recreational perspective, with its emphasis on public access, and Straub adopting a newer ecological interest in the integrity of the entire shoreline ecosystem.[74]

With the various options airing in the press, McCall's weekend helicopter tour of the coast was well-timed to intensify the emotional drama. The tour started at the lavish Salishan Beach resort and ended at Cannon Beach, where protesters carrying signs urged a boycott of the Surfsands Motel. While the governor absorbed the controversy into his own environmental agenda, his oceanographers determined that an elevation line 16 feet above the mean high-tide line, marked by the upper limit of drift logs in many locations, was the most satisfactory means of delimiting the dry-sand areas subject to public recreational use. After another week of discussions, the committee reported out the revised bill using the 16-foot elevation line as a provisional definition of the public dry-sand area, pending a survey to establish a fixed permanent property line based on local conditions. The decision was predicated on recreational rather than ecological considerations. It was the intent of the committee, as a member explained, to "provide for an orderly development" of the beaches as recreation areas. "We do not wish to make our beaches into a wilderness; rather we want to encourage the public use and enjoyment thereof."[75]

In late May the beach bill passed the house by a 57–3 vote. Although all principal parties in the controversy agreed to the final bill, it was McCall, architect of the 16-foot line, who appeared to be the winner in the battle of the beaches. The governor hailed the bill as the first great environmental victory of his administration and predicted it would be "one of the most far-reaching measures of its kind enacted by any legislative body in the nation." Others took credit as well. A Portland television station complimented the 30,000 citizens who "answered the alarm with letters, telegrams and telephone calls" and lauded its own diligence in insisting on a strong measure.[76]

Yet the bill, as revised at McCall's request, was in fact a compromise intended as much to define beachfront landowners' development rights as to establish public access to dry-sand areas. The bill's enactment was only the opening round of an extended and increasingly complicated controversy over the future of the Oregon Coast. Indeed, Jackson supported the beach bill partly as a means of limiting public recreational interest to the dry-sand areas, thus clearing the way for highway development in the vegetated zones. On July 7, 1967, the day after Governor McCall signed the beach bill, Jackson announced that the state would proceed with the Nestucca spit highway relocation.[77] But since public opinion was running heavily against the relocation, McCall, vowing to let popular sentiment determine the final decision, ordered Jackson to schedule another public hearing on the proposal.

Like Maine's site-permit hearings, the decision to open the Nestucca spit to public input galvanized Oregon environmental constituencies. An August meeting in Portland, attended by 700 activists, led to the formation of a new grassroots Committee to Save the Beaches (CSB), soon affiliated with Straub.[78] Lining up alongside SOS, CSB, and other statewide environmental groups were Pacific City locals concerned about increased traffic and noise in their quiet village and business owners worried about losing tourist development opportunities. Paul Hill, president of a local chamber of commerce, complained about a "high-speed highway going between us and the recreation area."[79]

Although each of these groups subscribed to the idea of the beach as public recreational space, they were polarized around the two major competitors for the mantle of Oregon's environmental leader. Governor McCall, seeking support as an environmental moderate, favored a modified beach route. Lining up with McCall were representatives from the Bureau of Public Roads, coastal real estate and tourist associations, and various chambers of commerce. McCall's supporters saw the project as a complement to the beaches' recreational purposes: "as it stands now few children will ever enjoy that beach because of the difficult access." Robert Straub, marshaling the "Willamette Valley forces," opposed the beach route, adding an ecological perspective to the debate by raising concerns about the vegetated zone.[80] Straub's Willamette Valley activists gathered 12,500 signatures opposing the highway and unrolled their petition at a rally in Portland the night before the hearing.[81] The next day they accompanied representatives of SOS, the Federation of Western Outdoor Clubs, the Sierra Club, and the Isaac Walton League to the coastal city of Tillamook, where 300 people jammed the small Elks Club hall for the two-day hearings.

The Tillamook hearings, intensely emotional and well publicized, widened the gap between recreationists and ecologists from Portland and civic leaders and development advocates from the coast. Representatives of the 230-member Pacific City Boosters expressed their resentment of the Portland activists for "sticking their noses in our business." The Boosters' Jan Byerlee described the spit as "an unused side pocket" that, without the road, would become "a ghetto for drunken fiascos ... hippy societies ... and lovenests." Lack of beach access, she suggested, would imperil the coast "in the event of [foreign] invasion or a tidal wave." Another relocation supporter quoted federal statistics showing that the most popular recreation activity in the nation was auto touring. With only 19 percent of Route 101 offering an unrestricted ocean view, the Nestucca spit highway would "provide the driver a feeling of closeness with the ocean."[82]

It was Straub, the man hung in effigy by local highway advocates the preceding year, who developed a formula for bridging this gap. Straub's appearance as the hearing's first witness touched off vigorous applause from inland preservationists and a round of boos from coastal highway advocates. Attempting to draw together the concerns of valley recreationists, ecologists, and local economic boosters, Straub raised the specter of California sprawl — an image around which all Oregonians might unify. Drawing attention to California's ruined coastline, with its speeding traffic, overcrowding, and overpopulation, Straub admonished all Oregonians — ecologists, day users, coastal residents — to "plan ... for the future." Oregon could avoid the mistakes made in California by preserving its beaches as a source of recreation and scenic beauty and "as places where the awe and the majesty of nature, the force and the reach of the ocean can be observed and contemplated without distraction and danger." Using the formula that had worked so well on earlier issues, he drew attention to a widely perceived "outside" enemy of both recreational and ecological ideals — "a few individuals and short-sighted organizations ... [looking for] ... a quick dollar" — and he raised the California bogey to generate a sense of crisis. Environmentalists and local business leaders followed with similar arguments, and the CSB presented its 12,500-signature petition against the spit route.[83]

The Tillamook hearing proved pivotal in the Nestucca spit controversy. Governor McCall had ordered the hearing as a means of extricating himself from the issue, and Straub, SOS, and CSB mustered enough opposition to allow him to reverse his position. In December, winter storm swells washed over the spit, leaving the proposed route littered with logs. Nature confirmed the claim that a highway would require elevating the right-of-way, creating an "absurd ridge running down the beach." Nestucca spit eventually became a state park, named for its principal savior, Robert Straub. Meanwhile, at Cannon Beach owners of the Surfsand Hotel began a drawn-out and ultimately unsuccessful lawsuit attempting to restrict public beach access in front of their motel.[84]

The political theater of Straub's Mother's Day march, McCall's helicopter ride, and the Tillamook hearing, plus the legislative hearings and the sustained media attention, helped popularize environmental ideas in Oregon. In November 1967 the State Highway Commission, with widespread media support, used the 16-foot limit to deny a permit for a beach road at Neskowin. Likewise, a proposed 100-home development at Seaside "suddenly became a matter of public interest" in Portland when the press published stories and pictures indicating potential damage to the beach. Secretary of State Clay Myers was compelled to inspect the $1 million development personally to help calm the "storm of unwanted publicity."[85] Acknowledging the powerful dystopian visions popularized by Straub during the Tillamook hearings, House Speaker F.F. Montgomery of Eugene, campaigning for secretary of state in spring 1968, vowed that "he would be the first to lead the fight against any attempt by California carpetbaggers to erect a picket line of hot dog stands along our shoreline." Like the hearings on the Maine coast, the beach controversy sensitized the public to the fragile environmental underpinnings of Oregon's new identity.[86]

Yet the beach controversy, like coastal preservation in Maine, revealed the fractures in Oregon's emerging environmental movement. Trying to create a middle ground were politicians such as Governor McCall and Commissioner Jackson, who, like Kenneth Curtis of Maine, sought to maintain links to both business and preservationist camps. In the shifting sands of public perception, they chose to draw legal lines between iconic regional landmarks and developable economic resources. In Oregon, these lines preserved the dry-sand area as a recreational asset — an essential ingredient in the Oregon way of life — but they fell short of protecting the living beach, which would have included the fragile spits, dunes, estuaries, and adjacent wetlands that defined coastal zone ecology.

As ecological consciousness gained support, this oversight divided Oregon environmentalists into two opposing camps. In 1968 the CSB split into two factions. Incorporating as Citizens to Save Oregon Beaches, one faction supported a statewide initiative giving landowners a single year to establish that the public did not have prescriptive rights to vegetated sections of the beaches. The other faction, incorporated as Beaches Forever, began work on a bill to raise gasoline taxes to pay for beach acquisitions landward of the state's new 16-foot elevation line, particularly sand dunes and spits. According to projections, a cent-per-gallon tax increase over four years would provide $30 million, enough to guarantee preservation of Oregon's vegetated beach zones. Moving beyond a traditional public-access approach, the organization hoped to save the fragile ecosystem from development pressures up and down the coast. With Straub behind Beaches Forever and McCall backing Citizens to Save Oregon Beaches, Oregon's two most prominent environmental advocates stood, once again, toe to toe.[87]

Only Beaches Forever succeeded in putting its initiative on the state ballot. Labeled Measure 6, it became a major subject of controversy during the 1968 general election. With bumper stickers, brochures, lawn signs, and schoolchildren passing out small bags of sand, organizers directed the campaign at the state's inland population centers, principally in the Willamette Valley. Sporting the slogan "Beware of Tricks in No. 6," the Family Highway Protection Committee, with substantial funding from oil companies, focused voter attention on the proposed higher gasoline taxes. Measure 6 also drew fire from groups like the Oregon Automobile Association and the Portland Chamber of Commerce.[88] While Straub reintroduced the specter of California sprawl, McCall argued that state purchase of vegetated beach sections would create a "multimillion-dollar real estate bonanza" for beachfront developers, who were, as McCall had insisted for years, "mainly from California."[89]

Despite an energetic campaign and strong support in cities like Portland and Eugene, Measure 6 lost. Some political diagnosticians felt that voters simply rejected the gas tax as an improper means of resolving the dry-sand question. According to the *Daily Astorian*, the aim was "a good one, but the voters ... didn't approve of the method." Others pointed to heavy spending by oil companies, particularly Standard Oil of California. Significantly, as Straub pointed out, "not one single Oregon contributor" was listed among the Family Highway Protection Committee's contributors. In any case, the defeat demonstrated that popular appeals to Oregon livability did not necessarily include the more ambiguous notion of ecological integrity, particularly when ecological preservation challenged private ownership.[90]

Following the defeat, a legislative interim committee, working with the governor's office, collected data, conducted legal research, and drafted a bill defining a survey line along the entire coastline. The 1969 legislature enacted the bill, and the Oregon Supreme Court confirmed its legal basis, public prescriptive rights, later that year.[91] The shore below the green vegetation line was dedicated as a "perpetual public park" belonging to all Oregonians.[92] The Oregon beaches were indeed saved, and at no cost to the taxpayers, but the settlement neglected the upland portions of the coastal ecosystem. And with the boundary line established at the dry-sand line, developers wasted no time in constructing bigger and newer motels, condominiums, restaurants, summer homes, roads, and related facilities — right up to the new boundary. Like efforts to establish parks elsewhere, saving the beaches as a recreational amenity only spurred nearby resort development.

The Maine and Oregon controversies fit into a broader pattern of public concern for the nation's coastline. In 1969 the federal Stratton Commission report, *Our Nation and the Sea*, called attention to the rising pressures on every coastal area in the country. Population, industry, and recreational facilities were gravitating to the shores. Coastal states contained 75 percent of the nation's population, and development along this narrow edge of the continent proceeded at double the national growth rate. The report concluded that land-use pressure had outrun the ability of local government to address it and suggested federal grants to foster new state planning and zoning agencies.

Subsequent Congressional hearings revealed other pressures: suburbanization, sea-lane traffic congestion caused by new jumbo tankers and ocean carriers, thermal pollution from nuclear power plants, burgeoning marine-based recreational demands, aquiculture developments, offshore oil drilling, and industrial and municipal pollution. In 1972 Congress passed the Coastal Zone Management Act, designed to "preserve, protect, and develop" the nation's coastline. The act defined scenic, ecological, and historic "areas of critical environmental concern" and encouraged the protection of those areas "of more than local significance" with grants to states for surveying and regulating use of these sensitive landscape features. But as the coastal campaigns in Maine and Oregon suggest, support for this type of preservation was tentative. In the federal law, private property remained "practically sacred"; Congress left any remedies to local planning officials, most of whom saw their mandate largely in terms of increasing tax revenues through development. The solutions embodied in the Coastal Zone Management Act were far from sweeping, and few states took advantage of its provisions.[93]

Thus Maine and Oregon, along with California, Delaware, and Hawaii, pioneered the politics of coastal preservation, while other states launched campaigns to protect mountain landscapes, wetlands, sand dunes, and other iconic natural features important to regional identity. In some instances proponents made their case for preservation on ecological grounds, but Maine and Oregon suggest that the primary impulse behind these early environmental crusades was recreational and aesthetic.

Acutely sensitive to issues affecting regional self-identity, Americans launched the environmental era on the strength of an old and enduring tradition of nature appreciation, modernized to accommodate recreational, scenic, and limited ecological values. The cross-pressures in these campaigns explain the difficulties inherent in passing effective preservationist legislation. In Oregon, preservation arguments prevailed, but the tension between recreational and ecological values drove a wedge into the environmental constituency. Likewise in Maine, the quest for a stable, scenic working rural landscape framed environmental activism. Only a few witnesses at Machiasport, Portland, and Searsport suggested the need to limit the use of petroleum products. Nor could environmentalists agree on whose backyard to sacrifice to protect ecological integrity along the rest of the coast. By accepting the consumption and recreational patterns of the Age of Affluence, environmentalists achieved a remarkable level of popularity; when ecological factors intruded, the consensus became more difficult to sustain.

Nevertheless, the environmental imagination gained permanent political standing in these divisive campaigns to defend regional nature icons. Coastlines, mountains, and other distinctive landscape features provided powerful symbolic antidotes to the multiple problems of urban life, and with new highways providing ready access to the countryside, urban vacationers and local inhabitants asserted their stewardship over these valued places. In the fight to secure pastoral and wilderness icons, the environmental imagination moved into mainstream American political discourse, where it achieved a place as permanent as America's love affair with rural places.

Willamette Valley, Oregon

"And a lovelier place to experience the change in the Earth's cycle cannot be found than here in the Willamette Valley of Oregon. Please—for those of us who care so much for this beautiful place—let's keep all of this ...."

—Letter from Nadine Harrang, Eugene, Oregon, to Oregon Governor Robert Straub

CHAPTER 5

NAVIGATING THE NATURAL STATE: NATURE, RECREATION, AND RIVER CORRIDOR PLANNING

In May 1975, as the nation brought to a close another war and ended a decade of civil strife, Nadine Harrang wrote to newly elected Oregon Governor Robert Straub from her home near Eugene. "Today," she informed the governor,

> the sun is shining — the kind of a day that is achingly beautiful — with the fuzzy outlines of new buds on the tips of slim branches, and muted green-yellow leaves of the oaks beginning to fill in the spaces between the hardy gnarled branches. The spirit fairly exults, and optimism and happiness repels [sic] anxious thoughts of the recent past. It's Spring. And a lovelier place to experience the change in the Earth's cycle cannot be found than here in the Willamette Valley of Oregon. Please — for those of us who care so much for this beautiful place — let's keep all of this loveliness. I realize the pressures must be great, but I trust the greater wisdom of preserving this country will remain a constant factor in your administration.[1]

Like Robert P. Tristram Coffin's essay, written nearly three decades earlier, Harrang's letter continued a time-honored custom of admiring the beauty of place in rural America. Unlike Coffin, who paid tribute to the land and to the people who shaped it, Harrang focused entirely on nature. Written amid an outpouring of ecological concern and wilderness appreciation in the mid-1970s, her letter suggests both continuity and change in the environmental imagination.

In the late 1960s the romantic view of the countryside shifted from the textured village-and-field landscape to the wilderness on the fringes of this imaginative construct. In Oregon, the vast solitudes of the eastern desert, the Cascade peaks, and the wild, wind-whipped coast beckoned as sources of freedom and authenticity. Maine travel writers rediscovered the "virtually limitless forests" inland from the coastal working landscape, a deep woods so silent and so vast "that those of us who spend our lives in the cities may have difficulty adjusting." In an article in *Natural History*, photo-essayist Eliot Porter rhapsodized about his summer home on Great Spruce Head Island in Penobscot Bay, where time brought "few physical changes." Oblivious to the concerns Tristram Coffin had raised two decades earlier about wiping the folk from the landscape, Porter suggested designating the islands a vast national park, set aside in perpetuity for bird and plant life. "Lobstermen and fishermen would still enjoy the freedom of the bay, and sailors and yachtsmen would still visit the same harbors in their cruises," but the islands themselves would become ageless and uninhabited natural sanctuaries — "an experiment in non-management."[2]

The stewardship encouraged by Harrang and Porter was sustained by an environmental imagination different from the Arcadian world of Robert P. Tristram Coffin and H.L. Davis. The veneration of pure, uninhabited nature dates from mid eighteenth-century romantic writers like Henry David Thoreau, George Catlin, and James Fenimore Cooper, continuing with John Muir. It was politicized by wilderness advocates like Robert Marshall and Howard Zahniser, but its popular appeal as a landscape ideal grew out of several postwar social and cultural trends. In the early postwar years, rural and urban Americans shared an essentially similar perception of the countryside as a world of simple country homes and neatly tilled fields set in a context of forest, mountain, and meadow. Nature and folk were inseparable, and contemplating this enduring interaction from the outside renewed one's sense of freedom and authenticity.[3] Although pastoralists romanticized rural work as a link between city and countryside, over time this connection was attenuated. As historian Lloyd Irland explains, the "ability of the urban [world] ... to import its corn from Iowa, its lumber from Canada, its oil from Iraq, and its steel from Japan led its citizens to believe that they had been liberated from dependence on natural resources."[4] When this functional link to the hinterland severed, city people began to idealize the countryside in different ways: as a natural landscape with no essential economic — or pastoral — identity.

At the same time, rural work was changing. Postwar demand for wood products accelerated the mechanization of forest activities, and a combination of corporate agricultural suppliers and distributers and federal and university research programs drove farmers in the direction of large-scale commercial operations. New ways of logging, ranching, and farming — heavier machinery, more petrochemical applications, larger landholdings — transformed rural and small-town America. E.B. White wrote of the new Maine farmer as "part industrialist, part mechanic, part chemist." Gone was the agreeable mix of garden, orchard, and field crops, barnyard fowl, and family cow. Commuting to the city for work, rural folk learned to "harvest the long, bright, weedless rows at the chain store, bringing home a carton of tomatoes with eye appeal and a package of instant potatoes."[5] Loss of self-sufficiency brought loss of rural innocence, while in the urban mind the rural inhabitant no longer sustained the idea of the folk in the pastoral landscape.

Political trends also undermined the idea of the folk. In the 1950s the American liberal establishment was taken aback by a surge of populist McCarthyite anti-intellectualism, and its members exchanged their faith in the masses for a cult of science and social science. The folk ideal lingered in popular politics in the familiar persona of Presidents Truman and Eisenhower, but it disappeared from presidential imagery with the election of John F. Kennedy. And when the New Left abandoned the Old Left's alliance with labor in the late 1960s, this further diminished the folk as a cultural ingredient in the vision of American reform.[6]

The rise of ecology as a popular political expression also helped marginalize rural folk in the urban imagination. The success of Rachel Carson's early works — *Under the Sea-Wind*, *The Edge of the Sea*, and *The Sea Around Us* — was a measure of her literary skills, but it also reflected intense popular interest in nature studies. Carson's books, like the Teals'

Life and Death of the Salt Marsh, Sally Carrighar's *One Day on Beetle Rock,* and other popular natural histories, encouraged a view of humans as foreign to the natural landscape, and of human intervention as disruptive or contaminating.[7] This new ecological perspective also took on aspects of the youth counterculture, with its own holistic "natural" approach. In every aspect of private life — clothing, grooming, food consumption, recreation — countercultural youth embraced pure nature as a protest against the sterile artificiality of modern merchandising, modern technology, and the modern city.

In addition to these natural lifestyle themes, the neo-Malthusian "limits to growth" discussion, centered around Paul Ehrlich and later Donella H. Meadows, added to the popular understanding that human activity was alien from, and antagonistic to, nature. "Within decades, the human race could resemble a colony of marooned ants scurrying frantically on a burning log," wrote one journalist after reading Meadows's Club of Rome report.[8] Redefining the rural world as a delicately harmonized natural system with powerful redemptive overtones, these social trends brought to fruition a new vision of the countryside, still valued in terms of freedom, authenticity, and permanence, but no longer as a harmonious blend of human work and natural process. Economically distanced from their own hinterland, and with more leisure and mobility than ever before, urban people rediscovered nature through play, and given the sway of popular ecology and wilderness enthusiasm, the wilder and more solitary the natural context, the more powerfully authentic and liberating the recreational activity became. Uncoupling folk and nature had enormous implications for environmentalism, changing the way Americans thought about, experienced, and protected natural landscapes.

These trends helped forge the environmental imagination into a powerful political ideology, but they also introduced a series of contradictions into environmental politics. By the 1970s three distinctive perceptions of nature and place commingled in the environmental imagination. The first was recreational. Although this perspective embraced wilderness appreciation, in most cases it accepted a degree of commercial mediation providing recreational convenience. A second strain emphasized ecological integrity. Less popular than the recreational perspective, ecology nonetheless gave environmentalism a moral intensity and scientific affirmation that preservation issues had not enjoyed since the turn of the century.[9]

The third strain, traditional pastoralism, had been marginalized by this time in the mainstream environmental movement, but it remained an important element in the agrarian ideology of rural landholders and commercial resource users, who accepted human manipulation of nature as long as it meshed with the popular notion of balance and harmony. Logging, for instance, fit into this traditional pastoral scheme, but clear-cutting sometimes did not.

A battle over Oregon's French Pete Valley, a recreational destination for Eugene–Springfield residents, illustrates the interdynamics of these three strains of thought. A tributary of the McKenzie River, French Pete Creek is bounded by a 10-mile-wide valley in the middle elevations of the Cascade range. During the 1960s the Forest Service developed a multiple-use plan for the valley that involved backcountry hiking trails, protected habitats, campgrounds, and logging operations — a blend of traditional pastoral, ecological, and recreational considerations. But when the agency announced its first timber sale in early 1968, a wave of protest swept through Eugene and Portland. To popular ecologists, the forest, with its high canopy and open understory, was a sanctuary for natural process; to commercial loggers, the aging trees represented decay, insect damage, and declining growth. Lost in this scuffle, according to conservation writer Anthony Netboy, was a "silent majority ... who want increasing [recreational] access to the great scenic areas of our country, and who view both sides in the French Pete issue with disdain." Less interested in pristine nature, these recreationists saw the mountain forests as a familiar setting in which to experience "the thrill of breathing clean mountain air, spend nights in chalets under moonlit skies, and for a few days or so live in god's country, far from the noise and smog of the polluted lowlands."[10]

In November 1969, as environmental protest escalated into civil disobedience, Sen. Robert Packwood introduced a bill to prevent logging and create an "intermediate recreation area" with a few primitive walk-in campsites, leaving most of the valley pristine for the hiker and backpacker. The solution, which pleased neither loggers nor ecologists, stabilized a familiar landscape and incorporated most of the recreational uses local people had come to expect. But this balance became increasingly difficult to sustain. In the late 1970s, after a quarter-century of struggle and compromise, the French Pete drainage was reclassified as part of the Three Sisters Wilderness, assuming the mantle of purity some activists sought but excluding many of the traditional pursuits that had been part of the compromise forged by Sen. Packwood in 1969.[11]

Unresolved tensions like these were especially evident in the development of river corridors and river greenbelts, management plans designed to extend the environmental gains of the clean-waters movement to adjacent riparian lands. The assertion of ecological values and recreational rights along these riverbanks drew the environmental movement into a more complex political arena, because here terms like natural, ecological, and pastoral were less precise. Rural landowners — farmers, ranchers, residents, resorters — saw the river as a traditional pastoral world, and largely in private terms, whereas recreationists and ecologists saw the river corridors as a form of commons. Protecting river waters from corporate despoilers drew together all those who appreciated nature in its various forms; saving the river banks did not. River corridors tested the limits of reconciling the divergent strains in the environmental imagination.[12]

As in coastal preservation, activists and administrators turned to state and regional planning as a way to mediate these conflicting interests. An array of prescriptive land-use controls, planning commissions, zoning programs, and techniques for managing critical habitat, they hoped, would reorder riparian land use in ways that balanced pastoral, wilderness recreational, and ecological imperatives. But this approach carried inherent biases, as planning theorist Frank J. Popper points out. Planning emerged from a history of attempts by urban elites to use nuisance laws, height limits, building codes, residential zoning, and other regulatory tools to segregate urban economic activity and neighborhood makeup in their own interests. Imposing these urban constructs on the countryside typically ignored or undervalued rural land-use traditions, employment needs, and recreational habits. Rural people were therefore generally skeptical about planning, even though, as Popper says, they cared as much as urbanites for the land.[13]

River corridor management thus presented a complicated challenge. On behalf of recreationists, planning would have to guarantee public access to these common waters. In the interest of rural property owners and resource users, it would have to ensure that new preservationist laws did not compromise older local prerogatives. And finally, in the interest of ecology, planning would have to protect sensitive natural processes, and in doing so limit both mass recreation and rural resource use. The accommodations that resulted seldom worked to everyone's satisfaction.

In Maine and Oregon environmentalists employed these elements of compromise — controlled public access, traditional use rights, ecological stability — as they pushed beyond flowing waters to protect the lands along the banks. But as the divergent meanings of nature crystalized in the 1970s, compromises like Sen. Packwood's became more difficult to sustain. Environmentalists, a rural landowner complained during a controversy over the North Umpqua River in 1974, "prefer to live in cities, but want the rural areas open for their recreation." Conversely, a Eugene couple told Governor Tom McCall that "from the time of Teddy Roosevelt to the present ... those living closest to areas in need of preservation have been its most vocal and active foes." In Maine, the *Kennebec Journal* likewise commented on the "unimaginative" nature of people who lived all their lives in the Allagash country: "perhaps only those who have lived for long periods in large, dirty, noisy cities can appreciate the true worth of unspoiled wilderness."[14] These unresolved conflicts help explain the emergence of anti-environmentalist property-rights movements in the mid 1970s on one hand, and the withdrawal of many environmentalists into personal solutions on the other. River corridor management was an important environmental accomplishment, but it also demonstrates the difficulties of consolidating environmental gains in the 1970s.

URBAN RIVER COMMONS: THE WILLAMETTE GREENWAY

In Oregon, the shift from reclaiming rivers to preserving riparian land began in the late 1960s when pollution abatement measures began to take effect along the lower Willamette River. A legacy of pollution and seasonal flooding had left the banks free of development, and most roads were located well above the river to avoid the meanders. As dams stabilized the flow, the floodplain grew a thick screen of blackberry, cottonwood, bigleaf maple, and Douglas fir, shielding even the downtown stretches from signs of civilization. A wilderness by default, the Willamette was seemingly untouched by the valley's million and a half people. With the waters once again safe for recreation, a small group of visionaries laid plans for a 225-mile greenway along its banks, claiming the floodplain as a public commons.

The greenway idea began as a reconciliation between ecological process and human use — an almost utopian pastoral ideal. Like the later stages of the water-pollution fight, it was an attempt to realign city and country at a time when mills and factories were moving to the margins of urban consciousness. A resident of Linnton, a Portland suburb "virtually eliminated" by highway expansion in 1962, sent a request to the governor that typified greenway support throughout the valley. The state, he insisted, could "put something back into the town" by turning its river frontage into a park. A recreational and aesthetic focus — a place to "enjoy the river, walk at its edge, and wade in the water" — would restore Linnton's shattered sense of community, he reasoned. Unbridled recreational access to the naturalizing river became a measure of urban quality of life.[15]

The greenway idea is at least as old as Ebenezer Howard's 1903 "garden cities" concept of a quasi-rural belt of land separating community from community. It drew from Frederick Law Olmsted's "emerald necklace" plan for Boston's park system, and it was incorporated into the motor parkway concept in the auto age.[16] Oregon's innovation, formulated in 1966 by University of Oregon Dean of Students Karl Onthank, was to link the greenway idea to recreational use of the recovering urban river. State Treasurer Robert Straub adopted the concept during his first gubernatorial campaign, calling for state acquisition of nonagricultural riparian lands between Eugene and Wilsonville, just south of Portland. Tom McCall, who defeated Straub in 1966 and again 1970, endorsed the proposal in his first inaugural address. Almost immediately after his election, McCall formed an advisory board to fashion a proposal, and without giving much thought to boundaries, approaches to acquisition, or rules of access, the board filed a report and prepared a bill for the 1967 legislative session. By this time, however, farmers and other riparian landowners were expressing concern about eminent domain, so when the planning team began refining its strategies, it adopted what McCall deemed a "soft-sell approach."[17]

Introducing the bill in the legislature, McCall asked that acquisitions be made "in a manner that injures no one and benefits all." This unlikely scenario could be achieved, he thought, through "imaginative yet wise planning and the cooperation of both citizenry and government." To address the concerns of rural property owners, the legislature abandoned the vision of a 225-mile greenway in favor of a Willamette River park system that would provide intermittent public access by purchasing some

150 miles of river frontage on a willing-seller basis. The Parks Division would provide technical help and funding, but local governments, more accessible to property owners, would make the land acquisitions. The bill won approval by a heavy margin, and the federal government offered $1.6 million to match state and municipal funding.[18]

Unveiled with a great deal of publicity, the greenway plan was, even in its modified form, highly visionary. Interior Secretary Stewart Udall found this totally new concept in river management breathtaking: "I had the feeling of being swept away by a new, dynamic force in conservation as I reviewed Oregon's plan." Thousands of acres of blackberry bushes, bigleaf maple, cottonwood, and Douglas fir in the heart of Oregon's fastest-growing urban corridor would become a public commons, preserving ecological treasures and encouraging multiple forms of recreation, with meandering parks, hiking trails, wildlife refuges, picnic areas, scenic turnouts, historical attractions, and camping spots for boaters — all widely accessible for free-ranging communion with nature, in some cases within blocks of the heart of the city.[19] The greenbelt was also an experiment in localism in a state where the federal government held over 50 percent of the land area. "There is no question that the federal government could have taken it over," an official explained. "But if we have any faith in people — local people — to do what is right, then we have an obligation ... to give it an honest chance for success." Widespread popular enthusiasm for the greenway boosted the administration's confidence in local initiative, and city and county governments, despite the complicated procedures, embraced the program.[20]

The enthusiasm that launched the greenway campaign dissipated in the ensuing debate over implementation. Farmers worried that recreational access would encourage trespass, theft, vandalism, and litter, and that public support for the program would encourage governments to take land by eminent domain. "The closer people get to a farm the more problems we have." Greenway supporters were equally dissatisfied with the slow pace of acquisitions. Under McCall's soft-sell approach, they complained, protecting the 150 miles identified in the plan would take forever.[21] And in fact, with less than 10 miles appropriated by 1970, the Greenway Commission became convinced that the 1967 law was "totally inadequate" and scaled the plan back to five parks and a series of thin protective buffers.

In 1971 Governor McCall received an additional $5 million in federal aid and sent some 30 right-of-way agents out to contact riverbank property owners. This time the agents went armed with condemnation rights for enforcing eminent domain. Using these aggressive tools the commission forged ahead with the plan, and by 1972 the public held more than 65 miles of river bank. Typical was a 1971 request from the city of Milwaukie for greenbelt funds to purchase waterfront land unsuitable for development yet of ecological and recreational value. State planner Greg Morley walked the property and observed that it needed "considerable modification," but with proper grading, a footpath through an old highway tunnel, and a pleasure-boat moorage, it would become a natural complement to the urban landscape, screened off from traffic noise by a high bank and by a "fine stand of large cottonwood and Douglas-fir." Morley envisioned a trail linking the tract to a large city park a few blocks away. Such acquisitions reinvigorated the program but also fanned the fears of eminent domain.[22]

Intense lobbying and emotional confrontations between greenway supporters and riverfront property owners in the 1973 legislative session sparked another revision in the greenway plan. A multitude of compromises based on public meetings throughout the valley resulted in a plan that would place the riverbanks under tightened zoning restrictions and limit the expansion of existing agricultural, commercial, and industrial land uses, with a generous provision for flexibility. The greenway itself would be confined to a narrow band of scenic easements, and officials would use condemnation proceedings only for the five state parks.[23] Farmers remained uneasy, but they were more comfortable with the new version.

With greenbelt diplomacy shifting from state acquisition to land-use stabilization through zoning regulation, the state commission began preparing a master plan by classifying existing riverbank uses in categories ranging from pristine to heavy industry. Again, however, the program languished. The Department of Transportation (DOT), in charge of the state's park system, found only two willing sellers of scenic easements among the 600 property owners it contacted between 1973 and 1975. One Corvallis-area farmer demanded $100 a month for a guarantee not to cut his riverfront trees; when the state refused the offer, he felled them. DOT, even with $5 million in federal funds at its disposal, was helpless to stop the land clearing.[24]

When Robert Straub, the greenway's major proponent, was elected governor in 1974, the program was at a standstill. In a private letter McCall
cautioned Straub to avoid changing the greenway law until the zoning plan
had been given a chance to work. Riparian owners, McCall pointed out,
were highly sensitive to condemnation; with a "sharp shift in the game,"
he warned, "there is every likelihood there would be an uprising of the
farming community." Straub agreed to "hold back the Greenway plan for
a short time to permit a testing of the water," but his more elaborate vision
beckoned, and the new governor sensed that a majority of the public
would support a stronger proposal.[25] A 1970 Oregon poll had indicated
75 percent support for the greenway, "friends" of various rivers were now
taking up similar causes across the state, and even elementary-school children were inundating the governor with letters about pollution and its
effect on "our friend the earth."[26]

After testing the waters for only two months, in March 1975 Straub
delivered a detailed message to the legislature outlining the changes he
thought necessary to restart the program. He acknowledged the landowners' concerns but reminded them that the valley population was projected
to grow by another million people over the next 20 years. "Already, people
are beginning to use the 85 percent of the Willamette River now in private
ownership, with or without permission." To speed up acquisitions, the
governor reintroduced the controversial condemnation issue, seeking eminent domain for a 500-foot strip along each riverbank, excluding farmland
and other developed land.[27] To assuage farmers' fears, House Speaker Phil
Land offered a series of amendments changing condemnation to a state
right of first refusal on property proposed for sale. Straub supported the
compromise, but wary riverfront owners lobbied against the measure and
the bill suffered a narrow defeat. Accepting the limits of public vision and
the fears of the property owners, Straub agreed to implement the plan
according to the 1973 guidelines, mostly through the zoning restrictions
with generous allowance for preexisting nonconforming uses.[28]

The legislative defeat was difficult for proponents to understand in light
of continued popular enthusiasm for the project. Some disappointed supporters embraced a lone-villain story centered on an unnamed state highway
employee who years earlier supposedly contacted farmers and convinced
them, in "heavy-handed fashion," that the state was about to take their property. A greenway supporter concluded that Oregon would be "paying for that

for a long time." Ken Johnson, one of the architects of the greenway, dismissed opponents as "vested interests" who simply wanted to "develop their land for profit." He viewed the greenway essentially as he viewed the clean-waters fight — as a struggle between the people and the profiteers.[29]

But Johnson and others like him failed to read the subtleties among the opposition. In fact, not everyone shared Straub's view of the river as an ecological and recreational commons. Farmers and other landowners saw the riparian landscape in traditional pastoral terms, as a natural place made better by ongoing human intercession. Farming, they insisted, preserved the banks "in the natural state that God and nature intended."[30] Where most urban environmentalists considered recreation the highest form of riparian land use, rural landowners typically saw recreation as essentially degrading. The greenbelt would bring "people pollution": beer cans, gum wrappers, dog dung, and other forms of urban clutter that clouded their vision of the pastoral order. Farmlands were "well kept"; if they were made public, "litter and destruction ... would result."[31] Some environmentalists, more worried about riparian ecology than recreational access, agreed. One criticized the park proposal, for instance, as an attempt to make "nice, clean rolling lawns of our river banks."[32]

Another element in this antigreenway coalition was the wealthy estate owner who saw riparian lands not as a public legacy but as a private sanctuary *from* the public. In a stinging letter of protest against Straub's initiative, ice-cream magnate Robert E. Farrell rejected the vision of the riverbanks as a commons. He was an "Eastern boy," as he put it, who moved west "with a dream to get away from the masses." Having escaped the twin scourges of the East — discarded beer cans and overbearing bureaucrats — he voiced the concerns of those who romanticized Oregon as a place to carve out private estates in the tradition of the western cattle baron. The idea of unlimited recreational access to a redemptive natural landscape struck Farrell as a dystopic "eastern" intrusion. "I have seen the areas that were kept up by private citizens like myself slowly deteriorate into a slum district ... because the great politicians thought they were going to remake the American dream and put everyone in the same boat, regardless of race, color, creed or economic ability."[33]

Although Straub assured distraught citizens that developed property would be "left alone," suspicions were deeply rooted.[34] "How about us ... having a hiking trail through your back pasture," farmers raged. "We might even pick some berries from Mrs. Straub's garden."[35] A West Linn landowner wrote that he had chanced upon a map of the proposed greenway system while in the local library: "much to my horror, I saw our home and river bank being eaten up by a park and a bike trail!" This sense of anxiety mixed with a host of specific reservations about the greenway. Some predicted a loss of agricultural jobs; others objected to purchasing floodplains that were unsuitable for development anyway. Many viewed cities and wilderness as a poor mix, reflecting a lingering impression of unimproved lands as dangerous spaces that attracted squatters, muggers, and rapists. Fields abandoned to the greenway were sure to grow back to bramble, making them useful only to "the few who ... are not allergic to poison oak, want a place to take their pets to run, and for murderers' hideouts."[36]

The most frustrating aspect of the greenway struggle was the fact that condemnation powers in the 1973 law were too weak to stop farmers from cutting trees along the river but real enough to encourage them to make their lands less attractive for greenway purposes. The dilemma mirrored the fight for a redwood national park just over the border in California: federal attempts to establish the park served as a signal to lumber companies to slash stands of virgin timber in a calculated effort to undermine the rationale for the park.[37] In March 1975 William G. Pearcy, an outdoor enthusiast and greenway supporter, wrote to Straub complaining that greenway politics accelerated the destruction of riparian forests. "One of the saddest moments for me was when I returned to a large woodland of big Douglas fir, bigleaf maple, and cottonwoods to find that it had been all cleared. The owner was tempted to leave several large fir but decided they made it look 'too much like a park,' so down they came." Since most of the land along the river was undevelopable floodplain, in Pearcy's opinion it did not need additional protection. The greenway project was "an ingrained threat to farmers and landowners, and when their hackles are up, our trees come down."[38]

Others agreed that the greenway vision had backfired. Contemplating the specter of more litter, more trails, and more woodland destruction, Irma Jean Wood conveyed to Straub her conflicted feelings about preserving a sense of wild nature along the Willamette:

> Some of my most pleasurable and peaceful moments are to walk down to the river and sit on the bank. I get such a sense of being at one with nature. It is as though there were not another human for miles ... I share your dream of

wanting to preserve this for our children and grandchildren and of making it possible for all the people of Oregon to enjoy. However ... I have reluctantly come to the conclusion that the most beneficial thing we can do to protect this beauty that we love so, is to forget the whole Greenway concept.[39]

Many environmentalists, too, felt that they could best achieve the alternate rationale for the greenway — ecological stability — by scaling back the earlier dream of unbridled recreational access to the river commons. House Minority Leader Roger E. Martin considered the greenway an agreeable idea, but unlimited public use left him apprehensive. "Neither one of us wants the banks of the river to look like Coney Island on the 4th of July."[40] Planners failed to comprehend in these critiques the complex mix of personal stewardship ideals, fears of wild nature, and ecological concerns about unbridled public access.

After 1975 the battle lines changed from saving the Willamette *for* the people to saving the river *from* the people. Local officials encouraged this shift by relaying reports of trespass and vandalism to the governor. "Careful, detailed planning," Straub replied, would "show just where and how public recreational use can be provided without damage to the river resource or the landowners." Intensive recreation would be confined to urban sections, while in rural areas access would be limited to "the relatively few who will venture to ... hike 5 or 20 miles along the bank." Public access would be balanced against habitat protection and "encroachment upon private properties adjacent to Greenway lands."[41] Straub's planners urged a program to "educate the public to confine its physical activities to those locations set aside for such use," with signs delineating public from private lands. The greenway was necessary, planners now assured property owners, to keep the growing ranks of recreationists off their property. It would "decrease indiscriminate trespass," as would the dense screen of blackberry, alder, and poison oak that would grow up along the banks under the plan's zoning restrictions.[42]

The compromise that brought together these different concerns – recreational access, property rights, ecological integrity – demonstrated the flexibility of the greenway vision. But it also demonstrated its limitations. The greenway idea began in 1967 as a way of fulfilling the western ideal of frontierlike free movement through a primitive natural landscape. Facing a backlash from landowners, planners increasingly emphasized ecological sanctions that limited mass recreational use and provided a buffer between public and private land. As one planning document pointed out, ecology would "necessitate that recreation take the back seat and only be allowed in a very limited and specific sense." A powerful expression of the recreational impulse in the environmental movement, the greenway was also, ironically, a kind of containment policy for exploding recreational use, appealing thus to both landowners and ecologists. In fall 1976, after several more public hearings — many of them still hostile — DOT finalized its planning and management program for the greenway, and DOT officials began erecting fences and posting signs delineating recreational land, private land, and nature's own domain.[43]

Carefully balancing recreational, proprietorial, and ecological demands, planners moved the program ahead under piecemeal land purchases, mostly in undevelopable floodplain tracts. Over the next several years, local governments, in concert with DOT, acquired lands and developed parks and provided other waterfront recreational access, while the state zoning authority oversaw public and private developments all along the river. In 1978 Governor Victor Atiyeh, far less the visionary than Straub, disbanded the Greenway Committee as part of an effort to achieve a "less expensive government." The program, Parks Superintendent David G. Talbot pointed out, had come through difficult times, but the era of problems — and perhaps of visions — was over.[44]

Once an urban blight, the Willamette, the nation's first legally designated river greenway, became the pride of the valley. Public and private stewardship, here on the Willamette, and later on other rivers across the nation, helped sustain the pastoral vision of a green heart for America's postindustrial cityscapes. Yet for all its successes, the Willamette River greenway left unfulfilled the utopian vision of a recreational and ecological commons without fences. For those who clung to the original vision of unbridled use in the bosom of nature, the compromise meant abandoning the public trust. Ernest Drapela, a greenway advisor, denounced the concessions to landowners: "had our forefathers succumbed to every anti-effort, Oregon would not have protected our Pacific Ocean beaches for all generations to enjoy."[45]

Drapela's frustrations express the difficulty of extending the environmental imagination to the private lands back from the water's edge, and also the difficulty of superimposing urban visions of uninhabited nature on an inhabited rural landscape. A utopian expression of Oregon's environmentalist aspirations, the original greenway dream was dashed on the rocks of public controversy, its fragments patched together into a collection of intermittent parks, easements, water access sites, and zoning regulations. A tribute to the popularity of environmental ideas, it was also a tribute to the growing divisions between private, ecological, and recreational components in its vision.

In Maine, where the pace of growth was slower and the political culture less aggressive, these divisions proved equally difficult to bridge.

URBAN RIVER COMMONS: THE KENNEBEC CORRIDOR

In Maine, the event that signaled the shift from protecting rivers to preserving riparian landscapes was a lawsuit that challenged a 200-year-old tradition of annual log drives on the rivers. As the heirs of this tradition, paper companies choked Maine's major rivers with millions of cords of pulpwood each spring and early summer, closing off recreational fishing and boating possibilities. By the late 1960s it was also clear that decomposing bark and wood left in the streambeds degraded river ecology.[46]

In November 1970 aquatic biology student Howard Trotzky and Augusta lawyer Jon Lund sued the companies that used the Kennebec River for log drives. Maine's attorney general joined the suit, and with additional federal suits pending the companies capitulated. Probably changing economics helped decide the issue, since overland wood-hauling was becoming more efficient with better trucks and better woods roads, but the well-publicized court case challenged the deference Maine people paid to paper company giants and was thus an important political victory for environmentalists. It also cleared the way for a successful whitewater rafting industry, broadening the constituency for the new river commons.

What made the lawsuit a turning point in river protection, however, was the controversy it stirred within the environmental community.[47] The Sierra Club and the Kennebec Valley Conservation Association endorsed Trotzky's action, but officials from the Natural Resources Council of Maine (NRCM) and Maine Audubon Society (MAS) objected that the lawsuit was a threat to their "working relationship" with the paper companies. More importantly, they felt that ending the log drives prematurely would encourage uncontrolled resort and recreational development along the river.

Environmentalists needed time to plan strategies for keeping the river in its "natural state."[48]

Like the Willamette greenbelt proponents, Maine environmentalists turned to the idea of riparian corridors on the Kennebec and other Maine rivers to protect recreational access, ensure ecological integrity, and sustain proprietary rights. Given the legacy of private ownership in Maine, the proposed river corridors were built on a somewhat different philosophy: the shoreline would remain largely in private hands, and existing land-use regulations would secure recreational rights and ecological integrity.[49] Reflecting Maine's minimalist political culture, the approach offered weaker protection than the Willamette Greenway, but advocates hoped it would be less divisive.

The Saco River, running through a relatively developed section of rural southern Maine and northern New Hampshire, was the first to receive protection under this corridor vision. Responding to growing recreational interest in the river, James G. Simonds of Center Conway, NH, hatched an imaginative scheme for a gigantic one-day cleanup to "sweep the Saco clear" of junk, scrap heaps, dumps, gravel pits, auto hulks, and trash.[50] On Save Our Saco day in July 1970, thousands of Boy and Girl scouts, camp owners, canoeists, and other valley residents flocked to the river with trash bags. Dam owners lowered the reservoirs, the National Guard used heavy equipment to cover over dump sites, and a Navy demolition team removed old pilings.

Building on this annual effort, citizen groups began lobbying for a protective corridor along the river, and in 1973 the state legislature created the Saco Corridor Commission, a regional body representing a variety of local interests and a "much needed common sense approach to land use control." For ecological reasons, the plan banned construction in the 100-year floodplain. For development purposes, it employed a flexible system of trade-offs between setback and frontage requirements: the closer a house was to the river, the wider the frontage requirement would be. This, commissioners felt, would avoid a repetitive, striplike housing pattern along the river without pricing the "average person" off the river by enforcing a large minimum lot size. And for recreational users, the plan offered a system of hiking, biking, and canoeing trails, public and private recreational facilities, scenic vistas, and water access points, coordinated by the state Bureau of Parks and Recreation.[51]

At about the same time, conservationists raised concerns about development on the Kennebec. In March 1970, with the Trotzky lawsuit pending, the NRCM helped form a Kennebec Valley Conservation Association, and former NRCM president Clinton B. Townsend inventoried the river's wilderness scenery, fishing holes, historic sites, critical habitat, access points, parks, and picnic areas. He also addressed the riverine area's liabilities: industrial intrusions, logging roads, dams, town dumps, pollution sources, gravel pits, and places where the "ubiquitous out-of-state land speculator" had already "reared his ugly head." With this inventory in hand, the association outlined an ambitious plan for preserving floodplains, slopes, and wetlands and guiding development to prevent "intermixing of incompatible uses."[52]

To encourage recreational use, the association recommended purchases for day-use parks, wildlife refuges, trails to waterfalls, boat launches, and canoe portages, but it ruled out larger land acquisitions; state legislators would be unwilling to bear the expense, and the corridor lacked the type of "nationally significant natural resources" that might warrant federal purchase. Behind this minimalist approach was a deeper logic related to a disconcerting rise in recreational use on the Allagash River after it received protective status. The association reasoned that large public purchases such as on the Allagash would generate too much media attention, inviting recreational use on a scale that would stress the riparian ecosystem.[53] Arriving early at the solution Willamette greenway planners eventually settled on, Townsend accepted existing property regimes as a buffer against uncontrolled public access. The Kennebec plan would reconcile ecological protection, private rights, and public access using an approach far less adventuresome, politically, than the Willamette greenway.

The upper corridor, owned in large parcels by paper companies, fell under existing wildland use rules that prevented development incompatible with wilderness values. In this remote and primitive area, logistical barriers, presumably, would prevent the crowds that plagued the Allagash Waterway. In urban sections along the lower river, planners made modest suggestions to expand public access to the river: a boardwalk along the gorge at Skowhegan, parks and walkways elsewhere, and a riverside park-and-trail system and restrictions on ridge-top development in the Waterville and Augusta areas.

Protecting the middle river, a mixed landscape of rural woodlands, farms, resort cabins, homes, and fields, was a tougher proposition under

Townsend's plan. Here, planners confronted small towns, multiple owners, diverse points of view, and entrenched local interests. Since much of this shore was unsuitable for building because of shallow soils over bedrock, few additional regulations, Townsend felt, were necessary. Dense forests and underbrush, steep cliffs, riparian wetlands, or private property would discourage mass recreation, limiting access to a few public launches. In many cases, recreational pressures would be managed "by what is probably the most democratic process, that of limiting ... access ... to those willing to make a certain amount of effort to reach it."[54] Townsend also recommended moderation in regulating privately owned land. The principal tools for limiting development would be voluntary conservation easements, municipal comprehensive plans, and existing state regulations.[55]

Townsend aired his Kennebec River plan publicly in 1971, drawing praise from those who saw it as a model for all Maine's major rivers. In the mid-1970s the corridor fell under jurisdiction of the Kennebec Regional Planning Commission, which turned it over to a citizens' Kennebec River Council composed of local business and conservation leaders. The council used the minimalist approach Townsend outlined in 1970 to map out a plan for conservation along the corridor. The group identified the various state agencies with potential authority over the corridor, and worked with conservation organizations to encourage landowners to consider voluntary management options such as land trusts, easements, mutual covenants, or deed restrictions.[56]

In 1975 the Southern Kennebec Valley Regional Planning Commission approved the river corridor plan. Citizens along the rural middle river, however, rejected it, despite a series of NRCM-sponsored conferences, workshops, tours, field trips, and planning sessions. Citizens living near the large reservoir behind the Kennebec's Wyman dam on the middle watershed expressed concern that a protective corridor would invite mass recreational use — "an invasion of out-of-staters and other persons" turning the reservoir "into a pleasure lake for all comers, with large boats and motors and other varieties of pollution." They preferred the region as it was, "a scenic spot with a minimum of activity." Responding to these anxieties about mass recreation, the council dropped plans for parks and public access to the reservoir. Nevertheless, the corridor idea continued to draw opposition.[57]

In 1977, after seven years of planning, meetings, and hearings, the council reported that the Kennebec River corridor was still far from achieving even minimal goals. The approach had worked best in the upper watershed, the council maintained, where large commercial timberland holdings simplified management strategies. In the rural middle sections, where ownership was fragmented, appeals to the conservation instincts of developers proved untenable in protecting the corridor commons. Since builders had no incentive for protecting the river against scenic degradation, they simply passed the social costs of development on to the public. A second line of defense, municipal planning agencies, proved equally ineffective. Local officials were better regarded by the public than those from state or federal agencies, but small-town officials — usually volunteers — lacked the experience, staff, motivation, and technical know-how to implement complex regulatory programs, and many distrusted "outside" professional advice or were reluctant to impose restrictions on fellow townspeople. Aware of the need to broaden the local tax base, these officials came under pressure to relax regulatory procedures.

The third strategy involved the multitude of existing federal, state, regional, and local regulations pertaining to various resources in the corridor. Since Maine already had planning mechanisms like shoreland zoning, subdivision laws, minimum lot-size laws, and industrial-site regulation, the commissioners saw little need for new regulations. But there were drawbacks to this approach as well. Land-use regulation was new in Maine, and landowners were sensitive about curbs on property rights and confused by the overlay of local, state, and federal regulations. Moreover, Maine's regulatory laws, although in some ways the most stringent in the nation, amounted to a collection rather than a system; overlapping jurisdictions discouraged comprehensive land-use planning. And finally, state agencies were too poorly staffed to conduct the onsite investigations needed to create the corridor, especially for a project on the periphery of their regulatory mandate.[58]

In 1981 a new and even less intrusive plan for the middle river failed passage, its opponents citing "over-regulation." New corridor regulations, they argued, were a moot point, given the existing corpus of environmental law, and even the plan's supporters admitted that there was "little need to establish a corridor commission to exercise powers that already reside with state and local governments."[59] The Kennebec Valley Conservation Association published a short recreational guide to point out attractions and access points and to demonstrate how to use the river "without harming it," but in general the regional commission and its Kennebec Council did little to highlight the river as a recreational resource. Commissioner John Szabo counseled against aggressive publicity; as the river was cleansed and the fish returned, recreational boating "would automatically return as well."[60] Indeed, designating a special protective corridor would generate unwanted recreational interest in this delicate riparian system. Thus on the Kennebec, as earlier on the Willamette, ecology trumped recreational access in resolving the tensions between private and public use of the river corridor.

As with the Willamette greenway, some saw the minimalist Kennebec corridor as a failure in environmental management. According to Eben Elwell, former chair of the Kennebec River Council, the approach allowed a complex of industrial, recreational, and commercial "special interests" to rule the river with significant veto power over the public will. But these multiple despoilers — recreationists, developers, transportation planners, dam builders, timberland owners — were less visible to the public than the mill owners who polluted the river a decade earlier. Thus protecting the river corridor failed to excite the popular imagination, and the question of preserving ecological integrity without inviting recreational abuse proved difficult to resolve. Nor was the sense of crisis all that clear, given relatively slow population growth in central Maine. The defense of the corridor would rest on the river's relative recreational obscurity and on the questionable premise that development pressures would remain low in this rural section. In 1989 the state created similar plans for all the major river basins in Maine.[61]

The Willamette planners began their campaign for a river commons by challenging rural property rights on behalf of recreational use and ecological integrity. They eventually compromised by stressing ecological integrity, thereby protecting both nature and property rights. Those in Maine began with the idea that property rights were paramount along the river and adjusted recreational goals and ecological protections to this existing regime. In the end, planners on both rivers achieved an awkward balance among ecological, recreational, and traditional pastoral demands, maintaining a fragile consensus as the environmental movement moved landward from the river commons.

RURAL RIVERS AND THE ILLUSION OF SOLITUDE

Controversy over rural and wild rivers demonstrates again the fissures that had developed in the environmental imagination by the 1970s. When Oregon's Rogue River was identified as a potential wild and scenic river in the mid-1960s, environmentalists sensed that the state should take similar measures to protect certain other rural rivers. Toward this end, state Rep. Don Wilner of Portland introduced a state scenic waterways proposal in the 1967 and 1969 legislatures. The measure failed, but Wilner and others collected signatures and launched a popular initiative, which passed in November 1970 by a two-to-one margin. Lands near designated portions of the Owyhee, Minam, John Day, Deschutes, Rogue, and Illinois rivers were classified into six use categories; dams were prohibited, and along a quarter-mile corridor back from bank only those activities compatible with specified scenic values were allowed. New buildings, for instance, were to be designed "in harmony with the natural beauty of the scenic waterway."[62] Management of these chosen rivers reflected the mix of recreational, pastoral, and ecological ideals that characterized the Willamette and Kennebec corridor proposals.

The scenic-rivers program was built on two premises: the public enjoyed prescriptive recreational rights to their rivers and the scenery around them; and ecological and scenic stability were key components of those rights. Recreation on a wild river, a state official noted, depended in good part on a feeling that "someone else, 50 years, or even 100 years, from now, will stand in the same place and be able to enjoy the same good fishing."[63] But ecologically minded environmentalists realized that the Scenic Rivers Act itself could undermine this stability by encouraging too much recreation. By declaring these rivers a commons, the act focused Oregon's burgeoning recreational traffic on a relatively few sections of the state's most remote and delicate river systems. As in Maine, at least some environmentalists worried that official designation carried the threat of excessive recreational use.

This point had implications for rural landowners as well. Given the invitation to public use, it was difficult to define boundaries for the new river commons. Recreationists, they found, assumed "the right to stop anywhere, camp anywhere and do anything they please." This distressed farmers and ranchers in remote sections of Oregon, who came face-to-face, often for the first time, with both urban recreationists and the state bureaucracy that supported their claims to a natural experience. "I sympathize," Governor Straub wrote to a colleague, "[these landowners] are a long way from Salem and light-years from Washington. Their perceptions of natural values and recreational needs are different from the urban proponents of river protection."[64] Glenn Jackson admitted that scenic designation would encourage public use, and this in turn would mean "litter, vandalism, demands to cross private lands, and increased disrespect for the rights of the property owners." Recognizing the contending ideas, he called attention to the need to protect the "adjacent property owners as well as the ecology of the river."[65]

Since neither ecological environmentalists nor pastoralist property owners were happy with Oregon's 1970 Scenic Rivers Act, in 1974 Highway Commission officials held hearings around the state to draw commentary on a more precisely defined plan for Oregon's scenic rivers. Although the original initiative had passed with a clear popular mandate, a new political climate, darkened by Watergate revelations and polarized by the Vietnam War, made preservation issues more volatile, especially where rapid residential growth and real-estate speculation pressed the bounds of weak local zoning ordinances.[66]

The hearings highlighted the growing tensions between urban and rural residents, the former generally concerned with scenic recreation and ecological integrity and the latter with property rights and a working rural landscape. Defending their commons, urban recreationists claimed the rivers by virtue of the "many happy hours ... spent in, and on, their waters," and used the hearings to encourage a shift from local to state decisionmaking in land-use issues.[67] Local boards and commissions, they felt, were simply tools of "greedy land developers," inadequate to protect Oregon's rivers. Levering control away from these purveyors of poor land-use decisions, vague zoning rules, and improper variances, the scenic-rivers program would lodge decisionmaking in the hands of state officials more accessible to urban constituents.[68]

Farmers and ranchers grounded their claims in an older pastoral vision. They saw the scenic waterways program as a scheme "dreamed up" by urban environmentalists and imposed on the countryside. Buildings that the ranch family lived in "365 days a year should not be forbidden because a river floater ... finds it to his dislike." At a deeper level, these

rural landowners defended their traditional multiple-purpose conservation ethic and the pastoral ideals upon which it rested. As Oregon's "true
conservationists," they sought to preserve the rivers as they "used to be,"
free of "people pollution" and litter.[69]

These differences were common in the nationwide debate over scenic
rivers. Although less likely to endorse the concept of an ecologically pristine river corridor, rural Americans prided themselves in their commitment to the land. Rural people, this testimony suggests, loved the land no
less than river recreationists, but they linked their ideals to a working rural
landscape rather than to recreational needs or ecological sensibilities. A
West Virginia representative, arguing against wild and scenic status for the
Shenandoah River, pointed out that West Virginia's rural inhabitants had
maintained the "conditions which the Act endeavors to restore or preserve" — farms, orchards, woods, and pastures — for over a hundred
years. A Wyoming ranch woman claimed that irrigated hay meadows, a
feature of the landscape for a half-century, expressed the "prudent uses of
all things that nature has given." The lush, green fields pleased the eye
more than "a dry, brown field through which runs a river lined with refuse
from campsites." Scenic designation and more water recreation would
only "disturb ... the delicate balance that we have worked rather hard to
arrive at." The ubiquity of such statements in congressional testimony
suggests a conservation ethic usually overlooked by urban environmentalists. As traditional pastoralists, rural people viewed the city as dystopic,
and felt that protective status would bring urban recreationists "carrying
off anything moveable; leaving gates open; cutting fences; and defacing
the land with all kinds of refuse."[70]

Oregon's scenic-rivers program failed to resolve the struggle over the
meaning of the countryside, but as on the Willamette and the Kennebec
administrators defused the issue when they struck a balance between
property rights and recreational use in the late 1970s. Over time, rural
owners and rural developers recognized that burgeoning recreational
traffic was inevitable, with or without scenic designation, as was
the accompanying litter, trespass, and negligence. During the later 1970s
a consensus emerged that the scenic-rivers program should "encourage
... but [also] control ... the development of recreational facilities and
related services."[71] Both the urban and rural public, Attorney General
Lee Johnson observed, endorsed the existing mix of private and public

"Households no longer plant gardens if they can avoid it.... Acting on advices from the TV screen, they harvest the long, bright, weedless rows at the chain store ..."
—E.B. White, writer

ownerships; both worried that rivers were being "nibbled to death" by the booming market for rural real estate, and both searched for ways to keep riparian land use stable. State officials responded to concerns about trespass, litter, and careless fires by designating public camping and picnicking sites to channel recreationists away from ecologically sensitive areas and keep them off private lands. Like the greenbelt, the scenic-rivers program balanced the concept of open access against the need to control recreational use in the interest of ecological stability and local landowners' sentiments.[72]

Regulating the largely public Rogue River corridor brought its own set of conflicting demands. Across the nation, whitewater recreation mushroomed as safer and more durable rafts, drift boats, kayaks, and canoes came on the market, and as traditional forms of commercial river guiding — highly personalized excursions with one or two wealthy anglers — gave way to large-scale sightseeing tours in fleets of rafts.[73] In 1972 more than 75,000 people rafted the Snake River, while annual traffic in the Grand Canyon leaped from a few hundred to 45,000. River runner Verne Huser, recalling better days on the Snake River when "you seldom saw another human being," complained in 1971 that the "almost steady stream of float trip traffic on the river" had to be carefully monitored to preserve the semblance of wilderness recreation. In 1970 the National Park Service placed a ceiling on the number of commercial operators on rivers in several western national parks and allowed access only on a special-permit basis. The Park Service and the Bureau of Land Management initiated detailed studies to determine the wilderness carrying capacity of river resources. In the Grand Canyon, the Park Service allotted 8 percent of its user days for private trips and divided the remaining 92 percent among commercial concessioners. Private users objected, commercial outfitters sued, and the Park Service became embroiled in debate over how to allocate wilderness carrying capacity.[74]

Corridor management on the Rogue River illustrates the impact of commercial use on these wild rivers. By the 1970s a veritable "boat parade" of downriver raft and drift-boat traffic confronted a fleet of jet boats churning upriver from Gold Beach. Jet-boat owners cleared and dredged the lower Rogue's bars and boulder fields using propeller wash, blasting, and bulldozing, steadily extending their reach into the wild canyons. In addition to rising commercial use, the river carried a growing burden of "drifting debris," as one official called them, people loose on the river in life rafts, inner tubes, and other floatable objects. In 1970, nearly 173,000 people visited the 84-mile section of the wild and scenic Rogue for angling, boating, hiking, and hunting.[75] If activity got any thicker, one journalist warned, it would be "necessary to install traffic signals."[76]

This rising commercial use altered the way Americans viewed their rivers as sources of freedom, authenticity, and permanence. A raft trip was neither a profession of environmental faith nor an act of preservation, but commercialization did introduce hundreds of thousands of Americans to nature, thereby broadening the constituency for environmental reform. Ironically, this rise in river traffic was accompanied by heightened popular perception of the river as a sanctuary from crowded urban environments. Civil strife and urban problems in the late 1960s accented the importance of nature as a personal sanctuary, and experiments with hallucinogenic drugs, eastern mysticism, counterculture living, and other forms of subjectivism reinforced the view of the natural, the primitive, and the unspoiled as a personal retreat from a world corrupted beyond redemption.[77]

Earlier descriptions had stressed the Rogue's masculine, confrontational nature. The river's tumbling waters, whirlpools, giant boulders, and thundering rapids defined the authenticating experience for the postwar generation: "handling a boat on the Rogue takes more than skill; in emergencies it calls for great strength," a river rafter warned in the late 1940s, emphasizing the larger-than-life character of the "thick chested and heavy muscled" guides. By the 1970s travel literature was more likely to emphasize the quiet, natural aspects of the scenery. Floating the Rogue, a *Better Homes and Gardens* article promised, you could "make peace with the world and yourself."[78] Thus the crowds of urban recreationists came chasing a vision of nature as solitude — "unpeopled bliss," as outdoor writer Wayne Thompson put it. Quiet water, Thompson emphasized, complemented the effect of the Rogue's legendary rapids. "There's an awful lot of silence in a couple of miles — a hush when the boatman rests his oars and permits the currents to gently carry the river craft." Away from the rapids, the wilderness experience became intensely personal, an illusion of "floating [in] ... the lap of nature ... virtually alone."[79] Travel descriptions began including critical notes on the number of human contacts made in passage on the Rogue, as "wild" came to mean essentially an escape from the human congestion that characterized so much of West Coast living.[80]

As a concession to the popular inclination to personalize the natural experience, in the early 1970s state and federal officials considered a controversial policy for restricting private and commercial traffic through the wild corridor. According to the 1968 act, managing agencies were to facilitate "river-oriented recreational opportunities" while protecting the river environment in a "natural, wild, and primitive condition essentially unaltered by the effects of man." Spiraling recreational traffic and the quest for solitude brought the contradictions in this mandate to the fore. "Lack of accessibility," according to one report, "is considered an asset since it helps to preserve the lower Rogue in its natural state, which is one reason for the area's popularity."[81]

The congressional mandate for wild rivers provided little specific insight on how to allocate the Rogue's "outstanding opportunities for solitude" among a growing number of users. The concept of wilderness carrying capacity was just emerging as a federal land-management tool, and managers knew little about those who used rivers as a wilderness experience.[82] Moreover, restricting public access went against the grain of federal recreation management principles, since agencies traditionally measured their success in terms of raw visitor-numbers rather than the quality of the experience.

The concept also went against the grain of local recreational use. While urban river runners were more at home with boundaries and interdictions introduced by distant governments, locals considered the river experience liberating because it was unencumbered by such rules. They "took care of things pretty well by themselves," viewing river recreation in terms of freedom from artificial restraints — a measure of their own independence. A river with regulations would be a "an entirely different 'animal.'"[83]

Local people also saw the river as a medium for social intercourse and relaxation in a familiar setting rather than as a wilderness experience. In small towns throughout the basin, outdoor recreation offered community consciousness and conviviality difficult for urban environmentalists to comprehend. Friends and family members worked together, drank together, worshiped together, hunted together, and fished and floated the river together.[84] Nature was an important component of both work and play, and a shared understanding of this working rural landscape — geography, lore, and flora and fauna — helped define community identity. River recreation affirmed the cultural commonalities among local people. Urban recreationists, in contrast, saw a wide separation between work, recreation, and residence, making it "seem normal to individualize leisure ... rather than engage in the kinds of activity that build reciprocal relationships of trust and interdependence." Finally, locals balanced their respect for the river's spiritual qualities against a realistic appraisal of the lower Rogue canyon as a cold, wet, unpredictable, and often dangerous place.[85] They saw the river more as part of a working rural landscape and less as a unique spiritual experience — a blessing, but at times a curse.

Having made their peace with the Rogue as part of a traditional pastoral commons — profane as well as sacred — locals grew anxious when its waters fell under the control of outsiders poised to change the rules of access and the very meaning of its wildness. They preferred to accommodate rising recreational pressures simply by using, as one river person put it, "common courtesy toward each other." What they feared most were not competitors for the river's solitude but rather federal restrictions that would enforce that solitude. "What is more scenic than people enjoying the Rogue River?" Regulations, they cautioned, should include "some provision ... for [the river's] human ... element."[86]

For administrators, the challenge was to blend the abstract vision of an unpeopled environment with the recreational traditions held dear by the river people themselves. Given similar concerns across the country, Congress crafted a flexible management system for wild and scenic rivers that tailored the national mandate for landscape preservation to local use traditions. "Like people," administrators explained, "rivers [each] have their own personalities and their own problems." Flexibility became the means by which the concept of wildness expanded to acknowledge rising commercial use, the urban quest for solitude, and the traditional pastoral perceptions held by local "river people."[87]

Key players in these negotiations included Oregon's whitewater guides, who were deeply torn by the twin necessities of boosting business and protecting the wilderness solitude upon which this business depended. With "a fishing boat in every hole" and jet boats passing by every few minutes, it was difficult to convince clients that they were enjoying a wilderness experience. According to old-timer Kenny King, the campsites were "badly congested," and he had been in the business long enough to know how competitive guides could be. "There will be much name calling, fist fights, and possible shootings," he admonished; "this is a poor way for a fishing party to finish their trip down a 'wild river.'"[88] Yet guides had other reasons to invoke the wilderness mandate. As administrators tightened commercial restrictions on other western waters, unrestricted Oregon rivers looked increasingly attractive to out-of-state outfitters. "Crowding of the Rogue was one thing," commented John Garret, a well-known river runner, "but out-of-state competition for some of the last unregulated commercial river in the West was something else." Oregon guides decided that the state "shouldn't wait too long" to begin protecting the "very fragile" waters classified under its own scenic waterways program.[89]

In a series of meetings between 1972 and 1976, commercial guides, environmentalists, local officials, lodge owners, and "river people" met with administrators to hammer out a policy that included a moratorium on new outfitters and a commitment to allocating a projected carrying capacity of 10,000 summer boaters among commercial and private parties. With many misgivings, the users agreed to an even split: 5,000 "starts" or launches for private, and 5,000 for commercial boaters. The policy capped the river traffic in general terms, but it left unresolved the questions of motorized boats in the lower corridor and peak-period traffic in the wild section. As Kenneth Mak of the Medford BLM office argued, the seasonal totals bore no relation to what was "happening out there daily" on busy weekends and holidays. During peak periods, Mak warned, the Rogue was "close to capacity."[90]

The problems of regulating people in an unpeopled landscape highlighted another thorny issue: who was responsible for the Rogue River? The corridor lay within the bounds of the Siskiyou National Forest and the BLM's public lands, but riverbed mining and gravel permits fell under the state Land Board, and the state Wildlife Commission determined policies for fish and wildlife. More important, the state Marine Board was responsible for river traffic, safety regulations, and boat licensing. Composed primarily of local and county officials, the Marine Board was skeptical of the restrictive policies issued by federal officials.[91] While urban recreationists pressured federal agencies for "wilderness with a capital 'W,'" local people, with greater access to the Marine Board, pressed a different viewpoint: "they enjoy seeing their friends on the river much like you'd enjoy seeing friends at a bowling alley." To better understand attitudes on the river, state and federal agencies held public meetings throughout Oregon and in San Francisco and commissioned Oregon State University to survey recreational boaters on the river itself.[92]

The public meetings brought little consensus on how to protect the Rogue against people pollution. Environmentalists criticized the access slots guaranteed to commercial guides; commercial users, they argued, would crowd out private boaters, overrun campsites, and degrade the wild river experience. As a way of appreciating the river, commercial rafting was less authentic, and it offered up the wilderness "to businesses and corporations." Raft and drift-boat guides defended their allocation and turned their sights on the power boats in the wild and scenic sections, with their "loud, chainsaw-like buzzing," pungent emissions, and destructive wakes. Long-time river residents, who needed motorized boats to get upstream, felt they had as much right to the river "as the boats filled with 'hippies'" coming down.[93] Tensions between Grants Pass residents, financially dependent on downriver drift traffic, and those in Gold Beach, staging ground for the upstream motorized traffic, complicated the decision.[94]

Consensus depended on balancing the commercial requirements of the outfitters, the traditions of the river people, and the wilderness ideals implicit in the federal mandate. The Marine Board and federal officials restricted power craft in the wild section but responded to local pressures by allowing camp and cabin owners upriver by permit, "in areas where their use is established and necessary."[95] The problem of drift-boat congestion proved more difficult to resolve. Three of the four river management agencies agreed with Ken Mak that daily peak-period limits were necessary to avoid the "OK Corral atmosphere" at popular campsites. But the Marine Board opposed the limits, since overall seasonal private use was still below the 5,000-launch ceiling in the master plan. Sensitive to constituent pressures, the board concluded that imposing access limits on local people was "a very touchy, dangerous situation."[96]

In December 1974 the Oregon State University study was published, showing that more than 80 percent of the 348 boaters interviewed supported limits to avoid overcrowding. One or two daily encounters with other parties, apparently, was "more enjoyable than none at all," but as numbers rose, "groups received less satisfaction from their river experience." With the OSU study in hand, in 1975 the BLM began charging a small camping fee in the wild section of the river, and in 1976 the agency limited overnight shoreside camping on BLM lands, which curtailed private trips.[97]

The most dramatic effect of the 1972–1976 regulations was to limit what some now called the "commercial exploitation of the Rogue." The BLM and Forest Service discouraged commercial expansion by placing a moratorium on new outfitters and restricting the size and number of weekly trips for those already on the river.[98] Commercial rafters gained assurance that the river would not become "easy for everyone to run," and out-of-state guides were held at bay because trip limitations kept profit margins thin. The remaining outfitters staggered their departures so that each party was "pleasantly oblivious to the activities of another."[99] Riverside property owners gained protection against burgeoning recreational traffic, yet maintained at least in moderation their right to use the river as they had before the federal designation. No one seemed particularly happy with the formula, but at last the agencies had devised a means of controlling commercial use and leveling out the daily peaks that transformed the Rogue from a wild river into a weekend carnival.[100]

Reflecting on the inevitable human manipulation that went into constructing the wilderness experience on the Rogue, Wayne Thompson surmised that some would view the river story as an environmental tragedy, "a spoil of man's desire to play, thus ruin, natural haunts." But, he noted ambiguously, since Jedediah Smith's sojourn through the valley in the 1820s, the Rogue had never been wilder than it was in 1975. And with a "few hundred bucks and an early reservation," he concluded, you could easily overlook this long history of human abuse while languishing in the "comparative privacy of your own tour group."[101]

CHANGES ON THE ALLAGASH

On Maine's Allagash River, rising recreational pressures brought similar problems. North-central Maine's new interstate highway and an expanded private woods-road system improved access to the wilderness river, and this, along with the rising popularity of river recreation and the national publicity resulting from the battle to preserve the Allagash, boosted canoe traffic from a few hundred people yearly to around 10,000. The Allagash Wilderness Waterway, implemented as a concession to federal preservationist pressures, was funded at a rate lower than a national park would have been, and access and use regulations were far less stringent than those for federal wild and scenic rivers. This minimalist, quasi-wilderness approach provided neither the facilities to accommodate mass recreation nor the administrative means to discourage it.[102]

In 1977 the *Kennebec Journal* ran an article titled "Allagash: An Outdoor Slum," pointing to the traffic, the overcrowded and littered campsites, and the encroaching logging activity that intruded on the wilderness traveler. The roar of chainsaws and skidders beyond the thin buffer and the parade of canoes reminded visitors that solitude had not been uppermost in the wilderness idea when the waterway was conceived. After finding 90 people in one primitive campsite, waterway Director Leigh Hoar resolved that his agency would have to devise some means to "keep the lid on use." Maine guide Sam Jalbert complained that the river had been "wilder" before it was designated a wilderness.[103]

Finding ways to protect the river from mass recreation was difficult because the meaning of wilderness had never been clearly resolved in the debate over protecting the waterway. Managers drafted rules regarding use of the "wild" river in the context of Maine's older pastoral tradition: recreational use of a working rural landscape. Leonard Pelletier, another guide, recalled the time when his canoe became immobilized on a rock. He walked back to Churchill dam, asked the warden to "turn off the water," and returned to find the canoe lying in a nearly dry riverbed. He and his party righted the craft, and when the water level rose, they were "once again on our way down the Allagash Wilderness Waterway." While some called for more restrictions on recreation and forestry activities, others suggested that the sound of chainsaws was simply part of the north woods experience.[104] Responding to public complaints, the Allagash Authority considered restricting access into the corridor from the expanding network of woods roads and mapping out line-of-sight "visual zones," with stricter standards for timber harvesting. But given its ambiguous mandate, the authority was unable to make headway on either goal. In fact, in response to local pressures, administrators expanded vehicle access over the next two decades from the original two points to fourteen.[105]

In 1974, amid growing controversy over the Allagash, regional author Lew Dietz reflected on river use and preservation all across the state. These waterways were "woefully vulnerable," he observed, and Maine needed a policy to shield them from loggers and developers. Yet the very act of preservation would encourage new forms of degradation, since the protected rivers would become magnets for mass recreational use.[106] Maine responded to this dilemma by applying the ambiguous Kennebec solution to its other wild rivers, using a combination of state and private management options that offered minimal protection against dams, development, and commercial loggers while avoiding high-profile federal wilderness status, which would draw more urban recreationists and encroach upon traditional pastoral liberties.

The risks in this strategy became evident in the mid-1970s during controversy over proposed federal wild and scenic designation for the upper Penobscot River. Advocates pointed out that the Penobscot corridor, linked to the Allagash Wilderness Waterway and to Maine's huge Baxter State Park, would create the largest block of protected wild country in the East.[107] Yet state officials balked at the concept of "another Allagash-type river system," and the major landowner, Great Northern Paper Company, offered recreational easements that ostensibly mitigated the need for federal protection. The federal proposal divided the environmental constituency. NRCM's Robert H. Gardiner endorsed the plan, but others worried that a federal "green strip ... across [the] map of northern Maine" would bring "too much publicity." Traffic on the Allagash had doubled since its wilderness designation, and similar status for the West Branch, they reasoned, would make that lightly used watercourse a "bumper to bumper" river. Local river users, meanwhile, feared that federal control would curb traditional forms of recreational use, resulting in a "loss of personal freedom" on the river.[108]

In the case of the Penobscot, these divergent values could not be reconciled. Burton Packwood of the Society to Protect the Penobscot and Kennebec Rivers spoke forcefully against the federal plan, and in 1976 thousands of Maine citizens signed petitions expressing the same opinion.[109] In 1977 the federal proposal was set aside, and almost immediately Great Northern announced plans for a dam at Big Ambejackmockamus Falls that would flood seventeen miles of Penobscot whitewater. Packwood fumed that Great Northern had used conservationists to help defeat the federal proposal, "so that [the company] ... could destroy the river itself." The struggle over the West Branch raged for eight years before the dam permit was finally turned down in 1985. More than any other issue, the "Big A" controversy highlighted the vulnerability of protecting rivers through conservation accords with private landowners.[110]

By the late 1970s new dam proposals were afoot for several Maine rivers, including the Kennebec Gorge, one of Maine's premier whitewater experiences, and a world-famous salmon pool on the lower Penobscot. Maine, the NRCM announced, needed a "rivers conservation strategy." In 1983 the legislature passed a Maine rivers bill protecting 1,100 miles of scenic river but encouraging hydropower development on other sections with a streamlined, one-stop permit process. Heralding a "system of recreation rivers ... unparalleled in the nation," Governor Joseph E. Brennan predicted the comprehensive site-selection process would bring the era of contention over rivers to a close. Maine's Land Use Regulation Commission would apply protective river zoning to the pristine upper watersheds, and downriver towns would develop shoreland zoning ordinances to ensure orderly development along the lower sections. Again the strategy avoided federal designation and the threat of national exposure, but at some risk. Without federal protection, as Conservation Commissioner Richard Barringer pointed out, any Maine river could be dammed under the Federal Power Act.[111] The Maine Rivers Act ratified the difficult course Maine steered between potential nature despoilers, but the solution fell short of full protection against ecological disruption or recreational abuse.

In the end, Maine negotiated the demands for ecological integrity, recreational access, and pastoral privileges through a policy of wilderness management by obscurity. Distance from the city, logistical disadvantage, a formidable phalanx of mosquitoes and black flies, and the careful avoidance of high-profile federal designations would stabilize urban recreational pressures while Mainers continued a traditional form of recreating in a working rural landscape. Maine's wilderness rivers, aside from the Allagash, remained hidden away behind a confusion of poorly maintained state highways and dusty logging roads. Local use of private logging roads for river access posed a significant river management problem, but at least the state's wild rivers seemed safely beyond the Northeast's urban recreational reach.

While the complicated formulas for using and preserving rivers in Maine and Oregon produced some inequities and no small amount of hard feelings, they managed in piecemeal fashion to blend together the divergent strains in the environmental imagination: the pastoral, the recreational, and the ecological. Thus they provided models when other cities and states began experimenting with river greenways and controlled access to wild rivers.

River greenways became an important part of the urban environmental movement in the 1970s, as community leaders realized their multiple benefits in providing recreational opportunities, mitigating habitat fragmentation, providing edge ecologies, preventing flood damage, cleansing waters, recharging aquifers, removing excess nutrients, and settling out sediments. As recreational and educational space, greenways drew city dwellers to the natural world and fostered a sense of stewardship. Denver, for instance, built a 10-mile greenway along the South Platte River from city limit to city limit after destructive floods in 1965 and 1973. Raleigh developed a similar greenway along the Neuse River, with "green fingers" up its feeder streams. The project, begun in 1975 with help from the local Sierra Club chapter, the League of Women Voters, and the city's garden clubs and homeowner groups, included floodplain management, zoning easements, stream-bank buffer zones, erosion control measures, and a system of footpaths, bikeways, and canoe trails. Towns in New York and Pennsylvania protected the upper Delaware River, and Athens, GA, created a 35-mile greenway along the north and middle forks of the Oconee River. Scores of other communities followed these examples, reversing a longstanding practice of channeling streams, filling wetlands, polluting rivers, and hiding watercourses behind intrusive forms of land use.

Provisions in the 1972 Federal Water Pollution Control Act Amendments and the 1987 Clean Water Act, particularly the goal of "no net loss" for wetlands, prompted scores of states, including Maine and Oregon, to enact comprehensive river conservation programs based primarily on ecological goals, but with pastoral underpinnings. Minnesota's Wetlands Conservation Act and California's Urban Stream Restoration Program assisted communities in preserving waterways, and Missouri's "stream team" encouraged networks and alliances among conservationists, landowners, farmers, and public officials. As one study of the greenway movement put it, citizens were beginning to realize that "curved is better than straight, rough is better than smooth, slow moving is better than fast, and vegetated is better than concrete covered." Rivers, wetlands, lakes, bays, and streams once again became rallying points for local conservation work and sources of civic pride.[112]

Controlling recreational use of wild rivers became a focus of environmental attention as well, and here again compromises on the Willamette, the Rogue, and the Allagash provided assurance that management could accommodate divergent values. The Snake River's spectacular Hells Canyon, for instance, remained unrivaled as a wild-river experience, even though its biggest rapids had been stilled by dams. Idaho's Salmon was likewise protected, although "exceedingly active" as a recreational river. The Stanislaus, secure from dams, tumbled through a "pure wilderness canyon" in California's tourist-trampled Gold Country. A complex blend of hydro-engineering, political brokering, and access management preserved each of these rivers as "at least a clever simulation of a wilderness waterway."[113]

Still, the terms under which these compromises were forged had to be negotiated anew for each river. Thus the Oregon and Maine models provided no universal solution to the tensions that were fragmenting the environmental imagination. In this ongoing battle, the Willamette, the Rogue, and the Allagash offered lessons for defining the future of America's rivers, but each also underscored the difficulty of blending urban-based recreational and ecological demands with the needs of local people whose perspectives on use, enjoyment, and conservation were different, but no less valid. Outdoor recreation, wilderness, ecology, natural beauty, harmony with nature — the catch-phrases of the environmental imagination — are contested terms. As such they often reveal the distance among contending themes in the environmental imagination, reflecting the different meanings people attributed to nature. But as these river stories also demonstrate, terms such as these can be drafted to build consensus as well as destroy it.

Belfast Harbor, Maine

CHAPTER 6

BRIDGE TO ECOTOPIA: LOCAL CONTROL AND STATEWIDE LAND-USE PLANNING

I n 1971 the Maine State Planning Office issued a small booklet intended to interest people in the process of comprehensive planning. The publication, *Maine Coastal Resources Renewal*, offered a sweeping vision of controlled progress along the state's besieged coast, a future far different from the industrial promotions advanced by oil-port developers during these same years. As an example of creative utopian planning, the document described the development of a hypothetical nuclear power plant on the shores of Belfast Harbor in mid-coast Maine. Turning a potential problem into an environmental blessing, Belfast citizens would direct the plant's warm-water discharges into ducts and pipes to heat their homes, their businesses, and then the waters of a swimming beach fronting their new destination resort hotel. An array of aquaculture hatcheries would bask in warm waters near the harbor, and greenhouses would operate year-round. Power from the nuclear plant would attract other clean industries to the area and run an energy-efficient local public transportation system. All of these facilities would serve a population prudently limited to 11,000 permanent and 20,000 seasonal residents. When demand for accommodations reached the prescribed limit, new migrants and visitors would be diverted to another of the 30 nuclear-powered communities projected for the Maine coast by the year 2000.[1]

Although a difficult leap for coastal citizens renowned for their downeast practicality, the utopian Belfast scenario epitomized the dream of comprehensive planning: rather than address development issues "singularly and in isolation," planners offered a new process that would blend Maine's pastoral legacy and its urban development needs into a unified, harmonious landscape. Dystopic predictions of a coast scarred by industrial blight would evaporate in the clear, crisp air of a new utopian dawn.

Roughly a year later, in June 1972, Oregon Governor Tom McCall launched a similar campaign for his state. The governor hired San Francisco planning consultant Lawrence Halprin to produce an illustrated, portfolio-sized booklet, *The Willamette Valley: Choices for the Future*, to serve as a workbook for citizens recruited to local planning sessions. The book presented two scenarios. The first, based on straight-line projections from contemporary trends, traced the steady erosion of Oregon's distinctive pastoral landscapes, using images of California-type sprawl to accentuate the crisis that lay ahead. Over the next 30 years, an unsightly profusion of houses, buildings, roads, shopping malls, and parking lots would spread like a cancer outward from the cities along Interstate 5, the state's major transportation route. Indiscriminate pursuit of new industries would bring more pollution and more outsiders to Oregon, and the once-beautiful Willamette Valley would succumb to sprawl, urban decay, traffic congestion, and resource depletion. The second scenario depicted a utopian Oregon shaped by intelligent planning. Urban expansion would be contained and communities rejuvenated through cluster development. An elevated train, part of a unified statewide public transit system, would speed commuters between Portland and Eugene, while small electric cars and buses, moving sidewalks, and walking and bike paths would carry residents through the bustling urban scene. The valley's rich soils, protected from suburban sprawl, would provide jobs for full-time and part-time farmers; rivers would run free and clear; and a combination of hydro, solar, and nuclear power would offer cheap, pollution-free energy for homes and factories.[2]

Visions like these became a staple of the environmental imagination in the early 1970s, as state agencies, conservation organizations, novelists, and journalists popularized the idea of a world in which nature and culture were harmonized. These environmental utopias, or ecotopias, painted a future in which the problems of pollution, sprawl, congestion, and disorder were resolved; energy was conserved; and people labored, consumed, and socialized in harmony with nature. As an ultimate expression of the environmental imagination, ecotopian thought was an attempt to mediate the contradictions between freedom and social control, permanence and economic growth, and preindustrial authenticity and postindustrial leisure. This imaginative exercise forced environmentalists to rethink the scope of their concerns as they shifted attention from preserving natural beauty and recreational opportunities to addressing the needs of a complex and dynamic human society. Translated into politics and state planning, ecotopian ideas broadened the movement's philosophical underpinnings and its sense of purpose while also highlighting its internal contradictions.

The idea of Maine as ecotopia was popularized in a book titled *Maine Pilgrimage*, written by former *Time* correspondent Richard Saltonstall. Linking Maine's pastoral themes to the idea of an "American renaissance," Saltonstall gave the state national standing as an ecotopian model. In Saltonstall's expansive imagination, Portland would become a "radiant East Coast San Francisco," with a restored historic waterfront, a thriving maritime economy, and a new, unintrusive outer belt of clean industries. With vision provided by state planners, citizens in smaller towns would blend new small-scale technologies and traditional ways of farming, fishing, and producing crafts to renew the rural economy. "Sea farmers" would take advantage of countless islands, coves, inlets, river mouths, and tide flats to cultivate a "vast aquaculture industry." Paper producers, pressed by southern competition, would surrender the north woods, and the forests would regain their diversity, eventually yielding energy, chemicals, milled products, paper, and a prosperous four-season recreation industry. Maine would rebuild its public transportation system by offering a combination of intra- and inter-urban buses, coastal and island ferries, float planes, local auto rentals, and the Mainetrain, a fast turboelectric passenger line that would whisk tourists to well-planned, low-density resort destinations across the state, "keeping the beaches and scenic sites relatively free of cars."[3]

Oregon's ecotopian aspirations were popularized in Ernest Callenbach's *Ecotopia*, a novel published in 1975. Perhaps the most powerful futuristic statement of the decade, Callenbach's work depicted life in a secessionist nation composed of Washington, Oregon, and northern California during a future in which most of America had succumbed to environmental disaster. Ecotopia's premise was a sustainable economy maintained by accounting directly for all production costs, including those traditionally externalized as burdens to society: pollution, resource exhaustion, and landscape degradation. Ecotopians wishing to build with wood, for instance, were required to complete a stint in the forest service, planting seedlings to replace the trees they would cut for building. In Callenbach's novel, a skeptical reporter from the East infiltrates Ecotopia, bent on exposé. His skepticism wains as he observes and records the logic, the sense of community, and the human creativity fostered by the Ecotopian approach to life. Falling in love with the new ecocentric nation, and in particular with one of its female citizens, the reporter sends a last dispatch to the outside world: he plans to stay.[4]

Promoted by government planning agencies and environmental organizations and popularized by authors like Saltonstall and Callenbach, ecotopian ideas marked a departure for the environmental imagination. The environmental movement began largely as a defense of specific iconic natural features, each distinctive because it was perceived as a common resource important to regional identity. The 1970s ecotopian literature broadened this field of concern to take in larger, less public landscapes that included the land, its resources, its scenic configurations, its people and buildings, its dynamic capitalist economy. This gave the movement a more complex vision and helped dissipate its reputation as antimodernist and reactive; however, as a new and much more comprehensive arena for environmental politics, ecotopia was also far more contentious than the defense of specific rivers or iconic coastlines. The shift from conservation and preservation to proactive planning was not easy.

Regional ecotopian literature served to elevate a series of planning experiments begun in the 1960s, when planners and political leaders realized that issues like habitat loss and disorderly suburban and rural development had become regional rather than local. At the state level, Hawaii, Wisconsin, and Vermont pioneered the comprehensive land-use planning movement by adapting zoning controls originally developed for cities to regulate entire countrysides. In 1961 Hawaii created a Land Use Commission and divided the state into four zoning categories based on conservation, agricultural, rural, and urban land use. Local zoning boards oversaw urban develop-ment; the Land Use Commission governed agricultural and rural lands; and the Department of Natural Resources administered conservation lands. Passage of the act was facilitated by a few large agricultural landowners and spurred by visions of a sprawling post-statehood tourist industry. Wisconsin's 1966 Water Resources Act, provoked by fears of overdevelop-ment along the state's many lakes and rivers, required counties to pass shoreland protection plans. In Vermont, the legislature reacted to a 20,000-acre resort subdivision proposal by International Paper Co. by pass-ing the 1970 Environmental Control Act, aimed at regulating developers with subdivisions of more than 10 acres or 10 lots within a radius of five miles. The pioneering examples of Hawaii, Wisconsin, and Vermont were, according to one report, "just the tip of the iceberg." Late in the decade state legislative committees, task groups, and volunteer organizations across the country launched similar campaigns as one of the most significant innova-tions in state policy to emerge out of the environmental movement.[5]

Across the nation comprehensive planning advocates articulated sever-al common themes, all tinged with ecotopian thinking. First, the cam-paigns signaled a shift from viewing land as a private commodity to regarding it as a social resource. Earlier land-use laws were written prima-rily to enhance urban real-estate values and promote urban economic development. The land-use laws of the 1970s extended the urban reach into the countryside, to protect rural land as a repository of social values for the entire society, or, as Tom Bell of the *High Country News* put it, as part of a "living community which cannot be bartered in the marketplace."[6]

Second, comprehensive planning was linked to America's aspiration for regional coherence: it would move the nation toward a postindustrial future in which each region fostered a self-contained, mixed-sector econ-omy linked organically to the natural resources at hand. The planning vision was thus a vivid expression of U.S. regionalism, tinted by the exclu-sionist tendencies implicit or explicit in most ecotopian literature.

Third, these visionary experiments assumed a much more powerful, more professionalized state administration. According to the planning rhetoric, only centralized state agencies were capable of implementing the sweeping political innovations necessary to protect the integrated rural landscape.[7] But advocates also saw planning as a shared-power concept, requiring grassroots participation. Literary ecotopias were democratic in a tribal sense, and the planning visions they inspired rested — in theory — on a similar citizen standing. Successful comprehensive planning was inconceivable without local support, experts realized, and they employed a variety of public venues to involve citizens in implementing the process. This engagement, however, was not always successful, and it was here, at the juncture of centralized expertise and popular participation, that the strengths and weaknesses of planning were driven home.[8]

The conflict between local prerogatives and centralized planning was implicit in congressional debates over the proposed but never passed National Land Use Policy Act. This legislation was initiated in 1970 by Washington state Sen. Henry Jackson as a companion to the National Environmental Policy Act (NEPA) enacted earlier that year. Jackson's bill provided federal assistance to states willing to control development in areas "of more than local significance," particularly coastal zones, shore-lands, rare ecosystems, and scenic and historic sites. Whereas NEPA required federal agencies to consider the environment in their management decisions, Jackson hoped that his land-use act would eventually require state governments and private developers to do the same. The new federal Council on Environmental Quality (CEQ) drafted its own version of the law, and after initial hearings a bipartisan consensus materialized behind the Nixon administration's proposal. It appeared that a new National Land Use Policy Act would follow NEPA into federal law.[9]

Bipartisan support was partly due to the vibrancy of environmental initiatives in the early 1970s, but just as important were fears that local grass-roots resistance to growth and development was getting out of hand. As CEQ Chair Russell Train testified before the House of Representatives, circumscribing local "exclusionary regulations" that prevented new developments "of regional benefit" was just as important as protecting areas of critical environmental concern. In short, planning could support, as well as contain, development. The act would replace idiosyncratic local influences with a more rational, professional administration capable of balancing competing environmental and economic interests.[10] The bill passed the Senate by a vote of 64 to 24, but it was killed in the House Rules Committee owing to strong opposition and eventual lack of support from the Nixon administration.[11] The fate of comprehensive planning would rest with states like Maine and Oregon rather than with the federal government.

The failure of the National Land Use Policy Act highlighted tension between the perceived need for centralized planning on one hand and a deep suspicion of centralized state initiatives on the other. Sensitized by Rachel Carson's exposé of business–government collusion, many activists and environmental publicists associated centralized power with environmental dystopia. Jean Anne Pollard, a muckraking naturalist writing in the *Silent Spring* tradition, portrayed Maine under siege by construction, real-estate, and forestry interests, each with a state agency serving its own interests. Citizens, Pollard argued, should remain vigilant against government bureaucrats of any sort.[12]

Yet the energy of the environmental movement, as the national debate suggested, was passing from the passion-driven grassroots organizations of the 1960s to the state planning agencies of the 1970s. Could these agencies retain the public confidence they gained during the river and coastal-protection campaigns? Could they implement the sweeping administrative changes needed to sustain the ecotopian revolution without igniting opposition from the rural people who would fall under the shadow of the new state apparatus? These questions were crucial because comprehensive planning raised the stakes in the political process by shifting the focus of environmental concern from discrete elements of the landscape to entire regions, and from simple protection of common resources to aggressively proactive regulation of private land use. This broader vision not only unnerved the corporations and developers who chafed under the new regime but also alienated many environmentalists, who found themselves uncomfortable in the company of state bureaucrats and other centralizers whose credentials as nature defenders were yet to be established.

A clear warning against the entreaties of state planners, technocrats, and ecotopians in general was published in 1971 by science-fiction writer Ursula LeGuin, at the time Oregon's most prolific author. *The Lathe of Heaven*, her novel about a future Portland, opened with a review of the multiple environmental crises in store for all U.S. cities. "Undernourishment, overcrowding, and pervading foulness of the environment were the norm.... It had always rained in western Oregon, but now it rained ceaselessly, steadily, tepidly. It was like living in a downpour of warm soup, forever." When young George Orr develops the ability to alter reality by dreaming, a psychiatrist, William Haber, seeks to control his thoughts to establish harmony, beauty, zero population growth, and all the other static indices of ecotopia. At this point, idealism begins to go awry. Haber has Orr dream away overpopulation, and 6 billion of the world's people are instantly swept off by a plague caused by carcinogenic chemical pollutants in the atmosphere. Haber turns out to be power mad, seeking an Orwellian world order under the planetary slogan "The Greatest Good for the Greatest Number." In the nick of time, the young dreamer saves the world, and we are returned to the present-day city. As the book ends, we see Portland, "half wrecked and half transformed, a jumble and mess of grandiose plans and incomplete memories, swarming like Bedlam."[13] Portland muddles on, the present imperfect world preferable to a world manipulated by those who grasp power through technology and misplaced idealism. Clearly America must address its environmental problems, LeGuin suggests, but through free and local choice, not bureaucratic central planning.

Maine and Oregon were among a small number of states that experimented with the new brand of ecotopian environmentalism in the late 1960s, and it is possible to see the strengths and weaknesses of the planning initiative as the movement passed into the hands of professional planners and their political allies. In both states the most significant support for planning came from a politically powerful rural landowning class to whom growth and development appeared threatening. Maine's coastal summer people and Oregon's Willamette Valley farmers were initially skeptical of state-directed land-use planning, but both held to pastoral traditions that spurred their participation in the search for new protective arrangements. Here, however, the story of planning in Maine and Oregon diverges. A state that rejected Franklin Delano Roosevelt and his New Deal activism in three successive elections, Maine harbored a parochial political culture that stressed minimal state intervention into land-use decisions. Oregon political culture was tinged with skepticism toward big government as well, but having benefited from New Deal and post–New Deal federal programs like hydroelectric power development, irrigation, and forestry, Oregonians were more likely than Mainers to accept the idea of regulation. More important, development pressures were intense in many regions of Oregon, while in Maine they were limited to coastal and some inland waterfront properties. And finally, Oregon's political leaders mounted a series of carefully crafted, professional publicity campaigns designed to convince the public that central planning would work;

Maine's political leaders were more content to leave the debate in the hands of contending interest groups. Thus the balance between professional expertise and popular participation played out differently in Maine and Oregon.

PLANNING ECOTOPIA IN MAINE

In Maine, support for statewide planning materialized during an unprecedented land boom in the mid-1960s, when general prosperity and a new interstate highway made Vacationland accessible to tens of thousands of urbanites living in the northeastern metropolitan corridor, many of whom were amazed to discover the bargains to be had for an old farmhouse with a water view in an underdeveloped area of the state. At first, Mainers were incredulous that nonresidents were willing to pay so much for land they themselves took for granted, but local speculators caught on quickly. Norman Langdon explained to a reporter the ease with which he became a small-time speculator: "I bought some land for darn near nothing, took over with nothing much to back up my credit, borrowed some money to buy more land, and kept on that way." By 1967, Langdon had seven active subdivisions in coastal Hancock and Washington counties, doing business under the name Paul Bunyon Shores.[14]

As stories of duping the "summer complaints" turned into stories of resales at triple and quadruple prices, Mainers began to realize that their state was no longer immune to the perils of prosperity. By the end of the decade, the opportunities for quick returns on southern and mid-coast lands were nearly exhausted. Priced out of their own dreams of a coastal or shorefront retreat, locals reacted by lambasting "the developers." Pointing to subdivisions appearing "like poison mushrooms on every side," Jean Anne Pollard lamented the collusion between local officials and "hucksters from away," and state official Donaldson Koons warned that "our available recreation property is being bought up largely by out-of-state people ... and there is not a hell of a lot we can do about it." But since participation in the land boom was so dispersed and widespread, the established environmental strategy of preserving particular natural areas was inadequate. Recognizing the need for a new approach, the Natural Resources Council of Maine (NRCM) organized a symposium on coastal development at Bowdoin College in Brunswick in October 1966. The conference was a pivotal moment in Maine's environmental movement.[15]

The Bowdoin event was organized by NRCM's Robert Patterson and Bowdoin College photo-essayist John McKee. Speakers included nationally known academics, government officials, lawyers, and planning consultants. In his introductory remarks, Patterson noted Maine's good fortune in having escaped the blight that engulfed so much of the eastern seaboard. To preserve Maine's options, Patterson promoted the idea of using comprehensive planning to isolate locations best suited for mass recreation and development — including those offering a "return on [high-value residential] investment" — from quieter refuges along the coast.[16] The rhetoric at the conference previewed a classic formulation: a crisis was just around the corner; Maine's heritage was threatened by self-serving developers; and environmental planning was the key to resolving the problem.

The conference demonstrated the importance of Maine's urban-based "summer folk" in this new exercise in environmental imagination. Issues were framed to protect their valuable resort investments, stressing the need for strategic points of public access to concentrate and contain the crowds and prevent local developers from despoiling the more sacred areas near existing summer colonies. Participants reasoned that because the coast was so deeply indented, relatively small access areas could provide "a feeling of untouched and open nature that out West would require many square miles." Beyond these sacrifice zones, the coast would remain inviolate, private, and secluded.[17] Harvard landscape professor emeritus Charles W. Eliot, whose family had summered in Maine for generations, advocated a series of preservation techniques: conservation easements, private land trusts, and federal and state reimbursement to local tax authorities for lands removed from development. As Eliot put it, these devices would reserve choice areas "for what the senior Olmsted called receptive re-creation" and protect residential areas "from invasion by the crowd." Preserving the coast as a timeless pastoral haven was a major theme at the conference.[18]

With this general assumption in the air, discussion centered on questions of political strategy. Eliot recognized that all those vaunted attributes of the downeast Yankee — self-reliance, independence, nonconformity, adaptability, make-do practicality — boded ill for the politics of planning. Nevertheless, with changes coming "much too fast," planners could not "wait for time to win these men and women over." He advised pursuing what amounted to a political shortcut: a centralized approach to planning and zoning wherever possible. The crisis was at hand, leaving precious little time to convince Maine's laconic locals of the necessity of the task.[19]

Eliot's alarmist approach helped set the tone for legislative battles in the years ahead; this sense of crisis would be used to justify the shift in power from local to state authorities. Yet others at the Bowdoin conference argued against such shortcuts. Municipal attorney Barnett Shur told the symposium that once Mainers vented their emotions, they would settle down and accept planning and zoning. The key to success, he advised, was to establish realistic zoning boundaries that confirmed existing uses. If planning advocates followed this simple rule, approached locals with reason and reserve, and endured the initial emotional attacks patiently, they would succeed at the local level.[20] Shur was familiar with Maine town-meeting politics, but his advice proved difficult to follow. Driven by a sense of crisis, planning advocates embraced the transfer of power from local to state authority.

The debate over strategies reflected a more fundamental issue: would Maine's coast and its forested interior remain cast in amber, its year-round residents locked into traditional occupations as authentic representatives of the nation's rural roots, or would they be allowed to follow the path to industrialization others had trod? Here again the symposium revealed a division in the planning and environmental community that would widen in the coming years. Some advocated federal or state funding to compensate rural Mainers who forfeited their development rights, combined with encouragements to preserve traditional occupations and architecture. Representing a new kind of national park, this proposal reflected a view of Maine as a timeless preindustrial haven set aside for postindustrial people, a place where urban vacationers could seek renewal and where a fortunate few could enter and never have to leave.[21] Others saw Maine as a developing region where the excesses of industrialism could still be avoided. Orlando Delogu, a law professor specializing in land-use planning, suggested a flexible and politically realistic system of zoning ordinances that would offer landowners a "meaningful range of uses" and a chance for economic returns on their land. A broad mix of zoning options could foster a dynamic economy that would incorporate the best of traditional Maine and postindustrial America.[22] The symposium concluded without resolving either set of tensions, and the split in underlying purposes continued to characterize Maine's planning movement.

The Bowdoin symposium helped direct the debate over the future of the Maine coast, where developers and preservers battled over a series of supertanker oil-port proposals. Some dismissed the idea that the coast could industrialize and still remain a pastoral haven, while others, like those who fashioned the Belfast ecotopian scenario, argued that Maine's working coast could adjust to a well-planned and dynamic industrial economy. Farther inland, the political and economic stakes were not as high. Maine's northern and western forest regions accommodated a few large seasonal resorts and a vast network of low-density recreational activities based primarily on public use of forest industry landholdings. Here development pressures were still a matter of conjecture, and the debate over planning and zoning remained speculative: activists predicted that industrial landowners would subdivide their lakeshore properties for recreational purposes, replicating the land boom already evident in Maine's coastal and southern inland areas.[23] The anticipation was not unreasonable. Vermont's path-breaking 1970 planning law, Act 250, came after a similar scenario in which International Paper Co., already ill-regarded because its New York mill had "turned the waters of Lake Champlain into a cesspool," planned to convert its commercial forest lands near Stratton into second-home developments. An investigation found that the company had virtually no plans for sewage treatment or open space. Under heavy criticism the company scrapped its plans.[24]

Acting to address this potential crisis, in 1967 state Sen. Horace Hildreth Jr. introduced a bill to create a Wildlands Use Regulation Commission. The commission would assume jurisdiction over some 10 million acres of northern forest, challenging a century or more of corporate domination. As legislator Sumner Pike pointed out, it was a type of legislation "completely new to us and to quite an extent ... new in the country because we don't have any Federal lands here." Hildreth's commission would, as one opponent had it, grant the seven-member board "more power than any other group in the State of Maine."[25] Landowners remained skeptical, so Hildreth compromised by limiting jurisdiction to land within 500 feet of roads and lakes, thereby leaving most of the commercial timberland unaffected, and he targeted potential recreational pressures rather than paper companies in arguing for the bill. Northern Maine, he cautioned, lay within a few hours drive of "nearly 40 million people, [who] each year have more and more spare time, who each year are earning more and more money, and who each year are looking for places to go." Despite Hildreth's insistence that the crisis was "absolutely inescapable," the measure failed to pass.[26]

Revived in 1969, the bill gained support from an emerging bipartisan environmental caucus which, eschewing Hildreth's circumspect approach, promoted it by reigniting public distrust of corporate wildland owners. Jon Lund of Augusta, for instance, announced that he was "tired of the big paper companies ... [taking] a big-brother-knows-best attitude." Company representatives reacted to this turn in the debate with predictable hostility, and under heavy lobbying the Senate Natural Resources Committee cut the bill's jurisdiction to land within 300 feet of "public" roads. The amendment rendered the law practically meaningless, since the state's unorganized townships had thousands of miles of private roads but virtually no public roads.[27] The full House and Senate passed the bill without scrutinizing this fatal flaw, which limited its jurisdiction to about two percent of the land area in the unorganized townships. As journalist Bob Cummings summarized, the new commission was "a fraud on the public with neither the power nor the funds to protect the wilderness."[28]

The vision of statewide planning and zoning was not laid to rest, however. The ongoing controversies over oil-port development focused public attention on comprehensive planning, and as the grassroots groups interested in coastal protection became more vocal, the likelihood of enacting legislation misunderstood by a majority of legislators disappeared. With the Machias Bay and Casco Bay oil-port proposals providing dramatic backdrop, in 1968 Governor Kenneth Curtis created an Office of State Planning, and in 1970 the legislature enacted the site location law by an overwhelming majority, giving the Environmental Improvement Commission (EIC) regulatory power over large industrial and subdivision developments.[29]

After only six months of operation, the EIC reported a backlog of 45,000 proposed lots, most of them speculations in recreational and shorefront developments. Sobered by the prospect of keeping track of thousands of miles of shoreline, EIC Chair Donaldson Koons approached

Curtis about a statewide zoning law that would force local governments to develop shoreland plans for themselves.[30] The 1971 Mandatory Shoreland Zoning Act, based on Wisconsin's pioneering work, gave organized municipalities two years to zone all lands within 250 feet of a pond, lake, river, stream or saltwater body. Failing this, the state would impose regulations. To justify the act, legislators fell back on a proven issue. Grass and algae, feeding on pollution from shoreside developments, were choking off sections of inland lakes, and clam flats along the ocean shore had been closed owing to septic conditions. The stress on pollution rather than planning reflected a lingering concern, raised earlier by Charles Eliot, that the downeast Yankee still considered planners, as one legislator put it, "starry-eyed environmentalists."[31]

A second 1971 land-use bill was a revival of the defective 1969 wildland planning law pertaining to the forested northern half of the state. Rather than focusing on the vision of a planned future, this time the bill's supporters raised the specter of an outside enemy: the "handful of big companies" that considered the forestlands their "sanctuaries immune from the law." The state's forests, according to the *Maine Sunday Telegram*, had been "granted, chartered, bought, sold, swapped, traded, burned, cut, re-cut, pillaged, conserved, and deeded" without much input from the people of the state. "Large national corporations now own much of the land," NRCM's Clinton B. Townsend editorialized. And demonstrably they had "no interest in Maine as such."[32] The new bill was "an opportunity to wrest from the absolute control of a small handful of land barons nearly half of this state ... placing its future development into the hands of the people, where it belongs."[33]

The bill's opponents skirted the planning vision as well, raising instead the specter of a different bogeyman "from away": the faceless, arbitrary state officials who would use the law to eliminate any form of land use unpalatable to Augusta bureaucrats. One opponent hypothesized that permits would be required for even minor alterations in land use, including those of his constituent "woodchucks, skunks, coons, foxes, beavers, and muskrats."[34] As far as these defenders of tradition were concerned, northwestern Maine was already a pastoral utopia, protected from change by its inaccessibility and distance from population centers. In the view of these critics, centralized planning would simply invite development by the wrong kind of people: urban vacationers attracted to an overpublicized resort area.[35] The Land Use Regulation Commission (LURC) Act passed by one vote, an outcome that guaranteed further controversy.

It was difficult to reconcile Maine's passion for local control with the newer enthusiasm for centralized planning. The argument behind the latter was laid out in a small volume titled *Maine Manifest*, published in 1971 by Harvard policy expert Richard Barringer under contract to the Allagash Group, Maine's first environmental think-tank.[36] A blueprint for statewide postindustrial development, *Maine Manifest* captured the aspirations of environmentalists seeking a balance between economic development and nature preservation. Maine was at a crossroads, Barringer announced. As the metropolitan economy pushed northward, the state would attract substantial investment, in good part because it *was* underdeveloped; its fresh air, clean waters, and unspoiled land were marketable commodities growing scarce elsewhere. This marginality was indeed Maine's greatest strength, leaving the state "unencumbered by many of the ills and rigidities of a society that has passed it by and now wants and needs what was left behind." The crux of the *Maine Manifest*, according to a *Science* magazine commentator, was that "local communities should capture a part of the benefits being realized, often by speculators, from the appreciating Maine property values."

But this was not something communities could accomplish independently, Barringer argued. Local governments were too vulnerable to development pressures, and too many Mainers still disdained zoning devices. Maine's only hope was a statewide apparatus that could implement a statewide growth strategy based on a broad mix of planning tools: zoning laws, tax policies, development corporations, access and conservation easements, public land purchases, and land trusteeships. Coordinated planning was Maine's only option, given the looming presence of metropolitan corporations and outside developers who were taking an interest in the state.[37]

The site location, shoreland zoning, and LURC laws conformed to Barringer's notion of an active state capable of planning an ecotopian future.[38] But these potentially powerful new laws were grafted onto Maine's town-meeting political culture with uneven success. Each was rooted in the argument that rapid development was inevitable, and thus central planning necessary to channel growth into appropriate places and desirable activities. Given Maine's locally centered political culture, the rationale for these imaginative new techniques depended on sustaining the specter of excessive development. In the long term, markets for northern Maine real estate proved less expansive than those for coastal property, and the wildlands remained on the margins of Maine's boom-and-bust real-estate industry. While speculative development did occur, it was seldom so extensive as to precipitate a widespread sense of crisis. These economic conditions set the parameters for conflicts between local control and statewide zoning in Maine.

Of the three new zoning initiatives, the site location law generated the least controversy. Other than the oil ports, the only proposed developments that fell under the law's jurisdiction were large-scale real-estate subdivisions, regulation of which sparked little public controversy. Moreover, with only two staff members to cover the entire state and a statutory requirement that they approve or deny applications within 14 days, permits were seldom contested. Of the first 136, only 4 were turned down. In addition, developers accepted the site location law with few complaints because it required no advance planning; it simply gave the EIC limited veto power over their development proposals.[39] With developers rather than planners retaining the initiative, the site location law did little to advance ecotopian ideas.

The Shoreland Zoning and Subdivision Control Act, which required proactive public planning by all organized municipalities, had a similarly limited effect. Operating on a thin budget, the State Planning Office could not provide adequate guidance for the state's part-time town officials, and when only 30 municipalities met the 1973 deadline for filing comprehensive plans, the legislature was forced to grant an extension. Meanwhile, as the *Portland Press Herald* put it, "the tendency by a developer to 'do it before somebody stops me' is given time to be put into action." By 1974, 235 municipalities had adopted shoreland ordinances; another 201 had ordinances imposed by the state.[40] Given the weak state oversight, this act too fell short of the ecotopian dreams that inspired it.

The Land Use Regulation Commission law remained a potent source of political controversy. Because development pressures were relatively weak in the north woods, this controversy was largely symbolic, involving various groups that imposed fiercely held cultural meanings on the north woods as a timeless natural paradise. The paper industry and its defenders held a traditional view of the woods as a working landscape open to various pastoral pursuits — hunting, fishing, camping, and, in a limited sense, wilderness recreation. In more accessible areas, resort owners and businesses dependent on tourism envisioned an expanding service economy predicated on these same traditional pursuits, with the working timberland serving as a scenic backdrop for their small-scale developments. Central state planning was anathema to both groups, as it challenged the traditional freedom of the woods. Environmentalists, meanwhile, supported planning on wilderness recreation and ecological terms. Professional planners, viewing the region as a vast empty land ready for innovative postindustrial planning, stepped into the breach to broker these claims, holding that development and nature preservation could coexist, but only if growth was carefully directed and controlled according to well-conceived land-use maps and related interventionist arrangements.

Much of the rancor over the north woods involved LURC's flamboyant executive director, James Haskell, a planner with a "rare capacity to speak his mind." Inspired by a vision of orderly development in the region, in 1971 Haskell suggested a moratorium on building permits pending completion of a comprehensive plan for the unorganized territory. Given LURC's low funding and the area's vastness — half of Maine and a quarter of New England's total land area — developers and paper company officials objected strenuously. The commission decided against the moratorium, but the ensuing controversy spawned a well-organized counterattack on LURC's planning efforts.[41] When LURC held hearings on its proposed interim guidelines in July 1972, a crowd of over 400 people offered an emotional spectacle reminiscent of the earlier oil-port meetings. Clinton B. Townsend of the NRCM castigated the "out-of-state corporations that still think they can run Maine like a colony," while the president of the Maine Forest Products Council testified that his foresters needed no "professors or professional jobholders" to tell them how to manage their property. When LURC announced its interim guidelines in October, it had scaled back its proposed timber-harvest regulations considerably, but this failed to mollify landowners who now dismissed any form of regulation as contrary to their interests and north woods tradition.[42]

In December, as a first step in imposing controls over the unorganized townships, LURC tackled the biggest zoning project in Maine's history: a half-million acre tract of western mountains targeted for ski-resort development. The commissioners held a number of meetings and hearings and debated zoning provisions for protected lands, lands appropriate for agriculture or forestry, and those suitable for small-scale business and industry, but it failed to reach consensus on large-scale resort development in the area. Rather than challenge LURC on its specific zoning proposals, industry lobbyists countered with a bill in the 1973 legislature to repeal the LURC law altogether. Haskell responded by asking the legislature for more funding to finish what his planners had begun. He opined, when challenged on the issue of property rights, that "like it or not, the United States is [a nation of] democratic socialism." Governor Curtis smoothed the political waters by admitting that his chief planner had a "public relations problem," but Haskell's statement stood as confirmation of LURC opponents' worst fears.[43]

Haskell's notoriety was accompanied by a growing split within the environmental community itself. LURC Chair Clinton Townsend and commissioner John McKee sided with Haskell, hoping to delay all new development in the unorganized territory until the comprehensive plan was completed. Others, including Christopher Hutchins of LURC and Donaldson Koons and Orlando Delogu of EIC, felt that actual development was slow enough to move ahead with a permitting process while completing the plan. After a period of painfully public internal bickering, the majority of commissioners rejected the Haskell–Townsend growth moratorium and proceeded to approve development applications.[44]

The split in the commission brought a number of erratic decisions, the most controversial involving a huge four-season resort complex planned for a semiwilderness area on Bigelow Mountain in western Maine. Wilderness advocates viewed Bigelow as a major preservation goal, while ecotopian planners saw it as an opportunity to demonstrate how planning could direct appropriate development. The commission first placed the area in a development zone, then reversed itself, and the developers sued the agency. With the commission divided, Governor Curtis opted against the Townsend–Haskell faction, appointing the more pragmatic Koons to head the state's new Department of Conservation (DOC), which also made him chair of LURC.[45] When Koons halted work on the interim comprehensive plan, several staff members leaked their concerns to the press. According to one, the commission had become a "weeping board" bending to every "hardship story" told by developers. Frustrated by the growing controversy, Curtis removed Townsend and another commissioner and in June 1974 Haskell resigned, complaining that his work had been "intolerably frustrated by the constant political ploys and maneuvers of the chairman."[46]

Under Haskell, critics had denounced LURC's attempts to plan their future; with Koons in charge, another set of critics denounced the agency's lack of commitment to planning. More an ecologist than an ecotopian, Koons urged commissioners to do less "academic planning" and more administering. Koons saw LURC as a classic environmental regulatory agency set in place to protect fragile and critical natural areas. Decisions would depend on whether a specific development would harm a specific natural character, not on whether it would conform to an abstract comprehensive plan. Landowners, not the state, should initiate planning decisions. A new LURC staff, containing no professionally qualified land-use planners, produced a document that many environmentalists denounced as "shallow, empty, vague, and inadequate."[47]

The 1974 election of independent Governor James Longley was a watershed event in Maine's planning story, reflecting in part the changed public mood brought on by the latter stages of the Vietnam War, the Nixon Watergate revelations, and the OPEC energy crisis. A conservative businessman, Longley was predisposed against statewide planning initiatives, and in the growing mood of uncertainty about government's ability to solve economic and social problems, his skepticism gained support among the Maine electorate. Although environmentalists joined to defeat another effort to repeal the LURC law in 1975, the legislature narrowed LURC's mandate by eliminating the requirement that its executive director be a professional planner. The change suggested that the agency's primary function was, as Koons insisted, to administer land-use permits, not guide the state toward ecotopia.[48]

Still, the agency remained immersed in controversy. Vitriolic attacks on the principle of regulated growth continued, and in March four more LURC staff resigned, citing low morale. Flagstaff Corp., the Bigelow developer, lobbied for overturning the LURC decision on its project, and Governor Longley rejected LURC's comprehensive plan as "too unspecific." In September 1975 Koons himself resigned as head of DOC, having parted company with the Longley administration.[49] To replace Koons as LURC chair, Longley appointed Richard Barringer. The author of *Maine Manifest* and a guiding light in Maine ecotopian thought, Barringer gained backing from many environmentalists, and, despite some grumbling, even from some representatives of the forest products industry. The *Portland Press Herald* pointed out that the new commissioner was neither an environmentalist nor a tool of the paper companies. He was, instead, "a natural politician, a person who knows how to play the game and to take a risk only when he has it pretty well calculated that the chips are going to fall on his side."[50]

Barringer personified the shift from town-meeting politics and local control to professional planning. The "environmentalism of the 1960s," he maintained, was no longer relevant: "the issues we are confronted with now aren't going to be solved by demonstrations, marches, flagwaving, or platitudes.... They are going to be solved by people knowing what the hell they are talking about." But the stormy history of Haskell's LURC and Koons's DOC showed above all else that placing environmental decision-making in the hands of the professional did not end controversy. LURC was in shambles, a result of interminable controversy, chronic under-staffing, limited budgets, and resignations. The fate of the agency belied Barringer's facile assertion about the calming influence of professionalism and centralized planning.[51] Rather, it illustrated the difficulties of statewide planning in a state that prided itself as an exemplar of local control and town-meeting democracy.

Was Maine any closer to ecotopia in 1975 than when planners launched their initiatives in the late 1960s? In the Longley era it would not have appeared so. The conservative governor ordered a halt to the coastal planning effort begun in 1969, state planning official Alec Giffen explaining that resistance by local officials and developers "scuttled the idea." One Longley LURC appointee, highway contractor Kenneth Cianchette, defined this agency's new philosophy as an attempt to "please [the people] ... and not be hard."[52] Another LURC official pointed out that "a lot of these rules have changed substantially. Places where one once couldn't build are no longer restricted."[53]

Nevertheless, the state had come some distance with its planning rev-olution. The creation of LURC was itself a major symbolic achievement, and in some ways its less-strident tone was a compromise with mid-1970s realities. The pressures for wildland development abated as inflation drove up interest rates late in the decade, and as the sense of crisis dissi-pated, environmentalists accommodated LURC's role as a "relatively small forest management oriented organization ... designed to regulate modest development."[54] Governor Joseph E. Brennan, who replaced Longley in 1978, resumed Maine's commitment to comprehensive planning, although with less strident rhetoric and an emphasis on the role of municipalities. The founding principle remained: the wild character of the unorganized territories was to be protected, and new development was to occur only in or near previously developed areas, provided that the area's resources could support it. Moreover, LURC's hard-won acceptance by the forest industry confirmed, albeit tentatively, the "unwritten ... law [that] the way property is used should be reasonably consistent with public interest."[55]

However, like the Allagash Wilderness Waterway, the LURC law was still a product of Maine's longstanding suspicion of central planning. In the second half of the decade, the agency faced several new challenges, including high-intensity forestry practices like clearcutting, chemical herbicide and pesticide applications, whole-tree utilization, and extensive company road building. Debates over these issues made it clear that the basic questions that had sparked the planning effort — who should decide the future of the Maine woods? — had yet to be resolved. And LURC, as the debates also made clear, was not the agency to resolve them; much of its regulatory apparatus was, as House Speaker John Martin put it in 1982, "window dressing."[56]

Sensitive to the development issues raised by the planning debate and yet unprepared to accept a comprehensive master plan for the north woods, Maine citizens traveled a different path toward ecotopia in the post-Longley years. In the last dramatic grassroots environmental effort of the 1970s, Maine voted in referendum to purchase Bigelow Mountain, the area slated for development by the Flagstaff Corp. as the "Aspen of the East." The Bigelow Mountain wildland purchase was the first in a growing number of state and private acquisitions of conservation land, a trend that eventually superceded comprehensive planning as Maine's claim to leadership in environmental legislation. Unlike Oregon, where just over 50 percent of the land base was already in the federal domain, Maine stood to benefit from expanding its public recreational lands. Beginning in the 1840s, the state had divested millions of acres of public lands, and by 1970 the public land that remained was limited to a few small state parks and 200,000-acre Baxter State Park, a wilderness reserve donated to the

people of Maine by former Governor Percival P. Baxter. In 1975 the legislature established the Bureau of Public Lands to shepherd its public purchases, and in 1981, after a long court battle, the state regained control
over thousands of acres of wildland in the "public lots" — sections set
aside at statehood in each township for schools and churches but subsequently abandoned to the timberland owners who bought up the unorganized territories.

Later in the decade Maine established and funded through periodic
bond issues a Land for Maine's Future program, which expanded public
reserved lands to some 450,000 acres scattered through the state.[57]
Municipalities also established community land trusts and conservation
commissions, and private groups such as the Nature Conservancy, the
Landguard Trust in Freeport, and the Ecological Land Exchange Co. of
Hallowell did the same. The most ambitious of these private groups, the
Maine Coast Heritage Trust, founded by Mrs. David Rockefeller, Thomas
D. Cabot, Andrew Wyeth, and others, campaigned for voluntary conservation trusts that would "coalesce property owners and their allies into a
force ... to perpetuate the wildness and beauty of the coast."[58] Over time,
the land-acquisition movement expanded from protecting scenery and
recreational opportunities to preserving critical ecological areas and sustainable timber and food production. Another legacy of Maine's ecotopian idealism, the land-trust movement proved more amenable to Maine's
conservative sensibilities than central state planning.

Donaldson Koons, college teacher, environmental activist, and state
official, garnered perhaps the deepest understanding of the strengths and
weaknesses of statewide planning in Maine, having led the drafting teams
that wrote two of the three laws that set the planning idea in motion. In
1976, upon his retirement from politics, Koons wrote a letter to
University of Maine President Arthur Johnson admitting that although he
had once held great hopes for land-use planning, he had become apprehensive about the effort to impose an essentially urban concept on rural
Maine. "We are not going to re-invent Maine," he told Johnson.

In a similar letter to Richard Barringer he drew contrasts between protecting public waters from pollution and planning private lands to control development. In the 1960s, advocates "were confronted with private
use of a public resource — the waters of the State — in ways which were
clearly harmful and exclusive. This had to be changed, and it demanded

an adversary approach." The architects of Maine's land-use revolution failed to recognize the differences between this public commons, with its natural grassroots constituency, and the private domain, with its more conflicted and confusing coalitions of support and opposition. Thus they missed a vital shift in the quest for ecotopia. Water pollution was a public crisis threatening a key source of Maine identity; threats to the north woods were more diffuse and complex, and addressing them required careful cultivation of each constituency, particularly at the local level.[59] Instead of engaging these local constituents — landowners, resort developers, commercial foresters — planning advocates premised their strategies on the belief that a coming land-use crisis would rally Maine people behind their effort. In fact, the crisis never materialized. Distant from the metropolis and accessible mainly by private gravel haul roads, the north woods saw only modest development in the later 1970s, and LURC's limited mandate seemed sufficient to accommodate these diffuse pressures.

Koons's conclusions were tinged by his recent strained relations with contentious environmentalists, developers, and industrial forest owners, but his assessment of the strategic choices made by planners and environmentalists bears scrutiny. Clearly political leaders did not aggressively promote planning among the voting public; instead they allowed the publicity surrounding the controversy itself to shape public opinion. Centralized planning, premised on the crisis consciousness of the late 1960s and the preconception that downeast Yankees were unapproachable on the issue, bypassed a vital arena of grassroots organizing and constituency building that had been the hallmark of the clean-waters and coastal-preservation campaigns, even though public support was more critical in private land-use issues than in pollution control and coastal preservation.

Despite the limitations of the planning effort in Maine, the ecotopian idealism upon which it was founded provided a lasting legacy. Unlike other elements of the environmental imagination, ecotopian thought advanced the notion that growth, not stasis, was essential to living in harmony with nature. Adopting this premise, thoughtful environmentalists transcended the limits of romantic pastoralism and exclusionary wilderness advocacy to confront a more complex world of economic change. Conflicted though they were, Maine's ecotopian planning campaigns helped bridge the ideological gaps among environmentalists, government land-managers, and property owners, thereby creating zones of tolerance that served the environmental movement well in the future.

PRACTICAL PATHS TO ECOTOPIA IN OREGON

In Maine, the proponents of comprehensive planning largely ignored the task of grassroots organizing. Oregon's planners and planning advocates, in contrast, made impressive efforts to involve local communities in the planning process through professional media campaigns and by constantly publicizing the crisis in land use. This, along with stronger development pressures, a long history of federal resource-management initiatives, and a developing rural farm crisis in the Willamette Valley, prepared the way for centralized planning as a practical, if not an ecotopian, solution.

Willamette Valley farmers Lois and Cliff Kenegy were introduced to comprehensive planning in 1963 after they discovered developers breaking ground for an 11-unit housing project on land abutting their farm near Albany, OR. Over the next few years, growing ranks of subdivisions and ranchettes impinged on the future of farming in the valley, as new suburban neighbors began complaining about machinery clanking in the early morning hours, pesticide spray drifting across their property lines, and irrigation water spilling through their lawns and gardens. A Mennonite couple, the Kenegys learned to articulate the defense of their farm in both economic and religious terms. The land, according to Cliff Kenegy, "belongs to everybody.... God has given it to us for a purpose, and the purpose is to provide food."[60] Concerned about conserving the valley's fertile soils, the Kenegys and farmers like them began organizing petition drives, contacting extension agents, attending meetings, and badgering county planning officers. Over coffee, at Grange and agricultural-extension meetings, and in legislative hearings, farmers initiated a land-use regulation campaign that would dominate Oregon environmental politics in the early 1970s.

In Maine, planning gained a relatively narrow constituency and was based on a projected land-development crisis. In Oregon, the crisis was real. Because the balance between local initiative and state guidance evolved in a context of considerable development pressures, a wide range of interests concentrated in the state's heavily populated Willamette Valley supported the planning idea.

A scene along the Salem Urban Growth Boundary, Oregon.

Each of Oregon's 241 cities is surrounded by an "urban growth boundary" or UGB. The UGB is a line drawn on planning and zoning maps to show where a city expects to grow.

Oregon counties had been granted authority to plan and zone rural areas as early as 1947, but they rarely used this authority. Beginning in the mid-1960s, the legislature provided tax breaks for farmland on the condition that county officials establish some areas dedicated exclusively to farm production. This idea was promoted by professionals like agricultural-extension agent Ted Sidor, the state's first prominent advocate of farmland zoning, who developed a slide show on the effects of sprawl and showed it repeatedly to local farm groups and civic organizations.[61] When the tax-deferral law was infused with federal planning funds, Sidor and Wesley Kvarsten of the Mid–Willamette Valley Council of Governments (COG) promoted soil studies and other planning techniques as a basis for county zoning initiatives. Most farmers remained suspicious, but the repeated presentations and constant discussion in farming circles began to take hold. Hector Macpherson, a dairy farmer who watched the steady march of new homes and subdivisions up the road from nearby Corvallis, recalled that activist farmers learned to use the California example as a rallying point against suburban sprawl. "You can still scare people by telling them that Oregon is getting more and more like California."[62]

In 1967 the legislature ordered an interim committee to study problems affecting prime agricultural lands. The first witness before the subcommittee was Representative L.B. Day, a labor union official who, while representing food-processing workers, became interested in preserving Willamette Valley farmlands. Day presented a series of maps showing that 20 percent of the best agricultural soils in the northern valley had been lost to urbanization during the preceding decade, and he predicted the loss of another 20 percent in the decade to come. At ensuing hearings, planning and soils experts repeated these dire warnings, raising the specter of an agricultural land crisis in Oregon's pastoral paradise.[63] When legislators toured the state in December 1967 to hear the concerns of constituent farmers and ranchers, they found growing support for linking farmland tax deferrals with farmland zoning. A Jackson County farmer testified that he was not, in principle, an advocate of zoning, but when a new suburban development across the road boosted his tax assessments, he decided to put his land in the tax-deferral program. Given the alternatives — suburban encroachment and higher taxes — farmers warmed to the idea of rural planning.[64]

Despite the crisis mood, farmers and others remained mistrustful. Unlike the explosive beach controversy, the problem of suburban sprawl had to be carefully publicized before it became a basis for popular politics. In several counties, consultants had already written planning documents and county commissioners had passed zoning ordinances, but given the lack of general public support, officials granted variances readily. In this political climate, Sidor felt, it would be premature for the state to seize zoning authority from the counties without first investing in a public-awareness campaign orchestrated from the local and county level. Expressing similar concerns, the legislative interim committee recommended giving the counties three years to educate, plan, and zone against sprawl; if the task was not complete by the end of 1971, the state would step in.[65]

The committee's draft bills, presented in November 1968, laid out the compromises necessary to pass mandatory agricultural zoning. Addressing Oregon's ecotopian aspirations, the committee highlighted air and water quality, open space, and scenic and recreational resources; on a more practical level, the committee avoided potentially explosive topics such as forestland zoning and offered tax breaks for landowners and state funding for local oversight agencies.[66] The legislature consolidated the drafts into Senate Bill 10 and held hearings in spring 1969.

While planning advocates in the legislature left the difficult task of cultivating support to the counties, Governor Tom McCall, determined to implement statewide planning as a capstone to his administration, proposed a more centralized solution, similar to Maine's LURC, but applied to the entire state. At the beginning of the 1969 legislative session, McCall outlined a planning process to be tightly controlled by officials and planning professionals in his own administration. County zoning seemed feasible — 6 of Oregon's 36 counties already had plans in place — but counties lacked qualified planning professionals and staff, McCall reasoned, and the issues were too complicated to process without such resources. To coordinate local efforts, McCall proposed 14 regional planning districts corresponding to the COGs and answering to the state's new Department of Environmental Quality.[67]

An able rhetorician, McCall took his message on the road, reminding Oregonians, and Americans at large, of the need to act forcefully against speculators who made "hamburger of the land, chopping it up into little pieces and offering it for sale as perfect for whatever use the prospective buyer wants it to be." McCall played upon Oregon's longstanding concern about in-migration and urban growth, adding his own rhetorical snap and his memorable penchant for expansiveness: "the steady scatteration of unimaginative, mislocated urban development," he announced, was "introducing little cancerous cells of unmentionable ugliness into our rural landscape." Using heady pastoral symbolism and invoking a sense of crisis, he tuned the exclusionary tendencies in Oregon's environmental rhetoric to a fine pitch. All Oregonians bore the costs of California-type sprawl, he asserted, in rising pollution levels, loss of farmland and forestland, degradation of open spaces, and diminished options for economic diversification.[68] The crisis rhetoric masked a more subtle concern for the McCall administration. Poor planning, endemic to local politics, led to acrimonious outbursts of local environmentalist energy that contributed little in the broader scheme of things. McCall hoped to channel these energies into a central administrative apparatus staffed by professional planners. This system would both control development pressures and dampen the explosive environmental mix that had scarred the governor's tenure on more than one occasion.[69] McCall's strategy resonated with an organizational push by valley farmers and environmentalists, and the campaign bore dramatic results.

However, while almost everyone agreed with the sentiments expressed by the popular governor, implementation, as in Maine, created a storm of controversy. The planning ideal, borne of ecotopian aspirations and grassroots concerns over disappearing farmland, dissolved into a debate over power and process: should planning be lodged in a centralized state apparatus, or should it remain with county and municipal administrators? Concerned about McCall's top-down approach, the Senate Agricultural Committee wrote the governor an instructive note on political realities: experiences with rural zoning at the county level demonstrated the need for more citizen involvement and local control. Unless the affected people were involved from the beginning, planning seldom succeeded. This cautionary mood derived from an experience in Clackamas County, near Portland, when in 1967 citizens repealed the county zoning regulation, leaving a distinct impression that they could handle growth "with[out] the help of the Governor."[70]

McCall's own inquiries reinforced the legislature's warning. Some local constituents favored the idea of planning districts, but others were lukewarm or hostile, including many large landowners and timber and industrial interests and their lobbying arm, Associated Oregon Industries (AOI). McCall, in short, found land-use planning more divisive and far less popular, generally, than river cleanup and beach preservation. Indeed, the well-attended hearings on S.B. 10 in spring 1969 demonstrated the need for a cautious approach even while they suggested that the issue could no longer be ignored.[71]

Opinion at the hearings ranged from the reasoned to the colorful, the pragmatic to the paranoiac. The Willamette Valley Grange sent a resolution opposing zoning on principle, and the Zoning AdjustMent Organization, or ZAMO, a suburban Portland grassroots group opposed to zoning, made repeated appearances before legislative committees. Agnes James of Albany informed the Senate Agriculture Committee that "when we don't let a man use his own knowledge and ingenuity we are depriving that man of something that is God given and rightfully belongs to him."[72] Paul Ramsay of Bend considered S.B. 10 a federal plot, leading first to regional government, then to interstate and federal government, then to the First Purge: "if you scratch a professional planner ... you will find either a bureaucratic sociologist, an American Socialist, or an American Communist. I have yet to meet a practical minded planner."[73] In contrast to zoning opponents, who drew on a long history of opposition to state regulations of any sort, advocates were just getting organized. The League of Women Voters had been working on air and water pollution and suburban sprawl for years, but only recently did the organization begin to see statewide zoning as a means of addressing these problems. The Oregon Environmental Council (OEC) adopted land-use planning as a major initiative, but it, too, was relatively new to the scene.[74]

In light of these political ambiguities, the legislature rejected the governor's centralized-planning approach and compromised with local forces by leaving the planning initiative in the hands of county government. Failing to provide state reimbursement for county administration, the legislature sent a clear message that it was still ambivalent about the initiative. In more positive terms, S.B. 10 mandated extensive public participation in the county planning process, a provision designed to promote the idea of land-use planning as much as to obtain public input.

For reasons ranging from the expedient to the quixotic, Oregon's entire civic culture — farmers, ecologists, government officials, business owners, industrialists, civic leaders, planning professionals, and plain citizens — became absorbed in the merits and demerits of statewide planning and zoning. With the press and the governor publicizing the dangers of suburban sprawl and county governments hosting public forums, popular consciousness of these issues grew remarkably after passage of S.B. 10. The bill thus helped build a broader coalition around the core interests of Willamette Valley farmers, environmentalists, planners, and the governor's office.

Opponents of rural zoning gained strength as well, inspiring a citizen initiative to prohibit zoning regulations outside city limits. Appearing on the November 1970 ballot, Measure 11 triggered a spirited campaign that kept the issue of suburban sprawl before the public at a time when most county administrators were taking a "wait and see" attitude toward S.B. 10. To defeat Measure 11, the OEC drew together a diverse coalition of interests, including Governor McCall, State Treasurer Robert Straub, the Portland City Club, the AOI, the AFL-CIO, and the League of Women Voters. The campaign crystalized an image of ecotopia's enemies as greedy California developers who would inflict scenic degradation, pollution, sprawl, higher taxes, and lower property values on Oregon to further their own selfish aims.[75] The defeat of Measure 11, the first statewide polling on the planning idea, was an encouraging sign for environmentalists.

As a complement to S.B. 10, the 1971 legislature created the Oregon Coastal Conservation and Development Commission, a regional planning agency with a mandate to zone the entire coast by 1975. Unlike Maine's Shoreland Zoning Act, which mandated town-level zoning, this measure lodged control in a supra-county planning commission — an experiment that would test the political and administrative bases for centralized planning.[76] Hopes for coastal planning were soon dashed, however, when the planner McCall recruited to manage the program resigned from the commission, asserting that local representatives were engaged in "covert actions" to aid their developer–clients. With commissioners still operating in an atmosphere of "smoky filled rooms," the experiment cast doubt on the principle that regional planning would rise above the interest mongering that characterized much of county-based politics.[77]

Senate Bill 10 and the Coastal Conservation and Development Commission represented tentative steps toward state-directed land-use planning. Neither was effective, but these experiments, like Maine's first LURC law, raised public expectations, drew supporters into a powerful coalition, and established a framework for public input at the local and county level. As Oregon planners and environmentalists digested the weaknesses in these initial legislative attempts, they prepared for the next step in planning for ecotopia.

County commissioners achieved some successes as they moved toward S.B. 10's 1971 deadline, especially in the mid–Willamette Valley, where farmers carried considerable political clout. It was here in Benton County that Lois and Cliff Kenegy asserted their claims that God gave Oregon good land for farming, not for housing tracts. The passage of S.B. 10 had raised the Kenegys' hopes, but Benton County supervisors rejected "exclusive farm" zoning regulations under the authority of S.B. 10. Soon after, a Corvallis housing developer purchased three large farms next to the Kenegys and platted them as 5- and 10-acre ranchettes. "I got up that morning, and it was raining anyway," Cliff Kenegy related, "and I told Lois 'I'm going to start knocking on doors.'... The first farmer I came to, a dairyman down the road a ways, I told him what I had in mind, and he says, 'well, I told my wife I was going to do the same thing.'" With support from

Oregon State University in Corvallis, Benton County farmers succeeded in enacting exclusive farm zoning. According to the Kenegys, local environmental activists from Corvallis "joined us in our agricultural concerns, realizing the value of our farmland for its own sake [and for] ... the concept of greenbelts around the city."[78]

One of the farmers who met the Kenegys during their Benton County campaign was Hector Macpherson, a dairy owner from nearby Linn County. Macpherson was unsuccessful in promoting exclusive farm zoning in his home county, but following this experience he was elected to the legislature in 1971. After watching the governor's regional-planning scheme go down to defeat in his initial session, he asked the Senate leadership for an interim committee to renew work on reviving the notion. When the Senate failed to fund his request, Macpherson contacted the OEC's Martin Davis, who suggested a collaboration. Working closely with Davis and the OEC, Macpherson created an informal interim Land Use Planning Committee at his own expense, after clearing the idea with Senate leaders. The 14-member committee represented a variety of environmental, governmental, business, farming, and civic groups, plus several professionals in economics, engineering, and law. Government figures included fellow Sen. Ted Hallock and McCall representative Robert Logan.[79] Macpherson's committee was an unprecedented effort to bring together traditional adversaries. Unlike a formal legislative interim committee, Macpherson's group avoided the press and the public. Over a period of nine months, it formed subcommittees, commissioned legal reports and documents, debated ideas and strategies, and eventually produced a bill to replace S.B. 10.

A central question in Macpherson's committee was the designation of "critical areas" to be accorded zoning protection as undeveloped preserves, a concept drawn largely from the pending National Land Use Planning Act. Environmentalists, guided by the OEC, urged a long list of critical areas. Dean Brice of AOI proposed a much shorter list, to be designated by the state legislature and included in the bill.[80] Following one particularly heated debate between Davis and Brice, environmental groups wrote another draft of the critical-areas section, agreeing, finally, to a strategy of avoidance: "the more specific you are, the more likely to have someone that finds fault with it. And so we took refuge in being general."[81]

While Macpherson's group was hammering out a new version of S.B. 10, Governor McCall stepped up his own public-awareness campaign. On

June 1, 1972, he formed the Willamette Valley Environmental Protection and Development Council, consisting of the four valley COGs and himself, and launched Project Foresight. The result was *The Willamette Valley: Choices for the Future*, to be used in the related Project Feedback.[82] With its contrasting dystopic and ecotopic scenarios, *Choices* left the fate of Oregon in the hands of the people, but it also made clear that centralized planning was the path that leads to ecotopia. In the fall and winter of 1972–1973, the administration held a series of carefully managed public meetings at which professionals led citizens through the planning process, abandoning along the way, McCall hoped, the unending string of grassroots controversies that characterized Oregon's environmental movement. Drawn by a new slide show on sprawl and by Halprin's persuasive *Choices*, some 20,000–25,000 people attended Project Feedback sessions. Project director Ken Bonnem, a planning professional recently arrived from Illinois, commented that the essentially flat landscape of his native Midwest gave a sense of endlessness: "here in the [Willamette] Valley, the land is finite. You can understand its limits." Bonnem quickly learned to dramatize California's tract houses, its clogged highways, its polluting factories, and its endless commercial strips. Environmental reporters contributed newspaper articles on suburban sprawl in the eastern desert and along the coast, extending the land-use debate beyond the Willamette Valley.[83]

When the 1973 legislative session opened, the campaign to shift control over land use from local jurisdictions to the state was well under way. Macpherson's committee had hammered out the basic compromises for a new land-use bill; the governor, his planners, and the media had focused public attention on sprawl as Oregon's most pressing environmental problem; and governmental, business, civic, development, and environmental groups were coalescing around the issue.[84] In February 1973, overflow crowds attended Senate hearings on Macpherson's S.B. 100. Proponents of state-directed planning insisted that local officials were "impotent, inadequate, and negligent in dealing with the pressures of an expanding population and economy," and came armed with stories about a "virtual Sodom and Gomorrah of land use planning" in one Portland suburb. Invoking a sense of crisis, they warned that the specter of unplanned development was "soon to be visited on the rest of the Willamette Valley and the Oregon Coast."

As the critical vote on S.B. 100 vote neared, Governor McCall brought his rhetorical powers to bear. In the most widely reproduced statement of his career, McCall announced that "sagebrush subdivision, coastal 'condomania' and the ravenous rampage of suburbia in the Willamette Valley all threaten to mock Oregon's status as environmental model of the nation.... The interest of Oregon for today and in the future must be protected from the grasping wastrels of the land."[85] The publicity was effective; a state-commissioned survey found that more than two-thirds of all Oregonians now considered the problem of sprawl a crisis.

Yet citizens were divided about where control over land-use decisions should rest. Most favored a mix of state, regional, local, and landowner control, but environmentalists and some state administrators still hoped for McCall's centralized approach, lodging authority in the state's 14 COGs. These differences drove a wedge into the planning coalition, and by March it was clear that S.B. 100 was in trouble. Washington County Commissioner Eldon Hout predicted that "the monotonous subdivision of the 1950s" would be "replaced by the monotonous planned unit development of the 1970s."[86] The sharply divided Senate Environment and Land-Use Committee began a painstaking "point-by-point review," and to rescue the bill, Chair Ted Hallock ordered a special committee to draft another compromise, "in the straight political sense." When environmentalists labeled this a sellout, Hallock lashed out at his erstwhile allies for "attacking the few liberals left who supported their positions."[87]

In late March, after a series of acrimonious public hearings, James Moore, representing the League of Oregon Cities, submitted an alternate bill in committee that eliminated the COGs as mediating agencies and put the counties "in the driver's seat." The revised bill required Oregon's 36 counties to develop comprehensive plans based on state goals and input from local citizens. The plans would be submitted to a new seven-member state agency, the Land Conservation and Development Commission (LCDC), and after more public hearings the LCDC would adopt and approve the county plans by January 1, 1975. Institutionalized local citizen input would meliorate concerns about landowner rights. The real secret, according to L.B. Day, was the people: planning would "start from the bottom up, not from the top down." But with administrative power lodged in the counties, planners, and environmentalists would find no shortcuts to ecotopia. In similar committee resolutions, business interests won their point on the number and definition of "critical natural areas" as well.[88] The conflict over centralization was thus reconciled through a compromise hammered out behind closed doors in subcommittee.

City representatives, who saw control pass into the hands of county administrators, labeled the compromise a step backward. Environmentalists realized that planning issues would be contested on a county-by-county basis, and that extension of the critical-areas concept would come only through the slow process of renewed grassroots activism. Although deeply disappointed about the COGs, McCall backed the measure. With McCall on board and the planning coalition patched back together, the compromise between state-directed planning and local initiative passed the legislature, moving Oregon to the front ranks of environmentally friendly states. The controversies over planning were far from over, L.B. Day remarked, but now there was "a mechanism for dealing with them."[89]

S.B. 100 was Oregon's greatest contribution to American environmental politics, and more than anything else it is responsible for Oregon's reputation as a leader in environmental policy. Public discussion of land-use planning continued throughout the 1970s, leaving S.B. 100 Oregon's premier symbol of civic commitment to ecotopia. The strength of Oregon's environmental imagination, expanded to incorporate entire landscapes in places with pressing growth pressures, largely explains LCDC's continuing popularity. According to historian Carl Abbott, the Valley's "conservative and moralistic" political culture incorporated planning in order to protect an Edenic way of life: "Oregonians ... liked what they have, and they have wanted rather smugly to protect it against unwanted change."[90] Ironically, at the end of the 1970s fewer than one in three Willamette Valley residents had been born there, and nearly 40 percent of the valley's immigrants were from California. H. Jeffrey Leonard points out that these new arrivals "provided the core of a constituency committed to protecting agricultural land from the encroachments of urban development."[91] Oregon's new citizens were eager to affirm their choice of living space by participating in land-use politics.

Popular involvement in the planning process was also an important part of the LCDC's eventual success. Over 10,000 Oregonians attended the three rounds of public workshops and hearings held in the first year after passage of S.B. 100, building local constituencies with a stake in the planning process. As one group of workshop attendees put it, local people accepted state planning because they felt "a positive involvement in the mythology of the State of Oregon."[92] At another meeting in LaGrande, in eastern Oregon, the crowd applauded when a local judge joked that this was the first time a state agency had come to ask local opinions before making up its mind.[93]

Yet LCDC's enthusiasm for local input weakened as both planners and the public realized how complicated land-use planning had become. In response, the LCDC put growing emphasis on recruiting key local leaders, holding special "pre-workshop" meetings with selected business, labor, and civic groups, and designing a "tighter process" that would "provide less opportunity to criticize." The result was a decline in citizen participation, accompanied by rising frustration with the concept of planning by public acclaim. Salem newspaperman J. Wesley Sullivan wrote Commission Chair L.B. Day that it was unreasonable to expect the public "to absorb the background necessary to make an appraisal of this great amount of fine print or to attend the hearings en masse. As a result, the special interests and staff have a preponderance of input."[94]

Thus while planners continued to regale Oregon citizens with the benefits of land-use planning, they turned more and more to Oregon's developers and politicians for detailed input. Responding to the political realities of the situation, Governor McCall helped form a new kind of environmental organization that could serve as a legal watchdog for implementation of S.B. 100. Named 1000 Friends of Oregon, the organization recruited environmental, business, and civic leaders whose most common characteristic was their position as "insiders." Determined to reconcile economic development with environmental protection, the 1000 Friends garnered a reputation for hard-nosed, persistent, detailed work that helped assure the law's full implementation. Combining thorough research, backroom dealmaking, lobbying, and litigation, the organization helped convert an ecotopian vision into a model planning and zoning statute.[95]

Despite this shift in levels of organizational power, Oregonians remained steadfast in their commitment to land-use planning. A 1976 effort to repeal S.B. 100 was defeated by a margin of 57 to 43 percent, and two years later another repeal initiative was defeated by an even wider margin, with support emerging in areas newly besieged by population growth and tourism, such as the north coast, central Oregon, the Medford area, and even eastern cities such as Pendleton and Baker. Efforts to gut the law in the legislature met a similar fate. This continuing commitment ensured that Oregon would remain America's leading environmental state, even though its most innovative feature, zoned "urban growth limits," fell far short of the ecotopian vision outlined in *Choices for the Future.*[96] In thousands of local zoning debates, Oregonians tested the strengths and weaknesses of land-use planning, keeping the issue at the center of public life throughout the decade.[97]

In Maine, Oregon, and a handful of other states, comprehensive planning offered the solution to a host of complicated environmental problems: incompatible land uses, habitat loss, landscape degradation, ecological disruption. Although the results were uneven and the politics divisive, these states advanced a powerful defense of common landscape values and achieved a great deal in rationalizing the use of precious scenic and ecological resources. Moving beyond the isolated river, mountain, and coastline, state environmental organizations helped to codify the common values Americans found in their rural landscapes. In Maine, where development pressures were mostly latent, planning initiatives remained modest in scope and achievement. In Oregon, where more intense development pressures and a vigorous public campaign prepared the ground for grassroots support, activists and political leaders were far more successful.

State planning processes helped reconcile divergent interests in rural property and realize common aspirations for a permanent landscape. Maine and Oregon would look different today without administrative efforts reflecting the planning processes implemented in the early 1970s. But the process of building and empowering planning agencies widened the ideological gap between those who saw the state as a protector of their landscape values — freedom, authenticity, and permanence — and those whose saw the state as a threat to those same values. In subtle ways, the planning process changed the dialectic of environmental ideals and environmental politics. Having offered up their lofty ecotopian visions to the planning professional, environmentalists found them reduced to a series of problems to be handled through the regulatory process. With the link to ecotopia severed, environmentalists lost a powerful tool for mobilizing grassroots support.

That aside, citizens in Oregon, Maine, and elsewhere gained much from the planning process; they secured affirmation that they lived in, or near, permanent and authentic rural landscapes, and this affirmation helped invest American pastoral and wilderness myths with new meaning. Ecotopia did not disappear from the environmental imagination; rather, political and economic trends in the late 1970s generated a new vision of freedom, authenticity, and permanence, secured not in a mythical rural past or a remote realm of pure nature but in a contemporary postindustrial society. Americans took up a new quest for livability in the later 1970s, secure in the knowledge that their natural landscapes were planned and preserved from a dystopic modern world. To be sure, tensions remained, but the quest for ecotopian landscapes accommodated an astoundingly diverse constituency anchored to the idea of permanence, authenticity, and freedom in communion with nature and mythic place.

Crosswater, Sunriver Oregon

CHAPTER 7

A VIEW ACROSS THE
GOLF LINKS:
PASTORALISM IN AN ERA
OF DECLINING CIVIC
ENGAGEMENT

In January 1973 Oregon resort developer John Gray stood before an assemblage of members of the National Association of Home Builders in Houston and publicly endorsed the stringent new environmental laws passed in his home state. He admonished fellow builders to do likewise. Careful land stewardship, he told the crowd, would show "in the color of [the] ink" on the developer's balance sheet. Thus in the new Oregon, nature would "enjoy the loving solicitude of a vast majority" and not just the "few noisy ecological nuts whose efforts bring little lasting benefit."[1] Gray was not exactly sublimating his economic interest in this environmental rhetoric. Stabilized land use, after all, was a key component of the well-groomed landscape he envisioned as backdrop for his exclusive Oregon resorts. The nature he defended, in speeches like this and in the resorts he built, differed from the pastoral images of the 1950s and the wilderness ideals of the next decade, but his commercialized expression of natural harmony was shared by many in the 1970s.

Gray's brand of pastoralism, a sumptuous resort set in an unspoiled natural context, was not widely accepted among the movement's activists.

Iconoclastic environmentalist Edward Abbey articulated the concerns felt by many old-time rural westerners — and old-time environmentalists — as they watched familiar rural settings grow into crowded upscale nature-oriented resorts of the type Gray saw as the new pastoralism. Telluride, Colorado, which Abbey claimed to have discovered during a "picnic expedition" into the San Miguel Mountains in 1957, had been an "honest, decayed little mining town of about 300 souls" as late as 1974. Soon after, it was "Californicated," in Abbey's words, into a "bustling whore of a ski resort with a population of 1,500." Like John Gray, California developer Joseph T. Zoline promised that Telluride's growth would be "controlled and orderly ... with full environmental protection." But what Abbey found disquieting was not the loss of ecological balance but the loss of local color and Telluride's organic sense of place. "What some of us liked so much about Telluride was not the skiing but the quality of the town which Zoline and his developmental millions must necessarily and unavoidably take away: its rundown, raunchy, redneck, backwoods backwardness. That quality is one you cannot keep in a classy modern ski resort, no matter how much money is spent for preservation."[2]

The subtle class tensions generated by the new Telluride expressed an inner conflict in the environmental imagination even as nature appreciation reached unprecedented levels of popularity in U.S. society. Like wilderness advocacy, the new pastoralism, actualized through personal lifestyle and consumer choices, erased the folk as an authenticating ideal and replaced them with an urban consumer. As Leo Marx pointed out, traditional pastoralism was too naive and too escapist to serve as a guide for reforming an urbanized and industrialized society. In its new guise — abstract, private, and commercial — the romantic pastoral tradition continued to resonate with U.S. popular culture, and Americans sustained their quest for a better environment by engaging these new forms of popular passion. However, because the new pastoral themes easily commingled with modern commercial culture, they dulled the environmental movement's oppositional edge.

The new pastoral vision reflected the changed political culture of the late 1970s. In the previous decade, environmental activists had won their most significant victories in the arena of state politics, challenging corporations on a variety of issues and forging a new national political ethos in the process. But even as opinion polls continued to show strong support for environmental goals, popular participation in statewide environmental campaigns waned in the mid-1970s, as politics lost the common civic vision that had sustained grassroots activism through the previous decades.[3]

A number of historians attribute the movement's loss of grassroots vitality to the "Washingtonization" of the major environmental organizations. Daniel D. Chiras, a sympathetic critic, noted the symptoms: emphasis on mass-mail recruiting; a political base composed of "soft members" willing to give money in lieu of personal effort; and a narrow focus on "imageable" issues likely to appeal to ephemeral public interest. Engaged in a war for power and funding at the national level, environmental organizations lost touch with ordinary people.[4] Richard N.L. Andrews notes that after the first wave of victories in the 1960s, "the politics of environmental protection shifted rapidly ... from mass mobilization and demonstrations to interest-group ... lobbying for tougher environmental laws, regulations, and enforcement," largely at the federal level. According to Andrews, business leaders encouraged this shift, seeing "preemptive federal controls" as easier to shape than more diverse and sometimes tougher state laws. In response, environmental organizations established offices in Washington and began to channel popular interest into institutional representation, lobbying successfully for a notable run of laws that included the National Environmental Protection Act, the Clean Air and Clean Water acts, the Endangered Species Act, and the Toxic Substances Control Act — each "remarkably successful in achieving what [it was] designed to achieve." Historian Mark Dowie maintains that this shift from local to national politics was predictable: "at some point in the history of every great American movement a federal strategy is applied." But, he continues, "in no other movement ... has a federal strategy absorbed such an overwhelming portion of talents and resources as it has in mainstream environmentalism."[5]

The decline in grassroots environmental activism should be viewed in a broader context. The mid-1970s saw the end of an era of popular political participation unique in modern America. Burdened by post-Vietnam and post-Watergate cynicism and the lingering memory of the Kennedy, King, Kent State, and Jackson State assassinations, the political culture was less receptive to grassroots organizing. The inflationary spiral brought on by the war and the Arab oil embargo strained the economy, complicating political proposals that affected taxes and consumer prices. The war itself also fractured the liberal consensus, as did the centripetal forces in the antiwar, civil rights, and feminist movements. No longer certain of the southern, trade union, rural, or suburban ethnic vote, Democrats wavered in their support for interventionist strategies. As Stephanie S. Pincetl points out, this period saw a "deliberate depoliticization of politics" in major party platforms and ideologies, leading to a "lack of real analysis and of programs that could produce the fundamental changes people seemed to ready to embrace."

Women were also moving into the workforce in greater numbers and had less time to devote to civic associations, and the American workforce was becoming more mobile. Families with less residential stability and more private leisure options disregarded the community organizations that had launched the environmental movement two decades earlier. In any event, nationalized markets and institutions seemed to call for nationalized political strategies. Finally, the waning of the counterculture deprived grass-roots environmentalism of a rich source of visionary thinking.[6]

Grassroots activists also confronted more perplexing issues. Remaining state-level problems — non–point-source pollution, soil erosion, groundwater contamination, hazardous waste disposal, pesticide spraying, habitat loss, acid rain, energy conservation — had much higher populist flash points and often lacked the intense anticorporate focus that animated earlier campaigns. New kinds of suburban growth, for example, complicated what had once seemed a clear-cut battle between developers and preservers, and here, as Audrey Jackson of the Oregon League of Women Voters explained, arcane technical expertise counted more than grassroots mobilization.[7] According to Maine land-use planner Holly Dominie, the newest environmental threat was a gradual accretion of personal housing choices that threatened the rural landscape — "people pollution," in her words. The term had been introduced a few years earlier by Paul R. Ehrlich, whose neo-Malthusian calculations about population growth tinged the movement with misanthropism. Significantly, Dominie used a collective pronoun to identify this environmental foe: "by not paying attention to the overall patterns we are creating … we are indeed despoiling Maine's scenery and visual character."[8] Moving from anticorporate populism to the politics of self-recrimination, environmentalists found the public less receptive.

The Oregon Environmental Council's experience in the later 1970s suggests the difficulty of organizing under these new conditions. OEC had played a key role in enacting the state's path-breaking S.B. 100 in 1972–1973 while pursuing a wide range of issues as the central coordinating body for over 80 environmental organizations. But once the technical tasks of the Land Conservation and Development Commission had been ironed out and city and county plans brought into compliance with statewide goals, land-use decisions were less open to the public controversy that had brought OEC into being a decade earlier. The "first flush" of popular activism was over, OEC leaders opined; the organization no longer stood on the front lines as the "brushfire organization of Oregon's environmental movement."[9]

OEC's activists and officers were more skilled and experienced by the second half of the decade, but defending hard-won legislative gains from the foes of regulation was less dramatic than crusading to pass S.B. 100, with fewer populist overtones. Important work remained: subdivision and suburban sprawl threatened the state's open lands; clearcutting spread like a blight across the Coastal and Cascade ranges; development pressures menaced the Oregon way of life; and Governor Victor Atiyeh showed none of his predecessors' caution about the politics of growth. Yet OEC seemed unable to reestablish its bearings in the post-S.B. 100 era.

Signs of slowing momentum appeared as early as OEC's fifth annual meeting in 1975. Guest speaker David Brower's focus on global and national issues and recognition of Hector Macpherson as environmentalist of the year for his 1972–1973 S.B. 100 fight suggest that OEC had fewer local accomplishments to celebrate in 1974–1975.[10] Indeed, the focus of the 1974 and 1975 OEC newsletters was largely national and international, highlighting such issues as a Japanese whaling boycott, the Alaska federal lands, nuclear radiation, national recycling and energy conservation, and predator eradication. The monthly newsletter became bimonthly shortly after, and when longtime Executive Director Lawrence Williams left OEC in 1978 for a position in Washington, DC, Oregon's flagship organization entered a period of uncertainty.[11]

On the tenth anniversary of Earth Day, OEC dedicated an issue of its newsletter to reflections on the 1970s and recommendations for the 1980s. Coming from the state's leading environmentalists, these ruminations reflect the choices and dilemmas that had divided the movement since the mid-1960s — working rural landscape or wilderness, recreational access or ecological purity, state oversight or local control. There was a new element of discord in these reflections, however. Most of Oregon's environmental leaders continued to endorse the idea of a politically managed defense of nature, but Randal O'Toole of the Oregon Student Public Interest Research Group urged instead a market-based strategy. The sad fact was that government ownership and regulation had failed, he thought; most environmental destruction was "either caused by or sanctioned by the government," through direct or hidden subsidies to timber companies, miners, ranchers, or corporate farmers. Fiscal conservatism would be the "watchword of the '80s," and environmentalists would do well to join the calls for reduced government and learn to use market forces to their advantage: they "must learn economics, and they must teach economics."[12]

Environmentalists faced some difficult choices in the later 1970s. They had achieved a great deal in curbing corporate plunder of the environment, but with their ideas now encoded in law, where was the movement headed? Organizations continued to negotiate their political demands in state and federal government, acting out a political drama that began back in the 1950s, but groups and individuals out of the mainstream movement pioneered a variety of new non-political solutions as well. Like O'Toole, they had learned to distrust centralized government, and they took the environmental imagination in new directions, inspired by a diverse set of highly decentralized personal lifestyle choices and consumer trends — new paths to ecotopia postulated on new ways of thinking about nature and place.

These changes were foreshadowed in a popular political treatise published in1970, the year passage of the National Environmental Protection Act signaled the movement's shift from the local to the national scene. In *The Greening of America*, Charles A. Reich recounted a pattern of denuded forests, polluted rivers, sterile cities, and rural lands "ripped by highways, high tension wires, and suburban swaths," and he attributed this devastation to a society that had become "more and more pointless and empty … plastic, vicarious, and false to our genuine needs." However, in assessing the causes of this crisis, Reich glossed over the sharp anticorporate analysis that had marshaled 1960s dissent and focused instead on the "American Colossus" — a combination of masterless technology, anonymous social forces, and a "growing atmosphere of unreality" in modern culture, all of which followed an economic imperative independent of any particular corporate villain.[13]

Dismissing the idea of a corporate despoiler, Reich, like Maine's Holly Dominie, traced the "American crisis" to diffused personal choices. Americans, he thought, were alienated from nature and place. Obsessed with mobility and change, they demolished and rebuilt neighborhoods before they could infuse these places with meaning and tradition; they allowed generic commercial considerations to dominate regional architecture. They ate food that bore no relation to the land, and lived and worked in "small, identical rectangles" with artificial air, weather, and sound, oblivious to the satisfactions of an intimate connection to place. Cut loose from the land, Americans experienced a deep sense of hopelessness. "In the spring, on the first soft, breezy, gently stirring day after a long winter, man feels a pang, an ineffable longing." They filled this void, according to Reich, through obsessive consumerism. A substitute for freedom, authenticity, and permanent contact with the land, consumption fit the growth imperatives of the American Colossus in ways the traditional work ethic could not.[14]

Reich's analysis of the culture of consumption was astute, but his search for solutions was problematic. Having obfuscated the antagonist in the American crisis, he sought remedies in an equally abstruse individual cognition. If indeed the villain was personal choice, then the solution would be personal as well. "Consciousness III," Reich's revolutionary reawakening of the soul, would reunite people with nature and place by forging a new sense of identity from the unique personality of each individual. Consciousness III would create a "dynamic togetherness," a constantly changing group interest that melded and dispersed according to immediate events: a rock festival or political protest, for example. After each group encounter, the Consciousness III citizen would reaffirm self-identity through intensely subjective experience: "he takes 'trips' out into nature; he might lie for two hours and simply stare up at the arching branches of a tree."[15]

Consciousness III would bond the individual to the community through a new form of consumerism: subversive, expressive, and grounded in real human needs. Bell-bottom pants, for example, offered a foundation for dynamic togetherness. "They give the ankles a special freedom as if to invite dancing right on the street.... A touch football game, if the players are wearing bell bottoms, is like a folk dance or a ballet. Bell bottoms … are happy and comic, rollicking. No one can take himself entirely seriously in bell bottoms." Consciousness III music linked audience and performer across race, ethnicity, and nationality; it "freely took over elements from Indian ragas, from jazz, from American country music … a protean music, capable of an almost limitless range of expression." Its technology — electric instruments and amplifiers — would liberate rather than encumber the soul.[16]

A somewhat bizarre *cri de coeur* for the 1970s, *The Greening of America* underscores the difficulties of organizing grassroots environmental initiatives in this decade. Reich's focus on alienation from nature and place drew attention to the rootlessness of the first generation of middle Americans to come of age in suburbia and hinted at the implications for their new pastoral ideal. Like Reich, suburbanites viewed the metropolis as an object of open hostility, thereby denying their own essential geography. This geographical alienation — the easy interchangeability of place in a mass-produced suburban world shielded from obvious signs of metropolitan infrastructure — fostered an equally interchangeable pastoral sensibility: a generic rather than organic sense of place.[17]

Reich's faith in subversive consumer choice also echoed the intensely personal perspective on social questions that marked the youth rebellion in its later stages. The politics of personal choice was a reaction to the dehumanizing effect of rationalism, industrialism, and mass culture, and thus an appropriate response to the bureaucratic structures that seemed to rule over a flawed society. But as historian David Burner points out, paradoxically, the personal nature of 1970s politics left these grass-roots movements vulnerable to the mainstream culture against which the rebellion is directed. Consciousness III, like the personal and moral tone of race, gender, and ethnic consciousness, of consciousness raising, and of conscientious objection to war, harbored a potential contradiction, in which, according to Burner, "from the exaltation of every spasm of the moral self, it is but a step to the celebration of every twitch of the hedonistic self."[18] For some, personal commitment to nature preservation could became an invitation to consume nature.

Reich's politics of self-recrimination also helps explain the waning of traditional environmental politics in an era without identifiable adversaries. As Murray Bookchin pointed out, this perspective places blame for environmental problems "not only on land speculators, construction barons, government bureaucrats, landlords, and corporate interests ... but rather flippantly, on the general public ... a target of insidious propaganda that demands concern, but denies the power of action to those who are most victimized."[19]

The Greening of America previewed the difficulties mainstream environmentalism would face in the 1970s. Because the old formulas for mobilizing popular concern — identifying a corporate foe and invoking the politics of place — were becoming less relevant, three new themes, each emphasizing personal choice and community politics, linked the new pastoralism with the defense of nature. First, some middle-class citizens acted on their environmental sensitivities by endorsing commodified symbols of nature and regional identity as a new consumption ideal. These commercial images, although they lacked the anticorporate edge that fueled earlier environmental politics, kept the passion for nature and place alive in a highly mobile suburbanized society.

Second, a demographic shift to the countryside — to exurbia — occurred in the 1970s, as young, two-income families abandoned the classic suburb for greener pastures, where they could learn to care for their adoptive neighborhoods and the surrounding environment. This, too, was deeply contradictory as a form of environmental consciousness, but it encouraged a preservationist instinct tied to a specific place.

Third, the second half of the decade saw experiments with new homesteading, urban gardening, and appropriate technology, all efforts that expressed a personal commitment to protecting nature by consuming less, using more efficient, sustainable technologies, and buying locally grown products. The latter movement was nonpolitical in the 1960s sense but still expressed traditional environmental values based on the idea of balance between people and nature. These variants of the pastoral ideal — dramatic departures from the ideological forms that motivated earlier environmental politics — infused the movement with an environmental imagination appropriate to an age of declining political engagement. The continuing popularity of environmentalism, and the strength of the ideology itself as a national political force, depended in good part on these conflicted ideals of the new pastoralism.

In an essay published in 1983, University of Oregon historian Richard Maxwell Brown linked these new pastoral images to a reassertion of regionalism, a movement surprising to cultural critics who assumed that mass media and mass-consumerism would completely homogenize American life. Regionalism would be revitalized, Brown argued, as a new set of myths replaced the national myths then in decline: the promise of American abundance, global destiny, democracy, classlessness. This powerful drive to reassert a sense of place, to counter the standardizing tendencies in mass consumption and modern technology, would help sustain the environmental imagination in a less political age. Environmental consciousness, Brown thought, would be manifest in regional magazines and preservation efforts, in regional lifestyle choices, and in the promotional and advertising materials devoted to regional living.[20] Brown's theory of commercialized regionalism was particularly appropriate in his home state, where a new urbane lifestyle could be fused to a powerful pastoral tradition that identified Oregon as a uniquely "natural" state. New resorts and real estate developments, new products and shopping facilities, and new kinds of journalism and advertising became the driving forces in America's new pastoral imagery. A view across the golf links — open-air recreation in an exclusive, commodified, and standardized version of nature — replaced the humble farm, wild nature, and the planned ecotopia that grassroots activists had defended with such vigor in previous years.

Commodification of nature had its roots in a postwar development historian James T. Farrell called the "Great Compromise," a concession in which American labor accepted degraded work in exchange for high wages and expanded consumer choice. The compromise, according to Farrell, "was good for the economy because it created an alternating current of production and consumption," and because it expedited the commodification of virtually every aspect of American life. In the next decade American youth rejected this commercial impulse, identifying nature and the organic as cultural counterpoints to the generic and synthetic world of their parents. This critique politicized nature images and provided a foundation for the environmental imagination. In the 1970s Americans redefined these nature images again. Once a refuge from commercialism, nature became an arena for commercial exchange.[21]

Commodified symbols of nature and place were part of a larger "lite green" consumer revolution that emerged in the later 1970s, particularly among a new suburban elite that adopted much of the environmentalist and counterculture apparatus of the late 1960s as a consumptive ideal. Journalist Joel Garreau found the ultimate expression of this new consumerism in suburban Marin County, CA:

> The unspeakably hip/chic suburban community north of the Golden Gate … thrives on tall redwoods, octagonal barns, alfalfa sprouts, walls made of planks nailed on the diagonal, hanging plants, and the highest achievement of the modern economic version of people taking in each other's laundry: crafts. In places like Marin … there is no end to the cozy shops featuring locally made pottery, woodworking, and leather-tooling. Much of it is marvelously innovative and of great technical quality, but that sometimes gets lost in the overpowering grooviness of it all. Like too much health food, it can make you sick.[22]

Green consumption was an expression of class as well as geographical identity; it set the new suburban elite apart from the ordinary suburbanite. In the class-conscious consumption of nature and place in suburban settings such as Marin County, we can see in bold relief the links forged between commercial capitalism and the environmental imagination.[23]

FROM OUTFITTER TO OUTLET: MERCHANDISING NATURE AND PLACE AT L.L. BEAN

The rise of the L.L. Bean Company of Freeport, Maine, highlights these consumer trends and suggests their impact on the environmental imagination. Leon L. Bean was born in 1872 in Maine's western hill country. Orphaned at age 12, he made his way working on farms and as a "Yankee salesman." Bean developed a passion for hunting and fishing, and with a partner in 1912 he opened a small outdoor clothing store in Freeport. Dissatisfied with the heavy, stiff hunting boots of the day, he fashioned a lightweight rubber version with leather uppers and began marketing what he called the Maine hunting shoe. Cautiously he added other outdoor sporting goods to his line, and by the 1920s his iron-clad guarantee of satisfaction, his common-sense approach to outdoor wear, and his engaging zest for the Maine woods had gained him a steadfast clientele.[24] In the half-century after World War I, when he managed the company virtually alone, Bean implanted himself "in the mythology of the Eastern sporting establishment" as an icon of old-style Maine pastoralism: the masculine engagement with the Maine woods through hunting, fishing, and camping. The Bean catalog, sprinkled with anecdotes about the Maine woods and "stitched together with ... pictures of L.L. and dead moose, L.L. and dead bear, L.L. and dead deer," was the premier representation of outdoor Maine.[25]

In 1954 Bean opened a Ladies Department, guided by the philosophy that "a man isn't fussy about what he looks like when he goes fishing [but when] ... a woman ... goes fishing, she wants to be dressed for fishing." The gender characterization appears spurious to the contemporary reader, but the introduction of style considerations in outdoor apparel was remarkably prescient. Bean continued to trade on the powerfully masculine image of outdoor Maine, but the company's role was changing. The merchandiser intuitively grasped the fact that shopping was America's fastest-growing recreational activity, and that nature shopping at his Freeport store and through his catalog married this new proclivity with the idea of the Maine woods, a combination that remained central to Bean's phenomenal success in the next decade.[26]

Leon Gorman, Bean's grandson and successor, joined the company in 1961 and became president in 1967, the year the store's founder died. Pushing the concept of nature as fashion even further, Gorman repositioned L.L. Bean at an opportune moment for catalog merchandising. Credit cards and telephone purchases were revolutionizing retail business; women entering the workforce had less time for conventional shopping; and traditional department stores were less likely to offer the onsite advantages of an expert, informed sales force. More leisure, more disposable income, and rising interest in outdoor recreation fueled a boom in camping gear and outdoor apparel. A new generation of customers — young, well-off, style-conscious — looked for quality in comfortable, elegant, sturdy outdoor gear and clothing. Gorman dumped Bean's old ad agency and promoted the company in *Grit, Mother Earth News, Parade,* and *Backpacker* magazine, marketing his outdoor wear with a close eye to fashion trends. A livelier catalog, a younger sales force, and a renovated retail establishment in Freeport catered to the new nature shopper. By 1975, with sales topping $29 million, Bean was doubling in size every three years.[27]

Like the traditional "sport" from the city, Bean's new customers were well-off and well-educated, but their understanding of nature was more generic. Catalog purchasing transcended geographical boundaries; consumers around the world adopted "Maine" outdoor apparel as a form of suburban leisure wear, much like the concurrent popularity of western wear outside cowboy country. According to a biographer, there were "more Maine Hunting Shoes on the patrons of the Lincoln, Massachusetts, Sanitary Land Fill" than one would "ever [see] ... in the woods of Maine." Gorman fully recognized the commercial potential of Bean's abstracted Maine woods image: "the whole idea of the state pervades our catalog, and I think when people buy the Maine hunting boot, they are buying a little of Maine too." Retaining the founder's emphasis on assured quality and his passion for the outdoors, Gorman generalized the Maine image to appeal to a national — indeed international — clientele.[28] Given the advertising message that consumers define ourselves through the commodities they buy, it was not difficult for some to declare their environmental allegiances — and their allegiances to place — through the L.L. Bean catalog.

This abstracted and generic attachment to Maine and its environment was part of a consumer revolution that popularized the appreciation of nature and place for a generation that found political engagement less effective as an avenue of personal expression. Sensitive to the connection between nature symbols and politics, *Audubon* editors began including lavish pictorial spreads in their magazine in the late 1960s, depicting the "artistry of nature" to highlight endangered species and threatened natural places. Consuming images of nature did not translate directly into campaigns to improve water-treatment facilities, reduce harmful pesticides, challenge federal predator-control programs, or even build nature centers — the National Audubon Society's focal campaigns in the 1970s — but the link between appealing images and nature preservation was important. Publications like *Audubon* blended with a broader array of abstracted images that infused the environmental imagination with new meaning.[29]

TWO POSTINDUSTRIAL CITIES: PORTLAND, OREGON, AND PORTLAND, MAINE

Popular appreciation of nature and place in a commercial medium, as well as an emphasis on regionalism, was evident in new urban forms as well as in the merchandising strategies of firms like L.L. Bean. Although mainstream environmentalism was predicated on a dystopic view of the city as dirty, dangerous, unnatural, and disordered, by the mid-1970s some Americans were beginning to view the city in a more positive light as a medium for regional identity. According to historian John M. Findlay, all new urban forms appear chaotic and threatening in their early stages; thus the sprawling melange of freeways, shopping centers, malls, housing tracts, and parking lots initially repelled more than it attracted. But as the postwar city and suburb matured, urbanites learned to appreciate the plasticity of the built environment, according to Findlay, just as they learned to appreciate the artificial landscapes of Disneyland, where mountains, lakes, rivers, waterfalls, flowers, and trees were all "rigidly distilled and controlled." The city could be molded into a landscape of livability with tools like zoning, downtown revitalization, and public transit.[30] Urban form began to make sense again.

Drawing inspiration from the city's artificiality was an approach in keeping with Daniel Burnham's turn-of-the-century City Beautiful concept, with its monumentalism and its international Beaux Arts forms. But in the mid-1970s urban redesign began to reflect a regional sensitivity as well. In Portland, Oregon, for instance, developer Ernest Hahn drew up plans for a huge mall complex in the southeastern section of the city. The project was, as a reporter put it, "your basic monolithic shopping center," but its generic northwest theme — a Cascade Court with natural rock formations, a Court of the Seasons, an Oregon Trail Court, and a Court of the Cedars, with forest foliage and cedar trunks carved into totem poles — glossed its commercial intrusion into the pastoral city. Hahn called his project Clackamas Town Center, a play on the small-town community life that the new pastoralism idealized, and he offered to host community college classes, theater, civic gatherings, ice-skating, and a branch public library. Opponents of sprawl raised the old California bogy, but the charges of "Californication" were dulled by Hahn's commercial idealization of themes that distinguished Oregon from California.[31]

Developers in Portland, Maine, likewise transformed a mix of old houses, tenements, and brick commercial buildings along the city's waterfront into a mid-city open mall with a generic colonial New England flavor, and lined the narrow cobble streets of the Old Port District with boutiques, bakeries, arts-and-crafts shops, specialty restaurants, gourmet and kitchenware stores, and fish markets. A scaled-down version of Boston's

Quincy Market, the Old Port adapted the city's obsolete industrial–commercial infrastructure to the romantic archetype of an old New England seaport, offering a regional-theme alternative to the huge Maine Mall built at the same time on the city's outskirts.[32] As in Oregon, the downtown renaissance, inspired by the plasticity of urban form, signaled a new appreciation for the city as an expression of regional identity.

Urbanites saw the countryside in a new light as well. Traditional pastoralists imagined rural America as an antipode to the city, a place of temporary refuge to be protected from all urban influences, with the presence of rural folk providing a sense of authenticity absent in the cities. Or they looked for ways to bring the countryside — the rural and the natural — into the city itself. Popular accommodation to the city in the mid-1970s blurred the distinction between urban and rural, and this allowed commercial developers to extend the forms of urbane livability outward to greener pastures beyond the city and the suburb. In Oregon, for instance, Portland's gentrification inspired a renaissance of regional theme resorts in the hinterland, ranging from restored classic railroad hotels to planned condominium and townhouse developments and fully renovated resort towns. Wine-tasting tours, guidebooks, restaurant reviews, specialty wine shops, and promotional pronouncements proclaiming a wine superior to California's cheap "screw-top" variety symbolized the reconciliation with California culture.[33]

Rural recreational activities like these fostered a different kind of appreciation of nature. Maine Department of Economic Development Commissioner James Keefe explained, after returning from a thought-provoking trip to Disney World, that "when you get right down to it people only want to *imagine* they're 'roughing it,' then come in at four o'clock, take a shower, [and] have a martini and a good dinner." Landscape historian John Hart pointed out that these country resorts routinely provided shops and activities for their patrons because Americans "become restless and uneasy if they are left alone with nothing to do but contemplate nature."[34]

The story of a ranch-resort development on the east slope of the Cascades illustrates the mix of elite consumerism and environmentalism in this pastoral perspective. Rancher Dayton Hyde, faced with new federal constrictions on his traditional way of life, converted his ranch into a resort featuring Oregon's natural landscape and its western heritage. Under the advice of his banker, Hyde wrote a book describing his decision in terms likely to draw attention from elite ecology-minded recreationists. In a carefully crafted narrative infused with environmental moralism, Hyde explained that a Forest Service official had shown him a map of the numerous logging roads planned for a nearby national forest, a prospect that would "destroy the land for all time." Hyde resolved that his private sanctuary would not become "another National Park, its beauty trampled by the multitude whose unconstrained mass destroys the very thing the park was set aside to preserve." With carefully controlled access, the ranch resort would be "managed on the quality basis soon to be lost forever from public [recreational] resources as the population burgeons." Hyde ended his book with a story aimed directly at the hearts of his target audience. He greets a fictitious owl, who "cocks his great shaggy head at me as though he asks a favor in return, as though he asks me to stay on and fight for his last domain." The thrust of Hyde's narrative was clear: "each one of us who cares [must] do ... his small ... thing in one last fight to save the land. I must not let [the owl] down."[35]

At the other end of the class spectrum, commodification of the natural experience involved an expanding use of motorized backcountry transport. For rural and largely local working-class individuals, four-wheel drive and all-terrain vehicles, snowmobiles, and later ski jets offered outdoor adventure at a "breakneck pace," as economist David Vail puts it. This, too, altered the terms of nature appreciation and divided it along class lines. Late in the decade, these tensions became manifest in efforts to ban motorized recreation from certain public lands, a campaign that helped cleave old-style populist impulses from the movement and give mainstream environmentalism an elitist veneer that anti-environmentalists learned to manipulate effectively.[36]

The trend toward "sentimentalizing" regional character — creating symbolic places to answer for empirical reality, as Lawrence Buell puts it — was epitomized in a new type of western destination resort village that emerged in the 1970s, mainly oriented around skiing. Pioneered by developers in Sun Valley, Idaho, and Aspen, Colorado, in the 1930s and 1940s, the western recreational–residential resort village modernized the turn-of-the-century country-club idea. As historian Robert Fishman remarks ironically, the "beautifully planted grounds of the older country clubs may be the most perfect realization of the cultural ideal of the picturesque ever created in the United States." These carefully maintained representations were open-air nature playgrounds, but they were also, Fishman suggests, elite private clubs, "institutionalized means to define the social boundaries of suburbia." Like the country club, the western ski resort made "nature ... the instrument of social snobbery."[37]

In 1960–1962 developers squeezed an ersatz Tyrolean village, ski facilities, and several subdivisions into a narrow, steep, semiwild valley of sage and oakbrush at Vail, Colorado, and in 1966–1969 the highway leading up to the resort was rebuilt as Interstate 70. Spectacularly successful, Vail became a model for heavily capitalized resorts anchored to recreational attractions like skiing, golf, and water sports appearing throughout the United States. Developers sited such resorts near rivers and lakes, on the coast, and in mid-elevation mountain valleys that offered isolation, regional-theme scenery, high-speed auto access, and open land suitable for subdivision. Around a series of ski lifts, golf courses, or other recreational facilities, developers assembled a mix of high-density condominium units, townhouses, patio houses, and mini-estates. A wildlife area, working ranch, or "live stream" routed through the community reinforced the regional experience.[38] These new regional-theme microtopias attracted a painfully self-conscious middle- and upper-middle-class clientele seeking temporary refuge from the city or suburb. Footloose, monied, and linked to metropolitan centers by air and express-highway travel, this new elite was learning the distance-business techniques that in the 1980s would allow its members to further attenuate their ties to the workplaces of downtown America.

In Oregon, the grandest of these new regional theme resorts appeared on the sunny eastern slopes of the Cascades and in the high desert, a land of old-growth ponderosa and lodgepole forests, quiet meadows and lakes, and desert rim-rock canyons. In the 1970s three cities emerged as popular destination resorts in this area. The smallest, Sisters, with a population of only 630, was a gateway for recreation in the nearby Cascade wilderness and a commercial hub for the opulent Black Butte Ranch resort area. Redmond, with 6,605 people, hosted Deschutes County's only commercial airport. The largest of the three was Bend, with nearly 18,000 people. Located on the eastern edge of the Cascade forests in central Oregon, Bend was once a bustling lumber-processing center. In the early 1960s a group of local skiers and business leaders relocated the area's downhill skiing site from a ridge near town to Mt. Bachelor, a 9,000-foot peak about 20 miles away. Originally developed for sun-starved Oregonians from west of the Cascades, by the end of the 1970s Mt. Bachelor attracted more than 500,000 skiers yearly from around the world.[39] Bend emerged from a lumber depression as Oregon's prime example of the West's new air-conditioned, resort-oriented Sunbelt cities.

In 1969 developers John Gray and Don McCallum unveiled plans for a 3,300-acre Sunriver resort complex near Bend, featuring lodges and condominiums, two giant pools, three championship golf courses, 28 outdoor tennis courts, 30 miles of paved bike and bridal paths, a skating rink, an airstrip, and nearby opportunities for fishing, hunting, horseback riding, sailing, rock hounding, canoeing, whitewater rafting, and, for the less intrepid, shopping and dining at the Sunriver Village mall. To offset the rootlessness of modern leisure life, Gray and McCallum simulated a sense of place by treating visitors to displays of Native artifacts and replicas of a pioneer past crowded with trappers, explorers, cattlemen, homesteaders, and celebrities such as John C. Fremont and Kit Carson.[40]

Promotional literature for developments like Sunriver repeated traditional pastoral themes: images of rivers flowing through peaceful meadows with grazing cattle; buildings set in juniper or pine woodlands; and perpetually sunny skies. However, unlike traditional pastoralists, promoters commodified these representations of nature and place as accessories to an array of marketable recreation activities, a "barely differentiated backdrop to modern leisure-life activities." Bend resorts offered two- and three-acre country-club home sites "developed in a natural setting which includes pine, juniper, clumps of bright green manzanita, lava outcrops, and the ever desirable views of the Cascades." Cluster homes were set in "park-like" landscapes secluded by juniper and pine and common open spaces.[41] On the site, nature was further abstracted as an ecosystem of artificial ponds, transplanted shrubs and trees, and winding paths through manicured lawns: the regional landscape reinvented as decor.

In Maine, as in Oregon, commodified nature assumed its most graphic form in the development of ski resorts. Sugarloaf Mountain, Maine's second-highest peak, followed national trends as skiing evolved from a "tough pioneer sport for rugged individualists into a lucrative, sophisticated industry catering to the relatively rich." The late 1960s brought recreational development to the Carrabassett Valley below Sugarloaf on a scale previously seen only along the Maine coast. Inns, condominiums,

motels, restaurants, night spots, A-frames, lodges, and clothing and equipment shops appeared along the winding access road. By the mid-1970s
Sugarloaf U.S.A. had become a full-service resort where "the skier would
never have to leave during a one or two week vacation."[42]

As in the West, the Maine ski resort inspired a new level of assertiveness
in shaping, packaging, and promoting nature. When a northern New
England consortium bought the Saddleback Mountain ski area near
Maine's Rangeley Lakes in 1972, the new owners announced plans for a
full-scale resort village with shops, condominium units, a country store, a
motel, a swimming pool, golf links, tennis courts, two trout ponds, and
pastoralization of nearby Saddleback Lake as a site for summer boating,
sailing, and swimming — all within walking distance of the housing complex. Saddleback developer John Christie called this a "sort of limited
wilderness" concept, involving carefully prescribed activities that shaded
imperceptibly from the wild to the suburban. Nature sports like skiing,
hiking, mountain climbing, boating, trout fishing, and horseback riding
would blend seamlessly with suburban cultural and recreational offerings
like tennis, theater, a convention center, a shopping center featuring
Maine crafts, and a business retreat.[43]

While Christie never completely fulfilled his ambitions at Saddleback,
others carried them out on a smaller scale at a variety of Maine locations.
At their most luxurious and exclusive, these resorts provided a total outdoor recreation and shopping experience virtually independent of the surrounding communities: the spirit of rural place free of encounters with
indigenous rural place-holders.[44] Consciously security prone, these microtopias were pastoral equivalents of the "edge city" complexes emerging at
the beltway intersections of cities like Houston, Los Angeles, and Denver.
Like these urban upper-class compounds, psychically sealed off from the
inner city and fully configured to the tastes of a new urban elite, the regional-theme resorts were self-sufficient, planned to the last detail, and selective in their patronage, offering access to nature for a class of individuals
too preoccupied to hike or backpack into it.[45]

As credit tightened in the mid-1970s, local and county planning commissions began rejecting proposals for large-scale resorts where they
seemed too disruptive or too far in advance of the area's needs. In
response, developers learned to accent the green themes and regional character in their development plans. In 1975 the pioneering Vail Associates

submitted plans for a new development west of Vail near the Holy Cross primitive area. The year-round "model environmental resort" was to be the "most environmentally sensitive major recreation facility ever built in the Rocky Mountain West." But in a notable instance of developers' dreams dashed by local skepticism, the plan was turned down. Newly elected Colorado Governor Dick Lamm praised Vail Associates as good "corporate citizens," but he recognized the validity of local concerns: the new development would "ignite explosive growth" in the open lands west of Vail and transform the access road into a "polluted, overcrowded, urban strip." The incident was significant not only because it demonstrated the region's willingness to "investigate the alternative of no development," but because it brought to light a new development strategy: folding environmental and place-sensitive rhetoric into proposals for high-intensity nature consumerism.

Developers discovered that tools like comprehensive rural planning and zoning, park-boundary buffers, and conservation easements, because they stabilized land use, could enhance the value of their resort complexes.[46] In Oregon, this trend was pioneered by Sunriver developer John Gray, who described his resort as an "ecological experiment, to determine if humans and wildlife could coexist in harmony within certain guidelines." The company employed a full-time professional ecologist to see that natural features on the property were inventoried, managed, and maintained. Retaining "as much of the natural, unspoiled beauty as possible," Sunriver planners harmonized nature preservation with the profit motive. Rather than build homes on land along the Deschutes River, for instance, Gray created a greenbelt commons, thereby enhancing the value of the more numerous cluster homes and condominiums further back from the river. By aligning ecology and regional character with his commercial goals, Gray shepherded several such resort projects through some of the "toughest environmental laws in nation," in the process discovering that environmental regulations, far from hampering development, were "a key element of healthy growth." With the number of Oregon homebuilding permits rising by 25 percent per annum compared with 14 percent nationwide, resorts like Sunriver needed protection.[47]

In Maine, John Christie, the developer whose grandiose plans for Saddleback set the pace for New England ski resorts in the early 1970s, led the campaign for green development. Christie noted that if old-style developers had been given rein, the roads leading to his resort would be lined with commercial buildings and the woods "dotted with A-frames." By contrast, Christie's complex was confined to 75 acres at the base of the mountain; parking lots were set back out of sight, and the living complex was linked by skiable connector trails and pedestrian walkways to the lifts. The corporation's remaining 1,625 acres were "marked only by ski trails in winter, and hiking and horseback riding trails in summer." The Saddleback Lake shoreline would remain "underdeveloped," and snow-mobile trails would be located carefully out of sight and sound of the ski trails. In the Carrabasset Valley below Sugarloaf Mountain, major landowners purchased and tore down unsightly camps and cabins to pro-tect the area's "natural assets."

Thus land-use regulations, the most contentious feature of state-level environmental politics in the 1970s, gained another layer of acceptance among large-scale, well-capitalized resort developers. Musing on Maine's land-use regulations, Christie pointed out that for a long time to come resort development in the western part of the state would be "clustered in an already developed region." Environmental regulations would rein in locals with property along the access highways, leaving the area pristine as a backdrop for vacationers at Saddleback. These regulations would also give developers like Christie an almost unassailable monopoly over large-scale development in Maine's western mountains.[48]

Although resorts like these helped sustain an appreciation for nature among their clientele, they widened the gap between these upper-middle-class green consumers and rural residents wedded to more traditional forms of pastoralism. In Maine, distrust of the "new element" in the Carrabasset Valley was pervasive. "Some fear the development of Sugarloaf ... will bring disaster to their utopian environment. Others are afraid that recreational activities will not be available at a price Maine people can afford. And, of course, the overriding question in the minds of many local entrepreneurs is: whatever happens, will I make money?" Uneasy about the lavish real-estate and recreational promotions that rede-fined their landscape as a suburbanite's dream, locals saw a familiar way of life altered at the hands of outside developers and Augusta bureaucrats, both espousing green philosophies. Rumors drifted through the valley about the "underworld types" behind the "ski establishment" and about state police looking the other way as flatlanders speeded to the resort.[49]

As the *Maine Times* pointed out, both locals and developers were environmentalists — each in their own way. "They all love the beauty of the mountains, river and woodlands. The people who preceded the ski industry want life to go on as always, but with some of that ski money in their pockets. The [developers] … also wish to keep the area semi-wilderness, in addition to expanding the facilities at the mountain."[50] Both groups hoped to preserve the land as it was, but the differences between the suburban dream of a sequestered nature experience and the local tradition of free-range hunting, fishing, hiking, snowmobiling, and camp-building exposed a growing gulf in the environmental constituency.

In an editorial titled "Sometimes You're Lucky," *Maine Times* editor Peter W. Cox reflected on the successful ballot initiative directing the state to purchase lands around Bigelow Mountain and protect them from massive resort development. Unlike so many environmental issues that emerged in the 1970s, the campaign against the Flagstaff Corp. offered Mainers a clear-cut choice: the natural state or the ultimate destination resort. Cutting across class lines, the decision to protect Bigelow was a monument to old-style preservationist thinking. And with Bigelow's destiny decided, Sugarloaf became the state's primary exemplar of the new pastoralism — a "mini-Vail," perhaps, but not the total recreation of nature envisioned for Bigelow. Located on the outer fringe of the Boston–New York vacation hinterland, Sugarloaf retained its local connections as a "Maine-based operation, which still caters to the Maine skier": Maine's compromise between the old pastoralism and the new.[51]

The effect of new consumer trends on traditional environmental politics was more dramatic in Oregon, where metropolitan pressures were intense and economic forces more dynamic. The evolution of *Oregon* magazine, published in Portland, suggests the pervasiveness of nature consumerism and the difficulty of holding to older environmental bearings in this new climate. The magazine was founded in 1971 as the *Oregon Times* by Phil Stanford, a young investigative journalist from Washington, DC. Modeled on the popular *I.F. Stone Weekly*, the journal was to be a "watchdog of Oregon's press and politics." Among its regular features was an "ecology" section edited by veteran environmental activist Betty Merten, and the first issue included a story by Samuel Werby titled "Death by Highrise" lamenting plans for at least 10 new apartment buildings in Portland's downtown. Four months after the magazine's founding the strident editor challenged OEC for "selling out" in connection with a licensing procedure for a nuclear power plant. At this point, even Portland's environmental leadership wondered whether the new magazine could sustain a following.[52] With advertising limited to book dealers, outdoor sports stores, and a few restaurants, and a subscription list still in its formative phase, the publisher turned the magazine over to a local political science professor who began offering reviews of fashionable Portland-area taverns and restaurants and articles by a broader selection of writers.

In its second year of publication, *Oregon Times* was still committed to environmental investigative reporting, but the arts and restaurant columns and the range of advertising reflected the new consumptive ideal emerging in the Portland area. Ads included personalized real-estate services, alternative dancing, herbal medicines, garden centers, craft instruction, concerts, and the "Indigene: A Restaurant Committed to Secession," inspired by Callenbach's *Ecotopia*. This year brought Betty Merten's final ecology column. Ending on a pessimistic note, she characterized the changing environmental scene in the words of an architect friend "who used to rant and rave at hearings" on air quality. "He doesn't go to them anymore. He explains, 'Why wait eight hours to testify when they've already made up their minds?'"[53]

The year 1974 brought a significant transition for the *Oregon Times*. Lifestyle advertising exploded. Vendors of crafts, beads, and garden supplies joined new entries for specialty furniture and foods, antiques, imported artwork, dinner theaters, film festivals, "Path of Total Awareness" services, and a new "Feeling Center." The magazine secured its first profitable year by publishing a compilation of reviews by popular restaurant critic Gloria Russakov. Holding to the magazine's original watchdog purpose, the new editor published an article complaining about Willamette River Greenway lands purchased in conjunction with speculative commercial proposals. But the contrast between the magazine's advertising, which celebrated the city's commercial development, and its traditional storylines, which castigated it, suggests the confusion of mid-1970s environmentalism.[54]

In 1975 the magazine paid its last respects to its leftist origins by publishing an interview with radical historian William Appleman Williams on the possibility that Oregon would secede from the United States to

join a new regional socialist commonwealth based on ecotopian principles. This was followed by an editorial extolling Ernest Callenbach's
Ecotopia and projecting a vision of Oregon as a stable state, complete with
legalized marijuana, monorails, and the worship of ecology.[55]

Another shift in ownership completed the transition. In a new
glossy-stock format, the editor announced in February 1976 that the magazine would "serve a more balanced diet of the good and the bad," pointing out that "after all, there's more to life in Oregon than politics and
humbug." Instead of reports about pollution and landscape degradation,
the magazine ran stories on the evolution of the valley's new wine industry and a boom in luxury resorts.[56] The reemergence of *Oregon Times* as
Oregon magazine, a "general interest consumer lifestyle magazine," paral-
leled a shift in Portland's popular political culture. Where Oregon environmental activists had lambasted California and the commercial culture
it represented, *Oregon* celebrated the hedonistic consumer lifestyle pioneered in the southern sister state. The generic expression of place in
Oregon magnified the state's regionalism and the pastoral principles upon
which this regionalism was based, but the approach had little in common
with the visions of nature and place that had launched the environmental movement two decades earlier.

The blurred boundaries between nature consumption and nature
preservation confused political activists and contributed to the decline in
general political activity at the state and grassroots level. Oregon environmentalists, for instance, showed curiously little concern for the natural
ecology obliterated by the new desert resorts. "Most Oregonians prefer ...
to leave the development (and exploitation) of dry and empty eastern
Oregon to those they considered too peculiar to live in the more hos-
pitable regions," an *Earthwatch Oregon* bulletin noted in 1978. At one
point, a high-desert study group researching critical land-management
issues in eastern Oregon identified mining, geothermal development,
leasing, grazing, and power-line corridors as areas of concern — while
saying little about resort and recreational development.

In the end the economy rather than the environmental movement set
limits on development. As interest rates soared late in the decade, the outlook for new resorts plummeted; the boom in high-desert home construction "burst like a hot-air balloon." In the 1980s Deschutes County recovered from the recession, but its subsequent growth was based on light

industry as much as on resort development, and this time it was guarded by a widely accepted county land-use plan backed by developers.[57]

Old-style environmental activists were uncertain about the place of new pastoralism in their movement. Many environmentalists hoped that low-impact recreation and leisure services would replace ecologically disruptive resource-extractive rural industries like forestry and mining. Like the grand Victorian-era summer hotels, these resorts concentrated tourist activity in a single location, relieving development pressures in more pristine areas. But to some, the elite vacationer's willingness to embrace this artificial nature seemed ironic. *High Country News* editor Tom Bell described Snowmass Village in Colorado as a "sort of place in which only Californians and John Denver fans could attend a conference on the environment and not feel hypocritical." Trees and mountains, he observed, were "gouged out to provide space for ski lifts, golf courses, softball fields, swimming pools, convention facilities, equestrian centers, and the rest of the prerequisites of wealth.... The place is a classic example of development that digs out all the hills, cuts down all the trees, drains all the water, and then sells subdivisions that tout the quality of their environment."[58]

Bell saw Rocky Mountain resort villages like Snowmass as popular fantasies of the West, "never-never lands" representing the last great enclave of open space left in the continental United States. Florida and California, he observed, had once offered authentic open space, but runaway growth had erased all but the memory. Disneyland and Disney World had become "opportunist death symbols," epitaphs to nature in areas now "too crowded and too cluttered to foster any grand dream of 'the great escape.'" Rocky Mountain resorts likewise catered to the memory of open space, Bell noted ominously, and Disney Enterprises held options on land here as well. It was "too early in the growth history" of the region for another fantasy land, but someday Mickey Mouse would replace the elk in the Rocky Mountain ecosystem.[59]

Clearly this was not the ecotopian vision most activists had been weaned on, and the class biases in this carefully crafted vision of nature alienated many. In a reflective article in *Audubon* magazine, Jack Hope noted that "ski facilities from Maine to Oregon are crammed to overcapacity," and he worried that developers were setting their sights on even more fragile areas. In the mountains of New England, few long-range vistas remained that did not include "the broad treeless swipes of ski trails radiating from mountaintops." The region's soils were thin, the slopes steep, and the growing season short; tree removal caused erosion and silted streams, and gasoline, oil, and salt runoff from roads and parking lots polluted small upland streams. Hope was impressed with the "ecological good sense" of some ski-area development; "still, the ultimate ecological sensibility would have been never to build these facilities in the first place."

Even more disconcerting to Hope was the obliviousness of the upper-class patrons of such resorts. Unconvinced by the argument that ski areas helped concentrate recreational pressures in a relatively small area, he concluded: "there is nothing in a heated outdoor pool, a helicopter chauffeuring service, a fancy boutique, or a $66 room that encourages humility or respect either toward our natural environment or toward ourselves. Rather, a week at a place like Snowbird only reinforces the belief that if you have the money, the tops of mountains will be lopped off to provide you with entertainment."[60] With the idea of nature confined to these carefully contained fantasy landscapes, place-oriented political activism waned among many sectors of the population.

EXURBAN MIGRATIONS

A pastoral construct keyed to the idea of a postindustrial countryside provided another source of new environmental awareness. This awareness rested on the belief that resource-extractive activities were no longer at the cutting edge of the rural economy, and that the countryside would become the seat of a new economic mix based on marketing, information processing, services, light industry, craft production, and, as historian Hal Rothman points out, the selling of images, emotions, and self-affirmation.[61]

Retired University of Maine president Winthrop C. Libby described what he saw as the ideal postindustrial future for the unspoiled but economically depressed eastern Maine coast. The area's local resources — deepwater ports, tidal power, fisheries, forests, blueberry barrens, scenery — would attract small, nonpolluting industries. Infusions of capital from these entrepreneurs would encourage downeasters to brighten their towns by clearing stream banks, beautifying public property, zoning out improper land uses, screening local eyesores, and improving roads, wharves, marinas, and recreation facilities. Magnificent natural scenery and a

Harbor at Stonington, Maine

brighter cultural light shining in the East would attract exurban migrants, retirees, eco-entrepreneurs, summer citizens, artists, craft workers, musicians, and resorters, who would broaden the tax base and facilitate further community revitalization.[62] This transition to postindustrial living would inspire a keener sense of place and a stronger defense of that place, but only if working-class locals and middle- and upper-class in-migrants agreed on certain premises regarding this somewhat suburbanized version of downeast culture.

Postindustrialism offered the prescriptive terms for an accelerated exurban migration as young, suburban families fanned out beyond the normal commuting distance to city jobs. Small towns in southern Maine and central Oregon became new growth centers feeding on the decentralization of work, the high-income, two-worker household, the limited-access commuter highway, and a growing preference for larger house-lot sizes. Late in the decade new telecommunications technology accelerated this trend, fostering footloose knowledge-based industries like pharmaceutical testing, computer software design, engineering, management and financial consulting, architecture, and subcontracted research, design, testing, and marketing. "Greenfielding" — the migration of entire firms to rural locations where wages, taxes, crime rates, and land values were lower — facilitated the spread of suburbia into the countryside.[63]

The postindustrial thrust of this suburban-to-rural migration was captured in a work titled *Regions of Opportunity* by writer and real-estate speculator Jack Lessinger, who labeled the new exurban landscape penturbia,
the fifth great demographic shift in U.S. history, following the transat-
lantic, westward, urban, and suburban movements. Moving to the world
beyond the suburb, Lessinger felt, brought a change in attitude not unlike
Reich's revolution in personal consciousness: suburbanites were
self-indulgent and insular; penturbanites would relearn community
bonding by "tak[ing] little from life and giv[ing] a lot." They would abandon mass consumption to became "Caring Conservers," seeking quality
of life in open spaces and simple natural surroundings. "Gifts bestowed
by nature" would "substitute for expensive goods and services made necessary by the artificialities of urban life." Mixing nature appreciation and
the postindustrial ideal into a new brand of pastoralism, penturbanites
would guard the livability of their adoptive communities, urging land-use
regulations to prevent runaway growth of the type they sought to escape
in the suburb.

Lessinger waxed ecotopian about his penturban world: "the genius of
the Caring Conserver thrives in the intimacy of small, rural areas where
people call each other by their first names, within jogging distance of
forests, mountains, and rivers ... and where creative, artistic, do-it-your-
self projects of all kinds ... win prestige from the community." Exchange
of skills would be commonplace; the new citizens would build their own
homes, sew their own clothes, and raise their own organic foods. They
would restore "decrepit nineteenth-century structures rich in nostalgia" in
lieu of building costly new homes and businesses. They would "volunteer
their loving labor as a substitute for expensive services in schools and hos-
pitals and in ministering to the young, the elderly, and the handicapped."
In Lessinger's vision, penturban life would transform middle-class values,
spawning a new form of environmental activism. "Intangible notions"
like peace of mind and unspoiled nature would take precedence over
rural extractive industries, and with the latter on the wane, rural areas
would benefit from low-impact service jobs, retail trade, finance, insur-
ance, real estate, education, and legal and other social services appearing
in the wake of penturban penetration. Wildland amenities would inspire
a shift from saving endangered species to preserving entire ecosystems as
habitat for plants, animals, and penturbanites.[64]

Like the new nature consumerism, Lessinger's celebration of penturbia struck some as specious. The concept of Caring Conserver, as Maine planner Evan Richert observed, came with a "built-in contradiction": it dismissed traditional suburbanites as "mass consumers," but glorified their upscale counterparts in sprawling exurbia "who want two-to-five-acre lots out near farms and along rivers and lakes." Although premised on a quest for rootedness, exurbia was in fact the most mobile of communities, according to Murray Bookchin. It was an entirely new social geography "based on the automobile, the suburban shopping center, and a high-income population that depends upon the city economically but is completely severed from it culturally." More "splatter" than sprawl, exurbia devoured farmland, wildlife habitat, open space, and recreational access at an alarming rate. In the West, exurbanites preempted precious streamside access. Where ranchers maintained a long tradition of friendly tolerance toward anglers and boaters, the new occupants prized exclusive private use, and their penchant for replacing streamside brush with landscaping threatened fish habitat and required large applications of water and weed killer. Mainstream environmentalists saw exurban migration as a planning disaster, bringing traffic, higher property taxes, lake and pond eutrophication, visual blight, and a "feeding frenzy" in rural real estate that forced many locals out of the market.[65]

But for all its liabilities, exurban migration was an expression of nature appreciation appropriate to a time of declining civic engagement. As historian Samuel Hays points out, environmentalism was an extension of the postwar consumer revolution, a collective and political search for amenities that could not be purchased in the marketplace: clean air, pure waters, unspoiled wilderness. For some, the difficulties in obtaining these amenities through politics heralded a new strategy. Having abandoned the collective quest for livability, exurbanites purchased these amenities individually, as localized or even privatized microtopias, surrounded by the small farms, pastures, woodlots, orchards, and other elements they prized in the pastoral landscape. They would defend this new world with stricter land-use planning and no-trespassing signs.[66]

The exurban movement generated a proliferation of community-related heritage, open space, and nature preservation projects, even though, like the new trends in nature consumption, its goals were difficult to reconcile with mainstream environmental activism. As nodes of the new exurban migration, isolated communities all across America were "wrenched into a strikingly different social and cultural world" as they were transformed by new resort and residential building and "commodified by new and ingenious methods."[67] Typical were developments in Cannon Beach, the resort town that sparked Oregon's popular environment movement back in 1967. The pastoral character of the Oregon coast depended in good part on the picture-book, small-town atmosphere of communities like Cannon Beach. As local author Murray Morgan wrote, the few towns that flourished along the coast were, "like the trees that take the first shock of the five-thousand-mile sweep of wind … tough and tenacious, but small."[68]

Cannon Beach struggled to retain this image as it grew from a sleepy summer resort into one of Portland's prime year-round resort destinations. Aware of the threats inherent in commercial growth, in the late 1960s town residents persuaded developers and homebuilders to adopt a special "Cannon Beach look," featuring use of local stone, natural colors, wood siding, and native landscaping. In a similar vein, downtown buildings, galleries, boutiques, restaurants, and motels were remodeled to create a "natural, aesthetically pleasing atmosphere," and special events like Sandcastle Day celebrated Cannon Beach's sense of place. The town welcomed the state's new Land Conservation and Development Commission (LCDC), becoming the first in Oregon to file its comprehensive plan. Building and landscaping practices like these were but one form of a larger cultural and psychic transformation of place in the 1970s — a preservation of old forms to serve new meanings.[69]

Adopting the preservationist mantle, local residents fought to protect the Cannon Beach look from development pressures. Activists rallied against a 175-unit condominium complex slated for the dunes north of Cannon Beach in 1971, for instance, gaining support from the LCDC and the Highway Department. After eight years of debate, the developer reduced the project to 71 units, created a public walkway to the beach, and left the westernmost dune front vacant. Still, although these accommodations helped address concerns over public access and ecological preservation, a set of buildings now loomed over what had previously been one of the largest natural dunes on the northern Oregon coast.

The outcome was not atypical; all along the coast developers were forced to compromise, but by abstracting local character and incorporat-

ing at least some ecological principles into their designs they gained public acceptance — and customers.[70] The subtle shift from organic regionalism to regional "look" — the postindustrial abstraction of place —
marked Cannon Beach's accommodation to the tourists and newcomers
who threatened to undermine community character, while infusing that
character with enormous commercial potential. Here and elsewhere,
town preservation and restoration borrowed from earlier campaigns to
save iconic landscapes, but these measures echoed the limitations of
1970s environmentalism: its essential commodity orientation and search
for private solutions to environmental problems.

Like Oregon's Cannon Beach, coastal communities in Maine strained
to meet the expectations of the consumption-conscious metropolitan
tourist while maintaining the pastoral character these visitors expected.
Freeport, a small, tree-shaded, colonial-revival town northeast of Portland
on the coast, was host to the expansive L.L. Bean Co. Bean's success and
Freeport's ideal location on the coastal tourist route attracted a series of
fashionable retail outlet shops along the town's main street, and the result
was an astounding 2.5 million shoppers besieging Freeport each summer.
As a quaint coastal town near urban amenities in Portland and Boston,
Freeport was also an ideal target for exurban relocation.

A recently arrived couple recounted the advantages of their old farmstead
near Freeport's Mast Landing in a 1970 *Maine Times* article. The couple's
postindustrial idyll included open fields, a quiet mill stream, and grassy
slopes with a grove of huge pines as a backdrop. With open space for jogging, picnicking, ice-skating, and cross-country skiing at the nearby Mast
Landing nature preserve, their neighborhood seemed a free and authentic
elaboration of the dream suburban backyard. A palpable sense of the area's
timelessness and permanence — stone walls, weather-worn barns, relics of
an old shipyard — added a sense of place for these uprooted exurbanites.
The feature article accented the farm's isolation — "only the birds and the
Mill Stream rushing over the rocks" broke the silence — but it also noted
the farm's convenience to the shopper's beat a few blocks up on Highway
1. The article ended with a brief but intimate restaurant guide to Freeport.[71]

Exurban postindustrial aspirations like these, coupled with commercial expansion following L.L. Bean's success, triggered strong preservationist sentiment in Freeport consisting of a complicated mix of exurban
gentrification, downeast exclusionism, and attempts at controlled com-

mercial development. Wary of the line of specialty outlets, boutique stores, and restaurants rippling outward from the L.L. Bean epicenter, newly settled Freeporters joined town elites to form a Merchants Association and initiated several "townscape projects" to rebuild sidewalks, plant trees, and protect the Freeport version of the Cannon Beach look.[72] Contemporary observers foresaw a period of cultural stress as exurbanite preservationists like these came face to face with the state's legendary Yankee individualism. A 1975 Bates College survey of residents in a nearby coastal town found that although income distribution for old-timers and newcomers was almost identical, older residents perceived an air of superiority among their new neighbors. Complaints about having to deal with "a lot of [new] people ... I could do without" were common. These attitudinal differences were not irreconcilable; in the Bates study, in fact, there was surprising agreement on issues involving jobs, local industry, and the environment.[73] But in Freeport, hothouse growth conditions coupled with intensive exurban penetration brought out certain class differences that fractured local preservationist opinion. Seeking a rural setting free of functional economic relations to the land, newcomers rejected the old Maine custom of neighborly coexistence and expressed their suburban "cultural paranoia," as one journalist put it, in a proliferation of new zoning initiatives.[74]

These tensions were manifest in Freeport as early as 1958, when a combination of wealthy locals and new residents, panicked by a growing number of house trailers and partially built homes, backed a comprehensive plan that banned outhouses, cellar living, and in effect a Maine tradition of building homes piecemeal to avoid mortgage commitments. Locals defeated this measure and a similar initiative a few years later, but growth in the 1970s widened such fissures. Some residents supported unbridled commercial expansion, particularly where land speculation remained in local hands. Others feared that Freeport was slipping out of their control. The new downtown retail outlets bore no relation to local needs, they complained, and they brought parking problems and rising levels of "obscenity, vandalism, and ... thievery."[75]

Thus issues like town preservation, hunting bans, forestry initiatives, and zoning ordinances opened seams in the constituency environmentalists forged in the classic campaigns against pollution and industrial development. According to one local, natives were "tuned into nature in a different way from out-of-staters." A Bath, Maine, shipyard worker deplored the new arrivals' propensity for "pick[ing] on the small people" while ignoring "big industry." For years, he noted, Maine hunters and anglers had fought the paper and power companies for control of Maine's rivers; "If these [newcomers] ... want to do something about pollution, why don't they go after the big companies instead of us?"[76]

In 1983 townspeople found common ground when they learned that a historic Main Street residence was to be torn down and replaced with a McDonald's restaurant. Natives and newcomers linked arms to protect their "quaint New England village," certain that opening the floodgates to fast-food purveyors would turn "Freeport into another piece of the 'plastic American scene.'" Armed with "Freeport Mac Attack" T-shirts, bumper stickers, buttons, lawyers, and petitions, townspeople forced the corporation to compromise by siting the restaurant in the existing house.

As at Cannon Beach, concessions to local character only eased the way for developers. McDonald's blended into Freeport's colonial revival character, but the effect was, like the Cannon Beach look, generic rather than organic. As a *Maine Times* writer put it, there was a "certain snobbiness to drawing the line between Cole-Haan [a fashionable shoe store] and McDonald's," but in the end Freeporters accepted the abstracted version of their village — if it drew a certain class of shopper.[77]

Towns along the Maine and Oregon coasts met the challenge of commercial development by abstracting regional identity or by protecting a few reminders of the past from the noise and glare of the well-beaten tourist path. A *Down East* magazine article assured tourists and natives alike that "although shopping centers and commercial developments are constantly closing in, the 'quaint village' still exists in Maine." Yet sense of place was subtly altered as tourists and exurbanites imposed new meaning on coastal Maine's shabby saltwater farms, old fishing wharves, and bobbing lobster boats. Retired executives tending their gardens, exurbanites restoring historic residences, and well-to-do shoppers hunting down antiques refashioned the Maine village. Native Mainers found themselves living in a familiar but strangely alien landscape[78]

The aspirations of those who occupied this new pastoral landscape set the scene for a conflicted form of environmentalism in the 1970s. A statewide lake conservation magazine in Michigan commented on a recreational boom similar to those in Maine and Oregon: as the "citizens of southern Michigan descend in a rush on weekends, our lakes are becoming so crowded with boats and water skiers that neighbor rises up against neighbor, and iron-clad organizations form overnight in order to protect their special use of the water." According to Evan Richert, "those leading the [preservationist] charge surely see themselves as 'Caring Conservers' — conserving neighborhoods and neighborhood character. But it also looks suspiciously like a ... modern-day territorial imperative."[79] The new pastoralism, and particularly its upper-class postindustrial predilections, offered little footing for the broad-based statewide activism evident in the clean-rivers, wild-rivers, and coastal-preservation campaigns.

POPULIST ENVIRONMENTALISM IN THE LATE 1970S

The new commercialized forms of environmental sensibility — nature merchandising, nature-oriented resort development, exurban preservationism — accented the class instincts of a broad section of the environmental constituency. In contrasting developments during this half-decade, grassroots environmentalists articulated a populist version of personal environmentalist activism.

Like the regionalist developers who gentrified downtowns in Portland, Oregon, and Portland, Maine, some environmental populists reengaged the subject of the pastoral city in the late 1970s, helping shift the mainstream movement from a fixation on preserving wild lands, wildlife, and wild rivers to a more democratic campaign for healthy, sustainable communities. If city building worked "against the form and characteristics of the natural environment," one urban planner admitted, "then the metropolitan area will no longer be embraced by nature, but will have marred, overrun, and devoured it."[80] Ecology-minded citizens advanced neighborhood "greening" projects — parks, walkways, bike trails, greenbelts, shade trees, public gardens, and other organic principles of design meant to "ruralize the city."

Others suggested more visionary approaches, promoting the pastoral city as a context for urban agriculture — community and private gardens grown in parks, on rooftops, in vacant lots, on land set aside by the city, in strips along the streets, in floating bioshelters in the harbors, and in old warehouses and unused factories. Urban agriculture would provide the city with fresh fruits and vegetables, fish, herbs, poultry, mushrooms, greens, vegetables, preserves, sauces, baked goods, and flowers, thereby increasing access to fresh produce, encouraging regional food independence, and reducing consumers' reliance on remote agribusiness corporations and their chemical fertilizers, pesticides, and fossil fuels. Others experimented with urban sources of cogeneration and renewable energy, and with urban orchards as a source of fresh fruit and purified air.[81]

Carrying this concept to the extreme, in 1974 Richard Britz of the University of Oregon's Urban Farm Project designed an "edible city," a series of agro-urban microtopias in which neighbors joined their back yards into common fields and cooperated in small-scale craft production. Urban agriculture would bring together the best of city and country, reawakening urbanites to the seasons and their cycles. Ultimately the "equilibrium between town and country" would be restored, "not as a sprawling suburb that mistakes a lawn … for nature, but as an interactive functional eco-community that unites industry with agriculture, mental work with physical, individuality with community." The new city would "increase our humanity rather than decrease it."[82] Although developments in "urbaculture" fell short of these ecotopian visions, the movement offered an exciting new direction in community-based urban environmentalism, maturing alongside the more commercialized and artificial trends in regional urban form as an expression of personal and community commitment in a time of declining political activism.

Another 1970s innovation in grassroots environmentalism was the back-to-the-land phenomenon. A revival of small-scale family farming, the new homesteading, as sociologist Jeffrey Jacob calls it, was sustained by disaffected suburban youth well acquainted with the problems outlined in *Greening of America*. Like Reich, the new homesteaders were motivated by an aversion to the city — a "feeling that the city was literally falling apart" — and by a romantic idealization of the countryside. New homesteading was founded on a Jeffersonian ideology updated by philosophers such as Wendell Berry and Helen and Scott Nearing and by magazines such as *Mother Earth News*.

But despite a common ideological underpinning, homesteading was a diverse phenomenon.[83] At one pole was the unabashedly commercial Sunburst Farms in Santa Barbara, California. A "self-created ecological utopia" with "psychic connections" to Eastern and Native American religion and to the "American pioneer spirit," Sunburst donated its $7 million yearly profits on whole-grain baked goods, organic restaurant produce, meats, and juices to Sunburst Communities, a tax-exempt religious corporation. Other forms of new pioneering included Oregon's Cerro Gordo, a quasi-communal experiment encompassing some 1,200 acres and involving 100 families.[84] The single-family subsistence farm, a more typical new homestead, stood at the other end of the continuum.

The most important models for this phenomenon were Helen and Scott Nearing, who, like the back-to-the-landers who followed them, had "gone through the big-city-industrial way of life" and were seeking something better. By the mid-1970s, when the Nearings's *Living the Good Life* became an environmentalist bestseller, the couple had spent nearly a half-century on a subsistence farm in Jamaica, Vermont, and after 1952 in Harborside, Maine. In their writings and their exemplary lives, they melded their Marxist perspectives with the folk ideals romanticized by regionalist writers like H.L. Davis and Robert P. Tristram Coffin to appeal to young idealists seeking authenticity and freedom in a traditional way of life. The Nearings lived their lives, in their own words, as "a part of the environment."[85]

A nationwide phenomenon, the back-to-the-land movement was strong in states like Maine and Oregon, where open land and an aura of environmentalism provided a sense of opportunity and new beginnings. Maine's mid-coast region was particularly attractive, offering relatively fertile soils, low property values, a strong rural heritage, and a scattering of operating farms to provide support and advice for newcomers. Here one Maine journalist discovered a symbiotic relation between old residents and new homesteaders. The inexperienced newcomers needed part-time jobs and practical advice, and in turn they helped fill a void in rural communities that had suffered dramatic population declines in the twentieth century. Their aspirations — peace of mind, self-reliance, autonomy, harmony with the land, a new foundation for the work ethic — were not all that different from those of their more settled neighbors. Although less political than mainstream environmental activists, new homesteaders were totally committed to their pastoral vision. Rejecting wilderness and commodified naturalism as sources of authenticity and freedom, they reworked older populist pastoral approaches to suit 1970s cultural conditions.[86]

Homesteading expressed both the strengths and the weaknesses of the new pastoralism. Committed to living lightly on the land, homesteaders pioneered the application of new "green" technologies like composting, passive solar construction, and solar, small-head hydropower, and wind energy and offered the self-sufficient farm as a working model for future society.[87] New homesteading popularized the idea of organically grown foods, helped bring Americans' consumption habits into critical focus, and challenged the logic of corporate agriculture. Perhaps more significantly, it helped tailor environmental ideals to the 1970s culture of personal choice. *High Country News* presented as an example John Mionczynski, a New York State native who with $72 began life over again in a solar-powered and wood-heated cabin in the Rocky Mountains. Although neither mainstream commercial farmers nor everyday consumers could hope to replicate Mionczynski's spartan lifestyle, the idea of combining traditional rural populist goals with new green technology signified that the problems of declining per-acre yields, high energy costs, soil toxicity, regional food dependency, and social anomie were not irresolvable. Rejecting the oppositional perspective of their 1960s predecessors, the new homesteaders opted for the politics of example. Although their search for a sustainable future seldom translated directly into political activism, their idealism was an important source of symbolic sustenance for a generation that found politics less satisfying as a form of personal expression.[88]

A third grassroots environmental trend emerging in the 1970s involved a proliferation of appropriate-technology products. Experiments with appropriate technology and simple living were inspired by rising fuel costs during the Arab oil embargo, by E.F. Schumacher's slogan "small is beautiful," and by Donella H. Meadows and her Club of Rome colleagues, who used innovative computer techniques to predict a frightening series of crises in resource availability and environmental conditions in the early twenty-first century.[89] *High Country News* publisher Tom Bell described America's energy options as he saw them in the mid-1970s: given the Arab stranglehold on global oil supplies — and the fact that the nation's best hydropower sites were already developed, the cost of coal was being driven upward by new federal strip-mining regulations, and public fears of nuclear power were mounting — Bell felt that the "old ways" were doomed, "whether we like it or not." The automobile would be the last great icon of "unrestrained exploitation of abundant [energy]

... resources, cheaply obtained at the expense of the environment." In the post-petroleum age, manufacturing would be small-scale, and growth would be reluctant — "grudgingly given, keenly observed," according to *Maine Times* editor John Cole. Drawing attention to the fact that natural resources are finite, writers like Schumacher and Meadows touched off a search for new "human-scale" technologies.[90]

The attraction of green technology depended in part on the populist, anticorporate symbolism pioneered by environmentalists in the 1960s. Experimenting with appropriate technology — in ideal terms, living "off the grid" — was a means of personal liberation from corporate hegemony and its emphasis on energy-intensive technology. But the appeal of green technology extended beyond this oppositional message. Even mainstream public figures like Laurance Rockefeller advocated environmental salvation through alternative transportation, energy conservation, and recycling.

Moreover, green technology had intrinsic consumer appeal. The *Whole Earth Catalog*, a primary medium for disseminating new ideas about green consumption, energy conservation, and other means for "creating a new society in the shell of the old," was remarkably successful, spreading the idea of simple living far beyond those who simply sought independence from corporate America. Journals like *Seriatim*, a northwestern environmental publication inspired by Callenbach's *Ecotopia*, focused heavily on the commercial trappings of alternate lifestyle choices: tools and equipment for solar energy systems, recycling, agronomy, bio-architecture, organic farming, and home health. Announcing the emergence of "an environmentally-attuned, stable-state society," *Seriatim* emphasized a less spartan — and perhaps more appealing — form of environmentalism made possible by purchasing green technology. "Once I had discovered solar and wind power," one essayist revealed, "I no longer was forced to think of life as a choice between technology and nature.... Wow — you could have your stereo, and clean air too."[91]

Appropriate technology rationalized the belief that environmental crises could be resolved through personal consumer choice, and this helped sustain a positive environmental message in the late 1970s. During these years, alternative energy sources and energy conservation became key themes in mainstream environmental publications, a focus that resonated with contemporary popular culture in the "era of limits."

Environmentalists promoted "soft energy paths" such as renewable sun, wind, and biomass energy sources to demonstrate the practicality of environmental solutions and project utopian enthusiasm. The Natural Resources Council of Maine, for instance, designated November 10 to December 10, 1976, "Energy conservation month in Maine," and asked television, radio, and newspaper broadcasters and school teachers to "remind Maine Citizens of our energy problems and the importance of energy conservation as a means to alleviate those problems." Bills aimed at energy conservation were prominent in the Maine legislature during these years, along with measures regulating electrical transmission lines and oil pipelines.[92] Like new homesteading, appropriate technology expressed both the strengths and weaknesses of 1970s environmentalism. Its approach — simple living — linked environmentalism to consumer capitalism and dulled the movement's adversarial edge, but these idealistic experiments with new "green" products challenged mainstream thinking about the inevitability of corporate-sponsored technology and, more important, gave substance to the visionary impulse in the environmental movement.

The path to ecotopia pioneered in the later 1970s infused the mainstream environmental movement with new ideas, new enthusiasm, and new ways of thinking about nature and place. Commercial regionalism, recycling, sustainable energy sources, organic foods, landscape preservation — the core innovations of late-1970s environmentalism — derived largely from a new pastoral vision: a harmonious blend of personal life choices and natural surroundings. But like the political approaches of the previous decade, these strategies left much undone. The pioneers of the new pastoralism asked both too much and too little of Americans. On one hand, like the rise of recreational nature appreciation in the previous decade, environmental consumerism assuaged feelings of guilt and fear without girding the nation to take on thornier environmental issues. On the other hand, simple living, subsistence homesteading, living off the grid, and constructing a self-sufficient urban household involved a total commitment that relatively few Americans were willing to make, even though they might adopt elements of this solution. The simple-living idea was forged in an era of apparent energy shortage, global food scarcity, and severe limits to growth. An end to the Arab oil embargo, new oil and gas discoveries, advances in genetics for food production, and economic recovery in the 1980s seemed to bear out the rhetoric of the Reagan administration, which was committed to disproving the dire predictions of the limits-to-growth community. Indeed, the declaration of limits was premature; rather than an era of global scarcity, where the vision of the good life would have to be radically adjusted, America entered an age of even more frenzied consumption. The simple life, country living, nature consumption, and commodified regionalism continued to inspire Americans, but the ecotopian strategies spun off from the mainstream environmental movement in the 1970s remained — at least for a while — out of step with the times.[93]

For a variety of cultural, political, and economic reasons, mainstream state-level environmentalists ended the 1970s in an ambivalent mood. Just as Maine and Oregon exemplified the emergence of state-level environmentalism in the 1960s, these two states demonstrated the difficulties of sustaining a grassroots defense of the environment in the 1970s. Operating in a national climate that discouraged visionary politics, activists found their own vision clouded by the complicated and contradictory abstractions of the new personal pastoralism. All too often, commercially constructed generic nature substituted for the real thing in the mind of the upper-middle-class suburbanite. This in turn changed the rules of environmental engagement. Unable to call up the old political formulas — crisis consciousness, populism, and the rhetoric of place — environmentalists abandoned state-level politics for the more promising field of Washington-based lobbying.

Mainstream environmental organizations continued to emphasize familiar issues: pollution, suburban sprawl, industrial expansion into critical natural habitat. To this they added a "new generation" of issues related to atmospheric carbon dioxide, changes in global climate, cancer-causing compounds in urban, suburban, and rural environments, and toxics in water supplies. In most cases mainstream organizations moved in the direction of greater ecological sensitivity and wilderness purity in the later 1970s, and these issues drew a passionate response from a core group of supporters. In Oregon, battles over the Hells Canyon Dam, various roadless areas in the federal domain, and the Oregon wilderness bill animated the environmental scene in the late 1970s.[94] But in a time of less intense political awareness and more complex environmental issues, these efforts were less likely to garner the broad public support generated during the clean-waters and coastal campaigns.

Still, when inflation and high interest rates discouraged "green" consumption and slowed resort development in the late 1970s, many Americans returned to the environmental defense that had traditionally united them across class and sectorial lines: crisis consciousness brought on by the Reagan administration's anti-environmentalist onslaught; a belief in the collective ownership of nature; and a strong regional identity, however conflicted that identity might be. In the 1980s the presence of toxics and other industrial hazards in rural, urban, and suburban neighborhoods also reignited grassroots protest based on the older formula of populist reaction to corporate abuse of the environment. Only then, and often belatedly, did environmentalists recall their roots in this yeasty tradition of local civic action.

Harrisburg, Oregon

Dairy farmer and citizen activist Edgar Grimes. His one-man antilitter campaign began after his dairy herd losses amounted to $50,000. The cows died, he says, when they ate broken glass mixed with their feed. The glass came from roadside litter.

CHAPTER 8

THE ENVIRONMENTAL IMAGINATION AND THE FUTURE OF THE ENVIRONMENTAL MOVEMENT

In this book we argue that Americans' understanding of nature and the natural changed in the years between 1945 and 1975 and that these changes sustained the environmental movement. In the late 1940s a fast-paced urban society looked to its hinterland for a sense of freedom, authenticity, and permanence gained through communion with nature and folk in a rural setting. This world of simple country homes and neatly tilled fields, set in a context of forest, mountain, and meadow, served as a repository for the values Americans left behind in their stormy passage to the modern age. Nature and folk formed a unified whole, and contemplating this union promised a sense of freedom and authenticity.

In the mid-1950s the contrast between this romanticized natural landscape and the realities of blight and water pollution elicited a political reaction that eventually gave rise to environmentalism. Challenging corporate control over America's polluted rivers required a popular imagination that transcended the utilitarian use of nature; clean-waters advocates promoted a new vision of urban rivers as vectors for reconnecting nature and the metropolis — for pastoralizing the urban environment. In the ensuing debate over the role of urban rivers in our lives, activists transformed pastoralism from an essentially nostalgic construct into a pragmatic and progressive vision of a new urban society linked to nature and the countryside by its pure and organically rich rivers.

Over time, this modernized pastoral vision gave way before the more compelling goals of wilderness preservation and ecological balance. Wild rivers, providing a chance for solitary recreation in an uninhabited natural landscape, epitomized these newer goals and the sometimes conflicted political campaigns they generated. In the older pastoral construction, nature was utilitarian social, and humanized; in the new wilderness vision nature became spiritual, personal, and uninhabited. The environmental imagination was in transition, and the battles over wild rivers involved a complex search for balance between these conflicting landscape traditions.

Building on these traditions — pastoral and wild — the environmental imagination became a mainstream political idea in the early 1970s. The formative battles of these years involved campaigns to defend iconic wild or pastoral landscapes, particularly those that defined regional identity: a coast, a mountain, a river, a wetlands, for instance. Acutely sensitive to issues affecting regional identity, Americans launched the environmental era on the strength of an old and enduring tradition of nature appreciation, modernized to accommodate new recreational, scenic, and ecological values. These campaigns politicized the environmental imagination, bringing together civic, scientific, and conservation groups, spreading the environmental ethic to new arenas, and giving the environmental imagination permanent standing in the American political lexicon.

In the early 1970s Americans emphasized a new set of concerns. Environmentalists and political leaders, realizing that issues like habitat loss and suburban sprawl had become regional rather than local, mapped out and popularized regional plans for an environment where nature and culture were harmonized. These planning utopias, or ecotopias, painted a future in which the problems of pollution, sprawl, congestion, and disorder were resolved by overarching regional planning. As an ultimate expression of the environmental imagination, ecotopian thought was an attempt to mediate the longstanding contradictions inherent in our relation with nature in America. This imaginative exercise forced environmentalists to rethink the scope of their concerns as they shifted attention from preserving specific natural features to addressing the needs of a complex and dynamic human society.

Subtle class tensions intruded into the environmental imagination even as the movement reached unprecedented levels of popularity. During the mid-1970s, in the years of waning grass-roots activism across the United States, the environmental imagination assumed a new cast, actualized through personal lifestyle and consumer choices. Commitment to protecting nature continued as a key aspect of American political culture, but in addition to old forms of environmental engagement, the political and social climate of the later 1970s generated new ideas and practices. Some Americans sustained the environmental imagination by engaging a new popular passion for consuming nature, while others experimented with personal forms of environmentalism through appropriate technology and alternative "green" lifestyles. Because the old 1960s formulas for mobilizing environmental concern — identifying a corporate foe and invoking the politics of place — were becoming less relevant, these new themes, emphasizing personal choice as an expression of environmental politics, were folded into the political defense of nature.

The environmental imagination was a creative blend of ideas involving, for instance, pastoralism, wilderness appreciation, popular ecology, redemptive recreation, planned development, and simple living, among others. While the relative importance of these constituent parts changed over time, each was an expression of affinity for nature and place. Together they made up the ideological framework for popular environmental politics. Gaining a better understanding of what Americans mean when they call themselves environmentalists, and how they understand the nature they hope to protect, helps us explain the persistence of the environmental movement and its impact on state, national, and even global politics.

"Why has environmentalism been able to avoid the fate of most short-lived movements, and how has it changed since the first Earth Day?" Riley E. Dunlap and Angela G. Mertig begin their influential *American Environmentalism: The U.S. Environmental Movement, 1970–1990* with these basic questions, and we have attempted to answer them in this volume. The postwar years brought to maturity the first generation of

Americans demographically more urban than rural. Their social memories, as Dunlap and Mertig point out, were acquired in a setting apart from nature, and yet it was this generation that touched off the most comprehensive grassroots defense of nature in U.S. history. Even more remarkable, these activists, publicists, and politicians managed to transmit the gospel of ecology to the next generation, sustaining, as Dunlap and Mertig suggest, the most prolonged reform movement of the twentieth century.

Most grassroots movements follow what Anthony Downs calls a "natural 'issue-attention cycle.'" They can be powerful agents of change, but they are more difficult to sustain than reforms orchestrated from above. Environmentalism seems to have transcended these limits. While we might debate its effectiveness in implementing specific policies, the fact remains that most Americans today agree with its basic premises. After several decades, the movement's key phrases — nature, ecology, wilderness, sustainability — still resonate, not only among a cadre of committed activists but across a broad cross-section of the American public.[1]

In explaining this popularity, some studies focus on the succession of crises and controversies that kept the environment in the public eye for three decades and more, while others highlight the structural economic trends that placed a premium on environmental quality as a basic consumer good. Between the emphases on political event and structural change falls the intriguing question of how Americans formulated a vision of nature and place worthy of such sustained political commitment. What, for instance, made Americans look at rivers in ways that precluded their use as sewers and dams? What made certain natural landscapes effective rallying points for environmental action?

State-level politics offers a compelling venue for answering these questions. In the 1950s and 1960s, civic activism aimed at state government led the way in formulating an environmental agenda, and throughout this period the rhetoric of state politics was closely linked to popular constructs of nature and place. Here issues like pollution, dams, development, and sprawl took on the texture of regional culture and life, and here we can see environmental activism emerging out of the landscapes that inspired it. From this record springs a sense of how Americans regarded their environment, and how their love of nature and place changed during the three decades between 1945 and 1975. While other forces drove environmental politics as well, this vision helped shape the movement's key arguments and allegiances. Tracing the popular regard for nature and place over time permits a better understanding of environmental politics — its ideological underpinnings, its dynamics, its pivot points, and its capacity for change.

This approach — scrutinizing the state-level record of environmental thought — has practical implications as well, in that it highlights the importance of place-centered politics in the environmental movement. Defending the natural and cultural fonts of regional identity can be a powerful means of reigniting grassroots enthusiasm and recruiting local leadership.[2] In an era of complex and confusing global problems, these regional issues bring environmental politics home. They teach individuals not to expect someone else to solve their problems, and they offer a venue for political activism in a familiar setting, where direct contact with legislative leaders and administrative officials is more likely to bring results.

This history, then, reminds us that state politics is still a productive arena for environmental activism, particularly in times of declining marginal effectiveness at the federal level. It may be, as Richard N. L. Andrews suggests, that current federal approaches — standard-setting, permitting, and enforcement — are "no longer sufficient to achieve significant further progress."[3] If so, state government, more flexible and responsive to collective citizen action, may offer new approaches. And as history shows, the defense of nature in state legislative battles can be a more democratic form of political participation. Understanding the substantial contribution states made between 1945 and 1975 helps us appreciate their role in forging new strategies for uncertain times. State politics, in short, should be taken more seriously as a foundation for the environmental movement as a whole.

Our second goal has been to distill from this state record a shared political vision: the idea of harmony between society and nature in its modern guise. The importance of this vision is underscored by its relative absence in today's environmental politics, where as Stephanie Pincetl suggests, statewide environmental organizations are often "embattled," fighting on a species-by-species, and development-by-development basis.[4] "What is missing from American environmental policy today," Richard N. L. Andrews says, "is a coherent vision of the common environmental good that is sufficiently compelling to generate sustained public support for government action to achieve it."[5] Such a vision inspired earlier reforms like the sanitation and City Beautiful campaigns of the late nineteenth century, the Progressive civic and conservation movements of the early twentieth century, and the New Deal idea of combining ecological, social, and economic recovery during the Depression.

Contemporary environmental politics seems to have lost track of this compelling vision of a world without pollution and despoliation. This imaginative construct is crucial to sustaining grassroots support for the movement. As this volume suggests, there are no set formulas for recreating the environmental imagination, but we also learned from the past that local and regional activities offer the richest possibilities for developing it. Environmentalists need to reconnect with place, and there they will find a firm footing on which to build the powerful ideological constructs so necessary to grass-roots support. If we engage in politics close to home, even while supporting the epic defense of nature under way in the halls of Congress and in the international arena, we will forge an environmental imagination appropriate to our times. Rekindling this imagination is of vital concern to the movement; it will inspire the next generation of environmentalists and encourage a more creative, more inclusive, and more coherent political strategy.

New visions, grounded in popular aspirations and popular culture, are crucial as the movement diversifies and reaches out to minority communities, religious groups, labor and trade associations, senior-citizen organizations, and educational and young people's institutions. The history of environmentalism suggests the need to weave coherent, positive goals into the rhetoric of specific environmental campaigns — protecting wilderness and wildlife, promoting environmental justice, developing appropriate technology, designing sustainable food production, conserving energy, ensuring healthy neighborhoods and workplaces, safeguarding air and water resources, controlling growth, and addressing global warming.[6] In these changing times, the crux of environmental activism remains the disparity between things as they are and things as they should be. We need to keep the latter before us.

NOTES

PREFACE

1. Margaret Lynn Brown, *The Wild East: A Biography of the Great Smoky Mountains* (Gainsville: University of Florida, 2000), p. 353.

2. J. Clarence Davies III and Barbara S. Davies, *The Politics of Pollution*, 2nd ed. (Indianapolis, IN: Pegasus, 1975 [c. 1970]), pp. 101–151; Richard N.L. Andrews, *Managing the Environment, Managing Ourselves: A History of American Environmental Policy* (New Haven, CT: Yale University Press, 1999), p. 202.

3. Telling a national story through local evidence is compelling and consonant with current trends in social and political history, but it is nevertheless difficult to achieve. One approach is to collect representative examples from many places across America and to weave these stories into a common theme. Another method is the single-location case study, which provides depth and narrative continuity but at the risk of losing national perspective. Samuel P. Hays's path-breaking *Beauty, Health, and Permanence: Environmental Politics in the United States, 1955–1985* (New York: Cambridge University Press, 1987) captures both the national sweep of the movement and its essentially local and regional character, and thus acknowledges the importance of local campaigns in the broader scope of environmental politics. But this attempt at comprehensiveness weakens the narrative thread. In fact, Hays's book is somewhat inaccessible to nonhistorians owing to its multiple themes and encyclopedic organization. One purpose in writing this volume is to reaffirm Hays's method while giving this approach a stronger and more unified narrative by focusing on two representative states.

CHAPTER 1: FOLK AND NATURE

1. Robert P. Tristram Coffin, *Yankee Coast* (New York: Macmillan, 1947), pp. 10–11, 46; Henry Beston, *Northern Farm: A Glorious Year on a Small Maine Farm* (New York: Henry Holt, 1976, c. 1948), pp. 4–5, 77, 236.

2. Beston, *Northern Farm*, p. 49.

3. Coffin, *Yankee Coast*, pp. 10–11, 172; Richard L. Neuberger, *Our Promised Land* (New York: Macmillan, 1938), p. 369.

4. Neuberger, *Our Promised Land*, p. 362; H.L. Davis, *Kettle of Fire* (New York: William Morrow, 1959), pp. 19, 21. See Gordon B. Dodds, *Oregon: A History* (New York: W.W. Norton, 1977), p. 222.

5. Davis, *Kettle of Fire*, p. 23.

6. Robert Ormond Case and Victoria Case, *Last Mountains: The Story of the Cascades* (Garden City, NY: Doubleday, Doran, 1945), pp. 8, 33–39. See Stewart Holbrook, "Oregon," in *The Pacific Northwest*, edited by Anthony Netboy (Garden City, NY: Doubleday, 1963).

7. Case and Case, *Last Mountains*, p. 200; *Medford Mail–Tribune*, October 26, 1970; E.B. White, *One Man's Meat* (New York: Harper and Brothers, 1942), p. 44.

8. Rene Dubos in Robert Burns, "Cultural Change, Resource Use, and the Forest Landscape: The Case of the Willamette National Forest," PhD dissertation (geography), University of Oregon, 1973, p. 205.

9. Case and Case, *Last Mountains*, pp. 16–17. See Stuart E. Jones, "Oregon's Many Faces," *National Geographic* 135 (January 1969): 81.

10. Leo A. Borah, "Oregon Finds New Riches," *National Geographic* 90 (December 1946): 705. See Robert Bunting, *The Pacific Raincoast: Environment and Culture in an American Eden, 1778–1900* (Lawrence: University of Kansas Press, 1997), pp. 87, 90–93; C. Hartley Grattan, "The Future of the Pacific Coast II: The Empire of the Northwest," *Harper's* 190 (April 1945): 402; Stewart H. Holbrook, "The State of Oregon," *American Mercury* 68 (March 1949): 351–56; Robert H. Babcock, "The Rise and Fall of Portland's Waterfront, 1850–1920," *Maine Historical Society Quarterly* 22 (Fall 1982): 79–85; Keith Montgomery Carr, "Changing Environmental Perceptions, Attitudes, and Values in Oregon's Willamette Valley: 1800 to 1978," master's thesis, University of Oregon, 1978, p. 117.

11. F. Wallace Patch, "Maine Becomes a State of Mind," *American Mercury* 91 (September 1960): 147. See Coffin, *Yankee Coast*, p. 46; Robert Cahn, "Oregon Dilemma," *Saturday Evening Post* 234 (October 14, 1967): 23; *Medford Mail–Tribune*, October 26, November 3, 7, 1970.

12. E.B. White, "Letter from the East," *New Yorker* 31 (December 24, 1955): 61; Case and Case, *Last Mountains*, p. 207. See Ian Tyrrell, *True Gardens of the Gods: Californian–Australian Environmental Reform, 1860–1930* (Berkeley: University of California Press, 1999), p. 13; John Warfield Simpson, *Visions of Paradise: Glimpses of Our Landscape's Legacy* (Berkeley: University of California Press, 1999), p. 25; Robert L. Dorman, *Revolt of the Provinces: The Regionalist Movement in America, 1920–1945* (Chapel Hill: University of North Carolina Press, 1993), pp. xii–xiii, 19, 124, 127.

13. Leo Marx, *The Machine in the Garden: Technology and the Pastoral Ideal in America* (New York: Oxford University Press, 1964), pp. 39, 141; Lawrence Buell, *The Environmental Imagination: Thoreau, Nature Writing, and the Formation of American Culture* (Cambridge, MA: Belknap Press, 1995), p. 31; Simpson, *Visions of Paradise*, p. 25.

14. Mark Luccarelli, *Lewis Mumford and the Ecological Region: The Politics of Planning* (New York: Guilford Press, 1995), pp. 102–03; Buell, *Environmental Imagination,* pp. 35, 44, 62. See Marx, *Machine in the Garden,* pp. 3, 13, 21, 28, 116.

15. Benton MacKaye in Luccarelli, *Lewis Mumford,* p. 79. See Murray Bookchin, *The Limits of the City* (New York: Harper Colophon, 1974), p. 6.

16. James L. Machor, *Pastoral Cities: Urban Ideals and the Symbolic Landscape of America* (Madison: University of Wisconsin Press, 1987), p. 14. See Elmer T. Peterson, "Cities Are Abnormal," in *Cities Are Abnormal,* edited by Peterson (Norman: University of Oklahoma Press, 1946), p. 20.

17. Henry L. Kamphoefner, "An Architect Protests," in *Cities Are Abnormal,* edited by Peterson, p. 128; Rutherford H. Platt, "From Commons to Commons," in *The Ecological City: Preserving and Restoring Urban Biodiversity,* edited by Platt, Rowan A. Rowntree, and Pamela C. Muick (Amherst: University of Massachusetts Press, 1994), pp. 31–32; Machor, *Pastoral Cities,* p. 213.

18. Machor, *Pastoral Cities,* pp. 6–7, 10, 13, 168; Lewis Mumford in Luccarelli, *Lewis Mumford,* pp. 28, 29, 48, 49, 51, 79.

19. Richard L. Neuberger, "My Home Town Is Good Enough for Me," *Saturday Evening Post* 223 (December 16, 1950): 30, 92, 93. See William G. Robbins, *Landscapes of Promise: The Oregon Story, 1800–1940* (Seattle: University of Washington Press, 1997), p. 294.

20. Jeffrey Jacob, *New Pioneers: The Back-to-the-Land Movement and the Search for a Sustainable Future* (University Park: Pennsylvania State University Press, 1997), p. 8.

21. Robbins, *Landscapes of Promise,* p. 239; Marx, *Machine in the Garden,* p. 176; Tyrrell, *True Gardens of the Gods,* p. 10. See Timothy O'Riordan, "The Third American Conservation Movement: New Implications for Public Policy," *Journal of American Studies* 5, 2 (1971): 156, 158; James E. Sherow, "Environmentalism and Agriculture in the American West," in *The Rural West since World War II,* edited by R. Douglas Hurt (Lawrence: University of Kansas Press, 1998), p. 59; Samuel P. Hays, "From Conservation to Environment: Environmental Politics in the United States Since World War II," in *Out of the Woods: Essays in Environmental History,* edited by Char Miller and Hal Rothman (Pittsburgh: University of Pittsburgh Press, 1997), p. 102.

22. O'Riordan, "Third American Conservation Movement," pp. 155–71; David Burner, *Making Peace with the 60s* (Princeton, NJ: Princeton University Press, 1996), pp. 6, 220–21; Warren Susman in Andrew Jamison and Ron Eyerman, *Seeds of the Sixties* (Berkeley: University of California Press, 1994), p. 8; Paul Lyons, *New Left, New Right, and the Legacy of the Sixties* (Philadelphia: Temple University Press, 1996), pp. 37–38; Raymond Williams in Buell, *Environmental Imagination,* p. 62. See Richard N.L. Andrews, *Managing the Environment, Managing Ourselves: A History of American Environmental Policy* (New Haven, CT: Yale University Press, 1999), p. xi.

23. Andrews, *Managing the Environment,* p. 201.

24. Bookchin, *Limits of the City,* p. 3; Michael Hough, "Design with City Nature: An Overview of Some Issues," in Platt, Rowntree, and Muick, eds., *Ecological City,* p. 41; Peterson, "Cities Are Abnormal," pp. 3, 5, 8, 16, 17, 21.

25. S.C. McConahey, "Economic Verities," in *Cities Are Abnormal,* edited by Peterson, pp. 159–60 (McConahey's emphasis). See Machor, *Pastoral Cities,* pp. 3–5; Kamphoefner, "Architect Protests," p. 139.

26. Peterson, "Cities Are Abnormal," pp. 3–4, 10.

27. Paul B. Sears, "The Ecology of City and Country," in *Cities Are Abnormal*, edited by Peterson, p. 44; Jonathan Forman, "Biological Truths and Public Health," in *Cities Are Abnormal*, edited by Peterson, pp. 96, 100–01, 108.

28. See Terry H. Anderson, *The Movement and the Sixties* (New York: Oxford University Press, 1995), p. 16; Jamison and Eyerman, *Seeds of the Sixties*, pp. 39–44, 55–60, 82–85.

29. Buell, *Environmental Imagination*, p. 52; Percival and Paul Goodman in Machor, *Pastoral Cities*, p. 213. See William Vitek, "Rediscovering the Landscape," in *Rooted in the Land: Essays on Community and Place*, edited by Vitek and Wes Jackson (New Haven: Yale University Press, 1996), p. 3.

30. Kenneth T. Jackson, *Crabgrass Frontier: The Suburbanization of the United States* (New York: Oxford University Press, 1985); Adam Rome, *The Bulldozer in the Countryside: Suburban Sprawl and the Rise of American Environmentalism* (New York: Cambridge University Press, 2001).

31. John M. Findlay, *Magic Lands: Western Cityscapes and American Culture after 1940* (Berkeley: University of California Press, 1992), p. 277; Earl Pomeroy, *The Pacific Slope: A History of California, Oregon, Washington, Idaho, Utah, and Nevada* (New York: Alfred A. Knopf, 1965), p. 388; Edward P. Morgan, *The 60s Experience: Hard Lessons about Modern America* (Philadelphia: Temple University Press, 1991).

32. William G. Robbins, *The Oregon Environment: Development vs. Preservation, 1905–1950* (Corvallis: Oregon State University, 1975), pp. 8–15, 32, 34–37; Robbins, *Landscapes of Promise*, p. 240; Bret Walth, *Fire at Eden's Gate: Tom McCall and the Oregon Story* (Portland: Oregon Historical Society Press, 1994), p. 88–89; Courtland Smith, *Public Participation in Willamette Valley Environmental Decisions* (Corvallis: Oregon State University, 1973), pp. 49, 51.

33. Neuberger, *Our Promised Land*, pp. 2, 24, 27–28, 90, 92. See Robbins, *Colony and Empire*, pp. 130–31, 148, 173, 180; Grattan, "Future of the Pacific Coast," pp. 404–05; Clayton W. Dumont, Jr., "The Demise of Community and Ecology in the Pacific Northwest: Historical Roots of the Ancient Forest Conflict," *Sociological Perspectives* 39, 2 (1996): 281–82.

34. Robbins, *Colony and Empire*, p. 130; Pomeroy, *Pacific Slope*, pp. 297–98; Carl Abbott, "The Metropolitan Region," in *The Twentieth-Century West: Historical Interpretations*, edited by Gerald D. Nash and Richard W. Etulain (Albuquerque: University of New Mexico Press, 1989), pp. 71–72; Miner H. Baker, "Economic Growth Patterns in Washington and Oregon," *Monthly Labor Review* 82 (May 1959): 502–03, 508.

35. Findlay, *Magic Lands*, p. 270. See Adam W. Rome, "Building on the Land: Toward an Environmental History of Residential Development in American Cities and Suburbs, 1870–1990," *Journal of Urban History* 20 (May 1994): 407–34.

36. First quote: Morgan, *60s Experience*, pp. 13–14; second quote: Findlay, *Magic Lands*, p. 277. See Pomeroy, *Pacific Slope*, p. 388.

37. "B–14–1 Nuisance complaints, A–P," Health Board, Sanitary Engineering Division General Files, 1936–1950, Box 29; "Selected Files, 1938–1961"; "Reports, 1936–1959"; "Reports (Quarterly) 1936–1947," Box 35, Oregon State Archives (hereafter OSA), Portland; Henry J. Vaux in Gordon A. Bradley, "The Urban/Forest Interface," in *Land Use and Forest Resources in a Changing Environment: The Urban/Forest Interface*, edited by Gordon A. Bradley (Seattle: University of Washington Press, 1984); Rome, *Bulldozer in the Countryside*, 2001.

38. "Quarterly Report of the Division of Sanitary Engineering: January, February, and March 1946," pp. 3–4, Health Board, Sanitary Engineering Division General Files, 1936–1950, B–14–1 (also file B–13–1), A–P; pp. 2, 5; "Memo," January 28, 1946, Report, Box 29, 2, OSA, Portland.

39. Peter James, "Ecotopia in Oregon?" *New Scientist* 81, 1135 (January 4, 1979): 29; Oregon Writers' Project, *Oregon: End of the Trail,* quoted in Robbins, *Landscapes of Promise,* p. 296. See Joel Garreau, *The Nine Nations of North America* (Boston: Houghton Mifflin, 1981), p. 263.

40. Neuberger, *Our Promised Land,* p.134; Holbrook, "State of Oregon," p.351; Jim Marshall, "Out Where the West Ends," *Collier's* 121 (April 24, 1948): 22; Davis, *Kettle of Fire,* p. 17, 23. See Borah, "Oregon Finds New Riches," p. 681.

41. Carr, "Changing Environmental Perceptions," pp. 134-35; *Medford Mail-Tribune* October 11, 1970; Stan Federman, "The 'Plumed Knight Rides Again in Oregon," *Smithsonian Magazine* 1 (September 1970): 37-38; *Oregonian,* April 14, 1977; Eugene Planning Commission, "Conference on Community Goals and Policies," typescript, May 25, 1966, n.p., Knight Library Oregon Collections, University of Oregon.

42. Charles DeDeurwaerder, "The Fight against Ticky-Tacky," *Oregonian Northwest Magazine,* November 16, 1969, p. 12. See Carr, "Changing Environmental Perceptions," pp. 109-110, 111, 113.

43. Kenneth P. Hayes, *As Envisioned by Maine People: A Report to the Commission on Maine's Future* (Orono, Maine: Social Science Research Institute, 1976), pp. 8, 9; Everett F. Greaton in *Lewiston Evening Journal,* April 5, 1954; Lewiston Planning Board, *The Comprehensive Plan: Lewiston, Maine* (Cambridge, MA: Planning Services Group, l962), p. 4; Vivian Grace Sampson, "A Study of the Textile Industries of Lewiston, Maine," bachelor's thesis, Bates College, l942; George Olmsted, Jr., in *Kennebec Journal,* April 1, 1963; William S. Ellis, "The Trouble with Maine," *Nation* 196 (June 22, 1963): 527-29; Richard Condon, "Maine Out of the Mainstream," in Richard W. Judd, Edwin A. Churchill, and Joel W. Eastman, *Maine: The Pine Tree State from Prehistory to the Present* (Orono: University of Maine Press, 1995), p. 531.

44. White, *One Man's Meat,* pp. 302-03, 307; "North by East," *Down East Magazine* 2 (February–March 1956): 11.

45. *Kennebec Journal,* January 10, February 28, 1963; Mary Ellen Chase, *A Goodly Heritage* (New York: Henry Holt, 1932), pp. 174-75; Beston, *Northern Farm,* p. 225.

46. Horace Hildreth, inaugural address, *Lewiston Evening Journal,* January 4, 1945. See Maine Development Commission in *Lewiston Journal,* January 17, 1945; Maine Development Commission, *Postwar Planning for the State of Maine* (Augusta, 1944), p. 22; Horace Hildreth in *Lewiston Journal,* January 17, 1945.

47. Hildreth, inaugural address; Geoffrey Faux, "Colonial New England," *New Republic* 167 (November 25, 1972): 16-19; Maine Development Commission, *Postwar Planning,* pp. 6, 8.

48. Maine Development Commission, *Industrial Maine: A Presentation of Facts Concerning the Advantages Which Maine Offers to Industry* (Augusta, ca. 1940), pp. 8-9, 15-16, 31, 32, 34, 36, 37.

49. William Dickson in *The Maine Coast: Prospects and Perspectives* (Brunswick, Maine: Center for Resource Studies, Bowdoin College, 1967), p. 45.

50. Coffin, *Yankee Coast,* 1947, pp. 157-58.

51. *Maine Times,* August 27 1976; Richard Pardo, "Something Up, Down East," *American Forests* 75 (May 1969): 14; *Lewiston Evening Journal,* January 17, 1945; "North by East," *Down East Magazine* 2 (July 1956): 19. See *Kennebec Journal,* January 10, 1963; *Portland Press Herald,* October 1, 1968.

52. John Fischer (*Harper's*) to Everett F. Greaton, November 7, 1955; Edmund S. Muskie to Alpheus D. Spiller, November 14, 1955; and the entire file, "Department of

Development of Industry and Commerce, folder 2," Box 525, Edmund S. Muskie Archives, Bates College, Lewiston, Maine.

53. John T. Rowland, "Maine Is Worth Saving," *Down East Magazine* 1 (May 1955): 30–31; Maine Legislature, *Legislative Record* (Augusta, 1959), pp. 1850, 1854–56, 2084, 2085. See Charles Callison, "The Billboard Battle Goes On," *Audubon* 65 (November–December 1963), p. 369; and (for Oregon) "Portland and Surprising Oregon," *Better Homes and Gardens* 45 (May 1967): 76–77.

54. Beston, *Northern Farm*, p. 182; Pardo, "Something Up," p. 43; Aimé Gauvin, "Raking the Muck to Mold the Future of Maine's Coast," *Audubon* 75 (July 1973): 132; "Maine Chance," *Time* 104 (August 26, 1974): 54.

55. *Kennebec Journal* August 29, 1956. See *Lewiston Evening Journal*, January 17, 1945.

56. Governor Kenneth Curtis in Andrew Hamilton, "Maine: Finding the Promised Land (without Losing the Wilderness)," *Science* 178 (November 10, 1972): 595–96; Richard Saltonstall, Jr., *Maine Pilgrimage: The Search for an American Way of Life* (Boston: Little, Brown, 1974), p. 102. See John N. Cole, "Letter from Maine: Oil and Water," *Harper's* 243 (November 1971): 48

57. White, "Letter from the East," p. 60; Cole, "Letter from Maine," p. 42.

58. *Maine Times*, May 17, June 7, October 18, 1974, April 1, 1977; Pardo, "Something Up," p. 44; Cole, "Letter from Maine," p. 42.

59. First quote: Gauvin, "Raking the Muck," p. 132; second quote: John Hay, "The Pace of Tides," *Audubon* 71 (March 1969): 27.

60. Cole, "Letter from Maine," p. 42.

61. Carr, "Changing Environmental Perceptions," p. 135.

62. Elizabeth Coatsworth, *Maine Ways* (New York: Macmillan, 1947): pp. 212–13; Coffin, *Yankee Coast*, p. 331; Case and Case, *Last Mountains*, pp. 5, 14–15.

63. Robbins, *Colony and Empire*, pp. 6, 61; Case and Case, *Last Mountains*, pp. 351–52, 356; Nancy Wilson Ross, *Farthest Reach: Oregon & Washington* (New York: Alfred A. Knopf, 1942), pp. 4–5. See Pomeroy, *Pacific Slope*, p. 166.

64. Coffin, *Yankee Coast*, pp. 248–49. See E.B. White, "Letter from the East," *New Yorker* 36 (December 3, 1960): 238.

65. "Fellow Americans: Keep Out!" *Forbes* 107 (June 15, 1971): 23.

66. Douglas R. Weiner, *Models of Nature: Ecology, Conservation, and Cultural Revolution in Soviet Russia* (Pittsburgh, PA: University of Pittsburgh Press, 2000), p. 229. See Tyrrell, *True Gardens of the Gods*, p. 13.

CHAPTER 2: POLITICIZING THE PASTORAL IDEAL

1. *Lewiston Evening Journal*, March 16, 1907; "Information by Frank I. Cowan ... State, v. Brown Company, Oxford Paper Company and International Paper Company, 1942," pp. 5–7, Androscoggin Pollution, Box 85, papers of the attorney general, Maine State Archives, Augusta (hereafter MSA).

2. "Preliminary Report ... Oregon State Sanitary Authority, January 5, 1945," pp. 3, 5–6, Stream Pollution Studies, 1944–1946, Box 4, RG–93A–18, Records of Governor Earl Snell, Oregon State Archives, Salem (hereafter OSA); George Abed, *Water Pollution Control Policy in the Willamette Basin* (Salem: Oregon Department of Commerce, 1965),

pp. 6, 21; Governor's Natural Resources Advisory Committee, *Water Pollution Control in the Willamette River Basin* (Salem: Oregon State Sanitary Authority, 1950), pp. 1–2.

3. "Preliminary Report Submitted to Governor's Special Committee, Oregon State Sanitary Authority, January 5, 1945," p. 3, Stream Pollution Studies, 1944–1946, Box 4, RG–93A–18, records of Governor Earl Snell, OSA; *Lewiston Evening Journal*, September 4, 1941. See Stuart E. DeRoche, *Fishery Management in the Androscoggin River* (Augusta: Maine Department of Inland Fisheries and Game, 1967), pp. 31–33; *Lewiston Evening Journal*, July 22, September 5, 1941; *Rumford Falls Times*, August 28, 1941.

4. Mabel (Mrs. Fred G.) Johnson to Straub, March 28, 1977, Administrative Correspondence, 1975–1978, 1120.2, Water, Box 94, RG 90A–30, Office of Governor Robert W. Straub Records (hereafter RWSR), OSA; George U. Benz and Margaret E. Benz to William H. Young, August 1, 3, 1977, ibid; Straub to Mrs. George U. Benz, August 17, 1977, ibid; Mrs. Glen Boehme to Straub, January 25, 1975, ibid; Straub to Mrs. Boehme, February 12, 1975, ibid.

5. *Lewiston Evening Journal*, November 8, 1940.

6. William Cronon, *Nature's Metropolis: Chicago and the Great West* (New York: W.W. Norton, 1991); Richard N.L. Andrews, *Managing the Environment, Managing Ourselves: A History of American Environmental Policy* (New Haven, CT: Yale University Press, 1999), p. 127.

7. John B. Jackson, *Discovering the Vernacular Landscape* (New Haven, CT: Yale University Press, 1984), p. 61; *Portland Sunday Telegram*, December 31, 1967.

8. See D.W. Meinig, "Symbolic Landscapes: Models of American Community," in *The Interpretation of Ordinary Landscapes: Geographical Essays*, edited by Meinig (New York: Oxford University Press, 1979); John T. Cumbler, *Reasonable Use: The People, the Environment, and the State, New England, 1790–1930* (New York: Oxford University Press, 2001), pp. 181–82; J. Clarence Davies III and Barbara S. Davies, *The Politics of Pollution*, 2d ed. (Indianapolis: Pegasus, 1975 [c. 1970]), p. 8.

9. Robert M. Paul in U.S. Congress, 87th Cong., 1st sess., House of Representatives, *Federal Water Pollution Control: Hearings before the Committee on Public Works* (Washington, DC: Government Printing Office, 1961), p. 300; M. James Gleason, ibid, p. 30; U.S. Department of Health, Education, and Welfare, Public Health Service, *Clean Water: A Challenge to the Nation: Highlights and Recommendations of the National Conference on Water Pollution* (Washington, DC: Government Printing Office, 1960), pp. 6, 8, 18; Andrews, *Managing the Environment*, p. 128.

10. First quote: Charles F. Schnee in U.S. Congress, 79th Cong., 1st sess., House of Representatives, Committee on Rivers and Harbors, *Pollution of Navigable Waters*, (Washington, DC: Government Printing Office, 1946), p. 105; second quote: Alfred LeFaber in ibid, p. 86.

11. Davies and Davies, *Politics of Pollution*, p. 28; Kenneth A. Reid in *Pollution of Navigable Waters* (1946), pp. 229, 231; ibid, p. 13; Holman Harvey, "Our Dead and Dying Rivers," *American City* 60 (September 1945): 106; U.S. Congress, 80th Cong., 1st sess., House of Representatives, *Water Pollution Control Hearings* (Washington, DC: Government Printing Office, 1947), pp. 188, 190; Howard Zahniser in ibid, p. 280. See Edward Connor in ibid, p. 109; J.D. Steele in *Pollution of Navigable Waters* (1946), p. 108; J.F. Parkinson in ibid, 108; Philip G. Platt in ibid, p. 157; Leroy E. Burney in Public Health Service, *Clean Water*, p. 8; Andrews, *Managing the Environment*, p. 203; *Water Pollution Control Hearings* (1947), p. 142; George A. Dondero and George Peabody in *Pollution of Navigable Waters* (1946), pp. 46–48, 111–16.

12. Sara Ebenreck, "Opening Pandora's Box: Imagination's Role in Environmental Ethics," *Environmental Ethics* 18 (Spring 1996): 13, 16; Daniel B. Botkin, *No Man's Garden: Thoreau and a New Vision for Civilization and Nature* (Washington, DC: Island Press/Shearwater Books, 2001), p. 10. See Mark Dowie, *Losing Ground: American Environmentalism at the Close of the Twentieth Century* (Cambridge, MA: MIT Press, 1995), p. 9; Stephanie S. Pincetl, *Transforming California: A Political History of Land Use and Development* (Baltimore, MD: Johns Hopkins University Press, 1999), p. xii; John Warfield Simpson, *Visions of Paradise: Glimpses of Our Landscape's Legacy* (Berkeley: University of California Press, 1999), p. 8.

13. *Kennebec Journal*, January 9, 1937. See *Portland Herald*, November 24, 1934.

14. *Portland Sunday Telegram*, December 31, 1967.

15. *Oregonian Northwest Magazine*, November 5, 1967, p. 14; David P. Willis and Jeffery A. Foran, *The Cuyahoga River Study* (Columbus: Ohio Wildlife Federation, 1986), in State Records, Ohio, Samuel P. and Barbara Hays Environmental Archives, University of Pittsburgh (hereafter SBHC).

16. Walter A. Lawrance, *A Twenty-Year Review of Androscoggin River Pollution Control Activities* (Lewiston, Maine: Androscoggin River Technical Committee, 1961), pp. 3–5; Joseph A. Warren et al., *Survey and Report of River and Stream Conditions in the State of Maine* (Augusta, 1930), pp. 1–3, 9, 12, 17, 18; Page Helm Jones, *Evolution of a Valley: The Androscoggin Story* (Canaan, NH: Phoenix Publishing, 1975), p. 143.

17. Leo R. Good, Cleas Lacombe interviews, Androscoggin Pollution: Brown Company et al., Information, Stipulation, Decree, etc., Box 83, Papers of the Attorney General, MSA. See also Mrs. Willard Andrews, E.N. Call, William H. Chaffers, M.J. Brooks in ibid; I.R. Lohnes, U.S. Department of Health, Education, and Welfare, *Conference in the Matter of Pollution of the Interstate Waters of the Androscoggin River, Proceedings* (Portland, September 24, 1962), p. 10; *Lewiston Evening Journal*, July 15, 25, 1941; Seth May to Attorney General Frank I. Cowan, July 25, 1941; Cowan to May, July 28, 1941; May to Cowan, August 4, 1942, Androscoggin Pollution: Brown Company et al.: Information, Stipulation, Decree, etc.; note under "Outline of Suggestions Growing Out of Conferences with the Pulp and Paper Section of the War Production Board, August 3–4, 1942"; "Frank I. Cowan, AG, v. Brown Company et al., Proof of Facts," Androscoggin Pollution: 1950, #2, Box 83, Papers of the Attorney General, MSA.

18. *Lewiston Evening Journal*, July 1, 15, 22, 25, 26, 31, August 8, 12, 13, 15, 18, 19, 21, 28, September 18, 30, 1941; *Rumford Falls Times*, August 28, 1941.

19. Elmer Campbell in *Lewiston Evening Journal*, September 5, 1941.

20. Rumford Falls Times, July 24, August 28, 1941; *Lewiston Evening Journal*, July 15, 22, 25, 1941; Maine Legislature, *Legislative Record* (Augusta, 1941), p. 856.

21. *Lewiston Evening Journal*, October 13, September 4, 5, 1941. See E. Sherman Chase, *Investigation of the Pollution of the Androscoggin River and Remedial Measures* (Augusta: Sanitary Water Board Bulletin no. 1, 1942), p. 20; *Rumford Falls Times*, August 28, 1941.

22. "Frank I. Cowan ... on Behalf of ... State vs. Brown Company ... Brief on Behalf of the State of Maine," Androscoggin Pollution: 1950, #2, Box 83, Papers of the Attorney General, MSA; John T. Mains in Maine Legislature, *Legislative Record* (Augusta, 1965), p. 603.

23. Brown Company to Seth May, October 21, 1942, Androscogin Pollution: Brown Company et al., Box 83, Papers of the Attorney General, MSA; Harold Schnurle in *Kennebec Journal*, March 9, 1954. See *Lewiston Evening Journal*, July 17, 22, 1941, January 7, 22, 1942, March 25, 1954; Chase, *Investigation of the Pollution*, pp. 16–17, 24; E.

Sherman Chase to Ralph W. Farris, October 26, 1950; "Memorandum of Conference on Androscoggin River Pollution ... November 13, 1943"; and "Conference, November 9, 1943 at Office of Metcalf and Eddy, Brown Co, Oxford, and International Paper," Androscoggin Pollution: 1950, #2, in ibid.

24. J. Elliott Hale, "New Stream Pollution Laws," *New England Townsman* 9 (July 1947): 6. See *Lewiston Evening Journal*, March 31, 1953; *Kennebec Journal*, March 31, 1953; *Rumford Falls Times*, June 19, August 21, 1947; Arno Karlen, "Muddled Waters: The Challenge to Maine," *Country Beautiful* (November 1962): 12; Maine Legislature, *Legislative Record* (Augusta, 1955), p. 2159; DeRoche, *Fishery Management in the Androscoggin River*, pp. 31–32; Hale, "The Stream Pollution Problem in the State of Maine," *New England Townsman* 7 (March 1945): 3; Hale, "New Stream Pollution Laws," *New England Townsman* 9 (July 1947): 6; *First Annual Report of the Maine Sanitary Water Board* (Augusta, 1945–1946), pp. 3, 17; ibid, *Third Annual Report* (1947–1948), pp. 2–3, 5, 7–8; Maine Department of Health and Welfare, Division of Sanitary Engineering, *Report on Water Pollution in the State of Maine* (Augusta, 1950), p. 7; *Report on Water Pollution in the State of Maine* (1950), p. 3; Memos, J. Elliott Hale, Technical Secretary, SWB, various dates, especially December 6, 1947, Sanitary Water Board correspondence binder, Department of Environmental Protection files, MSA.

25. Andrew Jamison and Ron Eyerman, *Seeds of the Sixties* (Berkeley: University of California Press, 1994), pp. 1, 5, 33–34; Kenneth P. Hayes, *As Envisioned by Maine People: A Report to the Commission on Maine's Future* (Orono, ME: Social Science Research Institute, 1976), p. 50; Terry H. Anderson, *The Movement and the Sixties* (New York: Oxford University Press, 1995), p. 22; Robert M. Collins, "Growth Liberalism in the Sixties: Great Societies at Home and Grand Designs Abroad," in *The Sixties: From Memory to History*, edited by David Farber (Chapel Hill: University of North Carolina Press, 1994), p. 11.

26. Robert P. Tristram Coffin, "The State of Maine," *American Mercury* 66 (January 1948): 70; "Maine's Ideal Man," *New Republic* 135 (September 10, 1956): 135; David Nevin, *Muskie of Maine* (New York: Random House, 1972), pp. 156–58; *New York Times*, January 10, June 20, September 5, 7, 1954; *Lewiston Evening Journal*, May 4, 1953; "Maine's Muskie Gets Build-Up," *Life* 37 (September 27, 1954): 26–27; Louis H. Bean, "Maine Casts Its Shadow," *New Republic* 139 (September 22, 1958): 8; "New Look in '58 Politics," *U.S. News and World Report* 45 (September 19, 1958): 37–50.

27. "Maine: Sorting the Straws," *Newsweek* 44 (September 27, 1954): 26–27; "Was It Really Democratic?" *Newsweek* 44 (September 27, 1954): 27; *New York Times*, September 14, 15, 1954; September 10, 1958; "Why Democrats Won in Maine," *U.S. News and World Report* 41 (September 21, 1956): 36–40; "Real Meaning of the Maine Election," *U.S. News and World Report* 41 (September 21, 1956): 31–35; "Change in Maine," *Time* 68 (November 26, 1956): 22–23.

28. Duane Lockard, *New England State Politics* (Princeton, NJ: Princeton University Press, 1959), p. 104, 107. See Maine Legislature, *Legislative Record* (Augusta, 1957), pp. 2314, 2336; 2510–13; *Kennebec Journal*, January 4, 7, February 8, 15, 23, March 10, 14, 1955, February 12, March 19, April 2, 11, 1957; *Lewiston Evening Journal*, January 26, 27, March 19, 1954, January 6, 26, 27, February 8, 22, 23, March 24, 1955; *New York Times*, February 16, 1958; "Department of Development of Industry and Commerce," Folder 2, Box 525, Edmund S. Muskie Archives, Bates College, Lewiston, Maine (hereafter ESMA); Jerome G. Daviau, *Maine's Life Blood* (Portland: House of Falmouth, 1958), p. 13.

29. Jamison and Eyerman, *Seeds of the Sixties*, p. 98; Vance Packard, *The Waste Makers* (New York: D. McKay Co, 1960); Andrews, *Managing the Environment*, pp. 210–24; Karlen,

"Muddled Waters," p. 13; Mrs. John B. Diamond, "Statement to the Committee on Natural Resources ... in Support of L.D. 245," n.d., ca. 1966, Water Pollution, vertical files, Maine Legislative and Law Library, Augusta (hereafter MLLL); Maine Legislature, *Legislative Record* (Augusta, 1955), pp. 1684–90, 1699, 2078; 2080; (1957), p. 1295–98; Lockard, *New England State Politics*, pp. 79, 80–82, 108; on the idea of a "third house," see also Pincetl, *Transforming California*, p. 98.

30. Alfred LeFaber in *Pollution of Navigable Waters* (1946), p. 86; Maine Legislature, *Legislative Record* (Augusta, 1955), p. 1691. See Kenneth A. Reid in ibid, p. 232; Senator Margaret Chase Smith, George Peabody, Rep. Karl E. Mundt, Sen. Wallace White, and Kenneth A. Reid in *Pollution of Navigable Waters* (1946), pp. 45, 47, 51, 77, 230.

31. Maine Legislature, *Legislative Record* (Augusta, 1955), pp. 1693–94. See ibid (1955), pp. 1685, 2156; Horace Hildreth, inaugural address, *Lewiston Evening Journal*, January 4, 1945; *New York Times*, June 23, 1956; July 21, 1963.

32. G.B. Davy to Edmund S. Muskie, August 11, 1955; Henry Gerrish to Muskie, August 1, 1955, Department of Development of Industry and Commerce, Folder 1, Box 525; Phil Brown to Muskie, n.d., ca. August 1955, Industrial Development, 1955, Box 531; news release, August 19, 1957, Department of Development of Industry and Commerce, Folder 1, Box 630, ESMA. See Maine Legislature, *Legislative Record* (Augusta, 1955), p. 1692; Clifford G. Chase to Edmund S. Muskie, November 24, 1955, Water Improvement Commission, 1955, Box 546, ESMA.

33. Bill Coffin, "A Discussion of Factors Confronting the Department for Development of Industry and Commerce in Attracting Industries to the State of Maine," n.d., ca. 1955, Department of Development of Industry and Commerce, Folder 1, Box 525, ESMA.

34. Sherman Corp., "Industrial Survey of Lewiston–Auburn" (Boston, for the Lewiston Chamber of Commerce, 1928), pp. 67–70, 115, 187–88; Emile Jacques in DeRoche, *Fishery Management*, p. 30. See Eva M. Leitman, "A Historical Perspective of the Franklin Company's Role in the Development of Early Lewiston, Maine," typescript, ca. 1969, Lewiston Public Library, pp. 15, 17; *Lewiston Evening Journal*, August 15, 1941, January 7, 1942, January 17, 1945, January 16, 27, 1954; Lewiston Planning Board, *The Comprehensive Plan, Lewiston, Maine, Program Report* (Cambridge, MA: Planning Services Group, 1962), pp. 4, 19, 21–24; Vivian Grace Sampson, "A Study of the Textile Industry of Lewiston, Maine," bachelor's thesis, Bates College, 1942, pp. 1, 3–4.

35. *Kennebec Journal*, January 7, 1955, January 4, 1957; *Lewiston Evening Journal*, January 6, February 22, 1955. See, for instance, Carl D. Land, "Islandomania," *Down East Magazine* 1 (September 1954): 9–13; Hodding Carter, "In the Buckboard Summer Time," 1 (September 1954): 14–16; "Margaret Hammel, "The One-Room School House," 1 (October 1954): 11–14.

36. Donald A. Fowler to Edmund S. Muskie, September 24, 1956, Inland Fisheries and Game, Folder 1, Box 590, ESMA; Robert H. Bradley, Jr., to Edmund S. Muskie, October 19, 1956, Pollution, Box 600; Hon. Frank Pierce to Muskie, December 27, 1955, Pollution, Box 541, ESMA. See Maine Legislature, *Legislative Record* (Augusta, 1961), p. 2141.

37. *Water Pollution Control Hearings* (1947), pp. 41, 61–62, 70, 73, 89, 185, 280. On labor, see John T. Curran in *Federal Water Pollution Control Hearings* (1961), p. 130.

38. Bill Devall, "The Eye of the Turtle: A Social History of the Northcoast Environmental Center, 1970–1978," typescript, draft copy, State Records California, Box 5, File: California, 1978, SBHC; J.W. Penfold in *Federal Water Pollution Control Hearings* (1961), p. 293. On women's organizations, see Mrs. John McEwan, p. 302; Dorothy Salisbury, p. 316, in ibid.

39. *American Environmentalism: The U.S. Environmental Movement, 1970–1990*, edited by Riley E. Dunlap and Angela Mertig (New York: Taylor & Francis, 1992), p. 27–37.

40. William M. Sullivan, "A Public Philosophy for Civic Culture," in William Vitek and Wes Jackson, *Rooted in the Land: Essays on Community and Place* (New Haven, CT: Yale University Press, 1996), p. 239.

41. *Kennebec Journal*, May 5, 1953. See *Lewiston Evening Journal*, February 13, 16, March 3, 31, April 16, 17, 1953, March 17, April 21, 1955; Mrs. Marion T. Weatherford to Edmund Muskie, June 29, 1955, Pollution, 541, ESMA; Karlen, "Muddled Waters," pp. 11–15; Maine Legislature, *Legislative Record* (Augusta, 1951), p. 2546; (1955), pp. 1701; (1866), p. 2446.

42. *Kennebec Journal*, February 25, 26, March 9,12, 31, April 1, 2, 6, 29, May 2, 1953; April 15, 16, May 6, 1955;

43. *Kennebec Journal*, March 9, 22, 1961. See Dunlap and Mertig, eds., *American Environmentalism*, p 21.

44. *Kennebec Journal*, January 5, 1955; Lockard, *New England State Politics*, pp. 93, 109; Maine Legislature, *Legislative Record* (Augusta, 1967), p. 2071; *Kennebec Journal*, February 18, March 24, April 12, 14, 15, 22, 1955, January 14, February 15, March 8, 22, 1957; *Lewiston Journal*, April 13, 1955; Maine Legislature, *Legislative Record* (Augusta, 1955), pp. 1688–89.

45. Maine Legislature, *Legislative Record* (Augusta, 1955), p. 1684; *Kennebec Journal*, April 14, 1955.

46. Maine Legislature, *Legislative Record* (Augusta, 1955), p. 1688. See *Lewiston Evening Journal*, April 16, 1953; January 27, March 25, April 27, 1954; April 21, 1955; *Kennebec Journal*, February 26, March 9, 12, April 2, 21, 1953, April 1–3, 1954, January 7, April 16, May 12, 1955, April 26, May 8–10, 1957, February 13, March 12, 26, April 18,1959; Maine Legislature, *Legislative Record* (Augusta, [1955], p. 1684, 1697; [1957], p. 1305; Paul W. Nutter to Warren Brockway and Milan Ross, March 22, 1955, Pollution, Box 541, ESMA.

47. Maine Legislature, *Legislative Record* (Augusta, 1955), pp. 1687, 1693, 1696; *Lewiston Evening Journal*, April 13, 1955; John T. Gould, "--And the Androscoggin Still STINKS!" pamphlet, Citizens for Conservation and Pollution Control, 1951, MLLL; *Lewiston Evening Journal*, March 25, 1954.

48. *Maine Sunday Telegram*, August 3, May 4, 1969, June 10, 1984; *Bangor Daily News*, June 8, 1984; Richard Pardo, "Something Up, Down East," *American Forests* 75 (May 1969): 44; *Waterville Sentinel*, December 14, 1967; *Portland Press Herald*, January 1, 1968; Neal R. Peirce, *The New England States: People, Politics and Power in the Six New England States* (New York: Norton, 1976), p. 400.

49. Governor's Natural Resources Advisory Committee, *Water Pollution Control in the Willamette River Basin*, pp. 1–2; William G. Robbins, *The Oregon Environment: Development vs. Preservation, 1905–1950* (Corvallis: Oregon State University, 1975), p. 21; George W. Gleeson, *The Return of a River: The Willamette River, Oregon* (Corvallis: Advisory Committee on Environmental Science and Technology and Water Resources Research Institute, Oregon State University, 1972), pp. 12, 49; Curtiss M. Everts, Jr., "Willamette Clean-Up" *Oregon State Game Commission Bulletin* 8 (July 1953): 2–3; "A Report of the Stream Purification Committee ... Oregon Division, Izaak Walton League of America, December 4, 1948," in Pollution, Water, Articles, 1969–1972, Oregon Environmental Council, Ms. 2386, Box 23, Oregon Historical Society, Portland (hereafter OHS); "Quarterly Report, September 30, 1946"; "Reports 1941–1954 (Misc.), Health

Board, Sanitary Engineering Division General Files, 1936–1950, A–P, Box 29, OSA; Harold F. Wendel to Earl Snell, February 16, 1943, Administrative Correspondence, 1943–1947, #5, Box 3, RG–93A–18, Records of Governor Earl Snell (1924–1927); "Preliminary Report Submitted to Governor's Special Committee, Oregon State Sanitary Authority, January 5, 1945," p. 3, Stream Pollution Studies, 1944–1946, Box 4, RG–93A–18, ibid; "Summary of Activities, April 1, 1946, to June 7, 1946," Health Board, Sanitary Engineering Division General Files, 1936–1950, A–P, Box 29; Memo to Dr. D.L. Inskeep, May 17, 1950; testimony of Kenneth H. Spies, October 8–9, 1965; testimony of Harold F. Wendel, October 8–9, 1965, Legislature, 53rd Session, Interim Committee on Public Health: Public Health Committee Files, 1965–1966, RG L6 68A–32, OSA; *Oregon Statesman*, September 30, 1965.

50. Henry Wallace, "Report from the Northwest," *New Republic* 116 (June 16, 1947): 12; Richard L. Neuberger, "Morse Versus Morse," *Nation* 170 (January 14, 1950): 29–30; Brent Walth, *Fire at Eden's Gate: Tom McCall and the Oregon Story* (Portland: Oregon Historical Society, 1994), 85–112, quote from p. 89.

51. Gordon B. Dodds, *The American Northwest: A History of Oregon and Washington* (Arlington Heights, IL: Forum Press, 1986), p. 332; Stewart Holbrook, "Oregon," in *The Pacific Northwest*, edited by Anthony Netboy (Garden City, NY: Doubleday, 1963); "Morse to the Right," *New Republic* 122 (June 5, 1950): 7–8; "Oregon: Pattern for '60?," *U.S. News and World Report* 42 (January 25, 1957): 83–84.

52. Earl Pomeroy, *The Pacific Slope: A History of California, Oregon, Washington, Idaho, Utah, and Nevada* (New York: Alfred A. Knopf, 1965), pp. 327–28; Robert Cahn, "Oregon Dilemma," *Saturday Evening Post* 234 (October 14, 1961): 26. See Donald G. Balmer, "The 1966 Election in Oregon," *Western Political Quarterly* 20 (June 1967): 593–94, 601; Riley E. Dunlap, "Legislative Voting on Environmental Issues: An Analysis of the Impact of Party Membership," PhD dissertation, University of Oregon, 1973, pp. 42–45.

53. Mark Hatfield in Cahn, "Oregon Dilemma," p. 21; *Oregonian*, August 27, 1965.

54. Wallace, "Report from the Northwest," p. 12. See Donald G. Balmer, "The 1962 Election in Oregon," *Western Political Quarterly* 16 (June 1963): 453–59; Balmer, "1966 Election in Oregon," pp. 593–601.

55. Holbrook, "Oregon," p. 60.

56. *Oregonian*, August 13, 1965. See *Medford Mail–Tribune*, February 2, 1963; *Oregon Journal*, February 28, March 4, 1966; Cahn, "Oregon Dilemma," p. 23; *Oregonian*, December 26, 1965.

57. *Oregonian*, January 10, 1967.

58. Bureau of Municipal Research and Service, *Procedure for Establishing County Planning and Zoning Under Chapter 537, Oregon Laws, 1947* (Corvallis: University of Oregon, 1947). See Paula M. Nelson, "Rural Life and Social Change in the Modern West," in *The Rural West since World War II*, edited by R. Douglas Hurt (Lawrence: University of Kansas Press, 1998), p. 38; Keith Montgomery Carr, "Changing Environmental Perceptions, Attitudes, and Values in Oregon's Willamette Valley: 1800 to 1978," master's thesis, University of Oregon, 1978; Horace Sutton, "Land of Lumber and Lakes," *Saturday Review of Literature* 31 (May 15, 1948): 30; "Quarterly Report of the Division of Sanitary Engineering, January, February & March, 1946," pp. 3–4, Health Board, Sanitary Engineering Division General Files, 1936–1950, B–14–1 (also file B–13–1), A–P, pp. 2, 5; "Memo," January 28, 1946, Report, Box 29, 2; L.A. Helgesson testimony to State Legislative Commission on Pollution, July 20, 1966, 1965–1967, Public Health Committee Minutes and Exhibits, Box 82, OSA; Adam W. Rome, *Bulldozer in the Countryside: Suburban Sprawl and the Rise of American Environmentalism* (New York: Cambridge University Press, 2001).

59. Governor's Natural Resources Advisory Committee, *Water Pollution Control in the Willamette River Basin*, pp. 4–9, 29, 32; Gleeson, *Return of a River*, pp. 12, 14, 17, 19–20, 21, 31, 59, 68; Robbins, *Oregon Environment* pp. 22, 25, 27; *Ninth Biennial Report of the Oregon State Sanitary Authority* (Portland, 1954–1956), p. 3; Courtland Smith, *Public Participation in Willamette Valley Environmental Decisions* (Corvallis: Oregon State University, 1973), p. 92.

60. Oregon Environmental Council, *Newsletter*, February 1969; *Maine Times*, April 17, 1970, May 10, 1974; John N. Cole, "Letter from Maine: Oil and Water," *Harper's* 243 (November 1971): 48.

61. *Maine Sunday Telegram*, February 8, 1970; Richard Condon, "The Tides of Change, 1967–1988," in *Maine: The Pine Tree State from Prehistory to the Present*, edited by Richard W. Judd, Edwin A. Churchill, and Joel W. Eastman (Orono: University of Maine Press, 1995), pp. 564–67; Peirce, *New England States*, pp. 404–05; Pomeroy, *Pacific Slope*, p. 333; Cahn, "Oregon Dilemma," p. 23; "Keep Oregon Livable," *The Fight to Save Oregon: An Environmental Progress Report* (Portland, 1972), n.p.; J. Herbert Stone to Pierre Kolisch, May 13, 1970, Governor's Committee for a Livable Oregon, Correspondence, 1970–1971, Ms. 2537, OHS.

62. Public Health Service, *Clean Water*, pp. 13–14; John D. Dingell in *Federal Water Pollution Control Hearings* (1961), p. 243.

63. David P. Willis and Jeffery A. Foran, *The Cuyahoga River Study* (Columbus: Ohio Wildlife Federation, 1986), State Records, Ohio, SBHC; Hal K. Rothman, *The Greening of a Nation? Environmentalism in the United States Since 1945* (Fort Worth, TX: Harcourt Brace, 1998), pp. 101, 115.

64. Sen. Robert Straub, Sen. William Grenfell, Stuart B. Mockford, to Rep. Haight, March 30, 1961, Exhibits, Legislative Assembly, 1961, House Local Government, Box 34, OSA; B.A. McPhillips, ca. March 29, 1961, Senate Bills 36 and 138 (Water Pollution), 1961 Senate Local Government Minutes and Exhibits, Box 44, OSA; *Twelfth Biennial Report on Water Pollution Control* (Portland: 1960–1962), p. 1; Bob Cooper, Vern Ayres, Doris Metcalf, Sen. William Grenfell, Alexander Brown, July 28–29, 1961, Minutes Interim Committee on Natural Resources, 1961–1962, Legislative Assembly, Exhibits, 1961 House Local Government, Boxes 34, 47, OSA; *Capital Journal*, March 30, 1961; *Oregon Statesman*, March 30, 1961.

65. C.C. Schenck to Clint Haight, March 29, 1961, Legislative Assembly, Exhibits, 1961 House Local Government, Box 34, OSA; *Oregonian*, February 18, 1956.

66. "Pollution in Paradise," in Walth, *Fire at Eden's Gate*, p. 143; George Laycock, "The Governor Is an Environmentalist: Tom McCall Fights for Oregon's Future," *Audubon* 75 (September 1973): 130. See Ethel A. Starbird, "A River Restored: Oregon's Willamette," *National Geographic* 141 (June 1972): 819; Charles Dubs, Ivan Congleton, Paul H. Weiland, Bob Gilchrist, B.A. McPhillips, Senate Bills 254 and 259, Minutes, Senate Local Government, file 2, March 1, 1963, Legislative Assembly, 1963 Senate Local Government Exhibits, OSA; Gleeson, *Return of a River*, p. 61; *Oregon Statesman*, September 26, 28, 1965; *Oregon Journal*, March 3, 1966.

67. John S. Wilson in 1965–1967 Public Health Committee Minutes and Exhibits, Box 82, OSA.

68. Stewart Udall in *Oregonian*, August 16, 26, 1965. See *Capital Journal*, August 16, 1965; *Oregon Journal*, February 28, 1966.

69. Farber, ed., *The Sixties*, p. 2; Stewart Udall in *Audubon* 66 (March–April 1964): 81. See Edward P. Morgan, *The 60s Experience: Hard Lessons about Modern America* (Philadelphia: Temple University Press, 1991), pp. 13, 16, 27–29, 173.

70. Jamison and Eyerman, *Seeds of the Sixties*, pp. 79–81, 92–6; Public Health Service, *Clean Water*, pp. 13–14; *Federal Water Pollution Control Hearings* (1961), p. 302; Robert W. Nero, "Detergents: Deadly Hazard to Water Birds" *Audubon* 66 (January–February 1964): 27.

71. Davies and Davies, *Politics of Pollution*, pp. 31, 39–40; Leonard Dworsky in Minutes Interim Committee on Natural Resources, 1961–1962, Legislative Assembly, Exhibits, 1961 House Local Government, Boxes 34, 47, OSA; "Conference of State Sanitary Engineers, April 28, 29, and 30, 1948," Health Board, Sanitary Engineering Division General Files, 1936–1950, A–P, Box 29, OSA; *Kennebec Journal*, January 4, 1957; New England Interstate Water Pollution Control Commission, *News Letter* 8 (February 1957): 10; "Municipal Pollution Control," *Maine Townsman* 18 (September 1956): 3; Marc Leggiero, "The Pollution Control Act a Year Later," ibid 20 (March 1958): 8–11; Edmund S. Muskie, "The Nation's Response to Urban Change," *Maine Townsman* 27 (May 1965): 4–5; Frank G. Chapman to Edmund S. Muskie, October 25, 1957, Pollution, Box 647, ESMA.

72. *Oregonian*, August 11, 1965. See ibid, August 16, 26, October 15, 1965; *Oregon Statesman*, September 25, 27, 29, 1965; *Thirteenth Biennial Report on Water Pollution Control* (Portland: Oregon State Sanitary Authority, 1962–1964), II: 8; Pomeroy, *Pacific Slope*, pp. 196–97, 199.

73. Irvin H. Luiten, 1965–1967 Public Health Committee Minutes and Exhibits, Box 82, RG L6 68A–32, OSA. See James B. Haas, Robert W. Straub, ibid; Kay Farkey, S. Wilson, Morris K. Crothers, J.O. Julson, and Irvin H. Luiten, ibid; Ted Hallock to Kay Farley, July 21, 1966, ibid; Alfred J. Kreft, Lloyd Dolby, William J. Bowerman in Interim Committee on Public Health: 1965–1966 Public Health Committee Files, Box 82, RG L6 68A–32, OSA; *Oregonian*, August 29, 1965; *Oregon Journal*, March 4, 1966; *Oregon Statesman*, September 30, 1965.

74. *Oregon Journal*, February 28, 1966; *Oregonian*, August 27, 1965; *Thirteenth Biennial Report on Water Pollution Control* (Portland: Oregon State Sanitary Authority, 1962–1964), p. 4; Kenneth H. Spies, Interim Committee on Public Health, 1965–1966 Public Health Committee Files, October 8–9, 1965, RG L6 68A–32, OSA; *Oregon Journal*, March 4, 1966; *Oregonian*, November 17, 1966. See Harold Wendell and Spies, Minutes Legislative Interim Committee on Public Health, Portland, Oregon, July 20–21, 1966, pp. 16–17, OSA.

75. *Portland Press Herald*, August 12, 17, 1958; Peter Partout, "Government by Lobby Killing Natural Resources," pamphlet, Citizens for Conservation and Pollution Control, 1953, MLLL. See J. Dennis Bruno, "Maine Leads N.E. in Filthy Waters," pamphlet, ibid; *Lewiston Evening Journal*, March 17, April 21, 1955, June 21, 1957; *Kennebec Journal*, March 9, 12, 1953, April 16, 1955; Maine Legislature, *Legislative Record* (Augusta, 1955), pp. 1684, 1700, 2448.

76. *Portland Press Herald*, August 12, 17, 1958; *Waterville Sentinel*, November 2, 1973; *Lewiston Evening Journal*, May 22, 1963.

77. *Bangor Daily News*, March 6–7, 24, 1965, April 11, 1969, February 10, 1970; *Lewiston Evening Journal*, March 4, 6, 1965; *Maine Sunday Telegram*, December 14, 1969; *Kennebec Journal*, March 20, 1965.

78. *Kennebec Journal*, February 17, 19, 1966; *Maine Times*, February 28, 1969; *Bangor Daily News*, December 14, 1966, July 9, 10, August 15, 20, 22, 1968, February 21, 1969; *Lewiston Evening Journal*, July 10, 11, 1968.

79. *Bangor Daily News*, July 25, August 3–4, 15, October 22, November 22, 1968; February 21, April 11, 1969, March 16, 1970; *Lewiston Evening Journal*, July 25, 1968; *Maine Times*,

January 17, April 11, May 10, 1969, January 30, 1970; "Maine: A Case of Sour Sugar," *Time* 95 (February 9, 1970): 12.

80. For Curtis, *Lewiston Evening Journal*, August 21, 1968; for Smith, *Maine Sunday Telegram*, January 25, 1970; for Muskie, *Maine Sunday Telegram*, February 15, 1970.

81. Aimé Gauvin, "Raking the Muck to Mold the Future of Maine's Coast," *Audubon* 75 (July 1973): 135; "Trying to Save Maine," *Time* 94 (October 31, 1969): 74; *Bangor Daily News* January 16, 1970.

82. Bob Cummings in *Maine Sunday Telegram*, May 20, 1973; second quote, *Waterville Sentinel*, February 14, 1981. See *Kennebec Journal*, November 24, 1972; *Waterville Sentinel*, August 15, 1970, May 26, 1973; *Maine Sunday Telegram*, August 9, 1970, June 6, 1971; *Portland Press Herald*, June 2, 1971.

83. Berry in *Maine Sunday Telegram*, May 27, 1973; Curtis in *Bangor Daily News*, June 4, 1973. See *Waterville Sentinel*, August 15, 1970, June 8, 1973, February 12, 1981; *Maine Sunday Telegram*, June 6, 1971, May 20, 27, 1973, September 29, 1974, July 30, 1978; *Kennebec Journal*, February 22, May 18, 1973; *Bangor Daily News*, June 4, 1973; *Portland Press Herald*, May 18, 1973.

84. *Maine Times*, October 24, 1968. See ibid, October 24, November 8, 1968; Morgan, *60s Experience*, p. 173; Laurance S. Rockefeller, "What Kind of America?" *Audubon* 66 (September–December 1964): 378.

85. Conference on Community Goals and Policies for Eugene, "Quality of Environment in Eugene," typescript, Eugene Planning Commission, May 25, 1966, Knight Library, University of Oregon, n.p. See Advisory Committee on Environmental Science and Technology, *Environmental Quality in Oregon, 1971: A Summary of Current and Future Problems* (Corvallis: Oregon State University, 1971), pp. ii, 3, 4; Lidia Shattuck; Frank Koch to Ted Hallock, February 23, 1963, Senate Local Government Exhibits, 1963, Senate Bills 254 & 259, OSA; Tom McCall, Interim Committee on Public Health: 1965–1966 Public Health Committee Files," RG L6 68A–32, item 2, OSA; *Medford Mail-Tribune*, February 2, 1963.

86. Laycock, "Governor Is an Environmentalist," p. 131. See Herbert Stoevener, House–Environment – Minutes, 1971, (February 12, 1971), Box 120, OSA; Conference on Community Goals and Policies for Eugene, "Quality of Environment in Eugene"; Carr, "Changing Environmental Perceptions," p. 116; *Oregon Statesman*, November 2, 1969; *Capital Journal*, February 3, 1970.

87. Bob Cummings, "The Late, Great State of Maine," *Maine Sunday Telegram*, August 30, 1970; Jean Anne Pollard, *Polluted Paradise: The Story of the Maine Rape* (Lewiston, Maine: Twin City Printery, 1972), pp. 110–11; *Maine Sunday Telegram*, November 5, 1972. See Peter N. Kyros in New England Regional Commission, *Public Hearings of the New England Regional Commission: Maine* (Augusta, November 2, 1967), p. 59; *New York Times*, September 13, 1960.

88. Robert W. Patterson, "Introduction," *The Maine Coast: Prospects and Perspectives* (Bowdoin College, Center for Resource Studies, October 20–22, 1966). See Bill Kovach, "As Maine Grows, So Grows the Statewide Debate over Ways to Preserve Its Environmental Resources," *New York Times*, September 13, 1970, p. 60; *Maine Times*, October 24, 1968, May 17, October 25, 1974; Cole, "Letter from Maine," p. 50.

89. Charles DeDeurwaerder, "The fight against Ticky-Tacky," *Oregonian Northwest Magazine*, November 16, 1969, p. 13. See Oral Bullard, "The View of the Artist" in ibid, pp. 8, 10; Columbia River Fishermen's Protective Union to Legislative Committee, July 17, 1966, 1965–1967 Public Health Committee Minutes and Exhibits," Box 82; Dan

Allen, Senate Bills 36 and 138, Senate Local Government Minutes and Exhibits, 1961, Box 44, OSA.

90. DeDeurwaerder, "Fight against Ticky-Tacky"; Carr, "Changing Environmental Perceptions," pp. 116; Gleeson, *Return of a River*, p. 8.

91. *Kennebec Journal*, Feburary 18, 1961, June 1, 1976; *Maine Times*, May 8, 1970, October 4, 1975; *Maine Sunday Telegram*, April 28, 1974; Kenneth Thompson, "The Hudson: Marcy to Manhattan" (and accompanying photo essay by Robert Perron)," *Audubon* 73 (March 1971). See Kovach, "As Maine Grows," p. 60; Frank Graham Jr., "Of Slime and the River" [the Androscoggin], *Audubon* 77 (March 1975): 9–15.

92. *Lewiston Evening Journal*, July 25, 1968; Mrs. Jane (Lloyd) Thomas, C.A. Fisher, John E. Jaqua, Esther Manseth, K.S. Lichty (August 25–26, 1966), 1965-1966 Interim Committee on Public Health: Public Health Committee Files, RG L6 68A–32, Item 2, OSA.

93. *Kennebec Journal*, October 17, 1975, July 15, 1978, March 18, 1981, August 31, 1982; *Waterville Sentinel*, October 3, 1981; *Maine Sunday Telegram*, January 7, 1968.

94. First quote: John S. Wilson, 1965-1967 Public Health Committee Minutes and Exhibits (July 20–21, 1966), Box 82; second quote: name obscured (Shady Cove, Oregon), to Sen. Ted Hallock, February 23, 1963, Senate Bills 254 & 259, 1963 Senate Local Government Exhibits, OSA; third quote: Eric W. Allen, Jr., in *Oregonian*, August 29, 30, 1965. See Dan Allen, Senate Bills 36 and 138, 1961 Senate Local Government Minutes and Exhibits, Box 44; Aaron Novick, Charles W. Swango, P.W. Schneider, Alfred J. Kreft, Aaron Novick, Neil Brown, Karl W. Onthank, Interim Committee on Public Health: 1965-1966 Public Health Committee Files, (August 25–26, 1966), RG L6 68A–32, Item 2, OSA; Charles Collins, House Local Government Exhibits 1961, Box 34; Oregon State Medical Society, "Resolution on Environmental Sanitation," 1961 Senate Local Government Minutes and Exhibits, Box 44; Robert McIlroy to Rep. Bill Bradley, March 25, 1961, HB 1635-1637, 1961 House Fish and Game Minutes and Exhibits, Box 31, OSA. On pastoral legacy, see *Oregonian*, August 16, 18, 29, 1965; *Oregon Journal*, February 28, 1966.

95. Samuel P. Hays, *Beauty, Health, and Permanence: Environmental Politics in the United States, 1955-1985* (New York: Cambridge University Press, 1987), p. 78; Andrews, *Managing the Environment*, p. 236.

96. *Kennebec Journal*, February 14, March 14, 31, April 3, June 8, 1961; *Maine Times*, October 10, 1969, May 10, 1974; *Portland Evening Express*, January 27, 1962; *Waterville Sentinel*, November 2, 1973. See *Lewiston Sun*, December 16, 1958.

97. *Kennebec Journal*, March 13, 1957; *Bangor Daily News*, March 24, 1965; *Maine Times*, April 26, 1974. See Carl Bloom, "The Penobscot River: A Study of Pollution Abatement Programs and Restoration Program," M.A. thesis, University of Maine, 1971, p. 2; Maine Water Improvement Commission, *Androscoggin River Classification Report* (Augusta, Maine, 1966), pp. 3, 5; "Pollution Control," *Maine Townsman* 25 (February 1963): 24; *Portland Press Herald*, September 6, 1965; *Lewiston Evening Journal*, November 22, 1966.

98. *Bangor Daily News*, April 1, 1970; *Maine Sunday Telegram*, August 3, 1969; *Maine Times*, February 20, 1970. See Pardo, "Something Up, Down East," p. 44; Andrew Hamilton, "Maine: Finding the Promised Land (without Losing the Wilderness)," *Science* 178 (November 10, 1972): 596.

99. *Kennebec Journal*, June 10, 1970. See *Maine Times*, January 17, February 21, November 14, 1969, February 20, 1970; *Bangor Daily News*, April 1, 1970; Legislative Research Committee, *Report on the Environmental Improvement Commission to the 105th Legislature* (Augusta, 1971), pp. 12, 13, 15.

100. *Maine Sunday Telegram*, May 20, 1973. See *Maine Times*, April 17, 1970, April 19, 26, 1974; *Maine Sunday Telegram*, January 7, 1968, March 8, March 1, August 2, October 11, 1970, May 28, 1972, September 15, 29, 1974, January 11, 1976; *Lewiston Evening Journal*, November 7, 1969, January 15, 1970; *Kennebec Journal*, June 7, 1961, June 13, 1982; *Waterville Sentinel*, October 2, 1974, February 14, 1981; Lawrance, "Twenty-Year Review of Androscoggin,'" pp. 5–6.

101. Tom McCall, Robert W. Straub, 53rd Session Interim Committee on Public Health: 1965–1966 Public Health Committee Files," RG L6 68A–32, Item 2, OSA; *Oregonian*, August 1, 13, 16, 21, October 9, 15, December 26, 1965; *Statesman*, October 16, 1965, June 11, August 11, 1966; Ken Johnson to Libby Loftin, May 4, 1977, 1150.3.1 Willamette Greenway, Box 96, RG 90A–30, RWSR; Balmer, "1966 Election in Oregon," p. 598; Joseph M. Allman, "The 1968 Elections in Oregon," *Western Political Quarterly* 22 (September 1969): 524; Balmer, "The 1964 Election in Oregon," *Western Political Quarterly* 18 (June 1965): 508.

102. "Governor McCall Radio Script Recording, 2/15/68," Governor's Committee for a Livable Oregon, Organization, November 1967–1968, Ms. 2537, OHS. See Laycock, "Governor Is an Environmentalist," p. 130; Robbins, *Oregon Environment*, p. 6; Advisory Committee on Environmental Science and Technology, *Environmental Quality in Oregon*, pp. ii, 3, 5; Steven E. Hungerford, "A Content Analysis of Environmental Reporting in Oregon Daily Newspapers for 1970," master's thesis, University of Oregon, 1972, p. 22; *Oregonian*, March 2, 18, 1970.

103. Dunlap, "Legislative Voting," p. 37; Floyd McKay, "The 'Kaffeeklatch' Constituency," *Nation* 208 (February 17, 1969): 205–06; Allman, "1968 Elections in Oregon," pp. 517, 521–22; "Oregon: Booming Dove," *Newsweek* 71 (May 27, 1968): 33A; "New Politics," *New Republic* 159 (December 7, 1968): 10.

104. Oregon Environmental Council, *Newsletter*, October 1969. "Governor McCall Radio Script Recording, 2/15/68," Ms. 2537, OHS; "McCall Administration Background Paper on Natural Resources and Environment," February 2, 1971, Correspondence, 1970–1971, Governor's Committee for a Livable Oregon, Ms. 2537, OHS.

105. *Medford Tribune*, September 16, 1967; *Oregonian*, October 13, 1967; Robert W. Straub, 1965–1967 Public Health Committee Minutes and Exhibits (July 20–21, 1966), Box 82, OSA.

106. First quote: *Medford Mail-Tribune*, February 27, 1967; second quote: R.E. Hatchard in ibid, May 8, 1967. See R.F. Poston, 1965–1967 Public Health Committee Minutes and Exhibits (July 20–21, 1966), Box 82, OSA; *Medford Mail-Tribune*, February 27, March 7, April 11, May 5, 1967; Robert W. Straub, Kenneth H. Spies, Harold F. Wendel, Interim Committee on Public Health: 1965–1966 Public Health Committee Files (October 8–9, 1965), RG L6 68A–32, item 2, OSA; Leonard Dworsky, Minutes, 1961–1962 Interim Committee on Natural Resources (March 30, 1962), Box 47 OSA.

107. *Lewiston Sun*, August 26, 1981, May 7, 1982, October 6, 1996; *Maine Times*, May 23, November 7, 1969, January 30, April 17, 1970, October 8, 1976; *Kennebec Journal*, August 18, 1979, April 21, 1979, July 11, 12, 1980, August 31, 1982, November 11, 1983, February 6, 1985, October 21, 1988, December 10, 1990, July 9, 1993; *Waterville Sentinel*, October 22, 1971, August 17, 1978, August 1, 1980; *Bangor Daily News*, July 16, 1968, August 15, 1972, February 12, 1981; *Maine Sunday Telegram*, August 19, 1979, September 11, November 27, 1983; Central Lane [County] Planning Council, *Crisis: Water: An Examination of Water Quality Conditions, Programs and Needs in Oregon's Willamette Valley* (Eugene, 1968), pp. 36–37, 39.

108.Harry F. Rivelli to Robert Straub, n.d., ca. May 13, 1977, Administrative Correspondence, 1975–1978, 1120.2 Water, Box 94, RG 90A–30, RWSR. See Michael McCloskey, "Twenty Years of Change in the Environmental Movement: An Insider's View," in Dunlap and Mertig, eds., *American Environmentalism*, p. 80.

109.*Lewiston Evening Journal*, September 4, 1941; *Kennebec Journal*, April 5, 1989. See *Kennebec Journal*, March 20, 21, June 6, 1990; *Waterville Sentinel*, December 6, 1989; *Bangor Daily News*, April 14, 1989; *Lewiston Sun*, January 4, 1990, January 15, 1991.

110. Orie L. Loucks, "Sustainability in Urban Ecosystems: Beyond an Object of Study," *The Ecological City: Preserving and Restoring Urban Biodiversity*, edited by Rutherford H. Platt, Rowan A. Rowntree, and Pamela C. Muick (Amherst: University of Massachusetts Press, 1994), p. 61; Paul Schneider, "Clear Progress: 25 Years of the Clean Water Act," *Audubon* 99 (September–October 1997): 36–47, 106–7.

111. *Recovering the Valley: An Environmental Status Report of the Connecticut River Basin, 1970–1983* (Easthampton, Massachusetts: Connecticut River Watershed Council, 1983), pp. 9, 11–15, 35, State Records, Connecticut, SBHC.

112.Schneider, "Clear Progress," pp. 44, 107.

CHAPTER 3: RIVERS, WILDERNESS, AND REDEMPTIVE PLAY

1. Andrew Jamison and Ron Eyerman, *Seeds of the Sixties* (Berkeley: University of California Press, 1994), pp. 71, 96; Mark W.T. Harvey, *A Symbol of Wilderness: Echo Park and the American Conservation Movement* (Albuquerque: University of New Mexico Press, 1994), p. 54; James A. Pritchard, *Preserving Yellowstone's Natural Conditions: Science and the Perception of Nature* (Lincoln: University of Nebraska Press, 1999), p. 198.

2. Aldo Leopold in Daniel B. Botkin, *No Man's Garden: Thoreau and a New Vision for Civilization and Nature* (Washington, DC: Island Press/Shearwater Books, 2001), p. 45.

3. Max Oeschlaeger, *The Idea of Wilderness from Prehistory to the Age of Ecology* (New Haven, CT: Yale University Press, 1991), p. 111.

4. "Rivers, Recreation, and You: Interview [Sen. Frank Church]," *Field and Stream* 71 (July 1966): 10; Senators Henry Jackson and John Saylor in John P. Saylor, "Once Along a Scenic River," *Parks and Recreation* 3 (August 1968): 58.

5. "Of Time and the Rivers," *Newsweek* 69 (June 19, 1967): 86.

6. Galen Rowell, "Fragile Nature, Fragile Man," *Sierra Club Bulletin* 58 (January 1973): 5.

7. Harvey, *Symbol of Wilderness*, pp. xiv, 53. See James E. Sherow, "Environmentalism and Agriculture in the American West," in *The Rural West Since World War II*, edited by R. Douglas Hurt (Lawrence: University of Kansas Press, 1998), p. 65.

8. Richard Starnes, "Shall Our Wild Rivers Be Eternally Dammed?" *Field and Stream* 69 (March 1965): 16. See Paul Mason Tilden, "Washington Newsletter," *Natural History* 74 (August 1965): 62; Kenneth Cmiel, "The Politics of Civility," *The Sixties: From Memory to History*, edited by David Farber (Chapel Hill: University of North Carolina Press, 1994), pp. 270–71; Luther J. Carter, "Dams and Wild Rivers: Looking Beyond the Pork Barrel," *Science* 158 (October 13, 1967): 233–36; Donald J. Pisani, "Federal Water Policy and the Rural West," *The Rural West since World War II*, edited by R. Douglas Hurt (Lawrence: University of Kansas Press, 1998), pp. 119, 122–123; Harvey, *Symbol of Wilderness*, pp. 25, 26, 51.

9. Lynton K. Caldwell, Lynton R. Hayes, Isabel M. MacWhirter, *Citizens and the Environment: Case Studies in Popular Action* (Bloomington: Indiana University Press, 1976), pp. 146–54.

10. William G. Robbins, *Colony and Empire: The Capitalist Transformation of the American West* (Lawrence: University Press of Kansas, 1994), p. 105; Don Holm, "The Rogue: A River for All Kinds of Boaters," *Oregonian Northwest Magazine*, April 17, 1977, p. 4.

11. *Oregonian*, May 23, 1948. See Robert Ormond Case and Victoria Case, *Last Mountains: The Story of the Cascades* (Garden City, NY: Doubleday, Doran, 1945), p. 299; Mae Eva Hopkins to Governor Douglas McKay, March 21, 1950, Lester S. Diehl to McKay, March 22, 1950, R.C. Buckley to McKay, March 22, 1950, D.E. Crabb to McKay, April 7, 1950, Rogue River Dam and Project, Jan. 1950–Feb. 14, 1950 (Letters from Public), Administrative Correspondence, Records of Governor Douglas McKay, Box 24, RG-57-98, Oregon State Archives, Salem (hereafter OSA); Dolan, "Rogue River Country," pp. 14–15, 45–48; *Oregonian*, January 12, 1967; LaLande, *Prehistory and History*, p. 182; Beckham, *Cultural Resource Overview*, p. 122.

12. Roger Keith Davis, "The Lower Rogue River Recreation Resources: Development and Potential," master's thesis, Oregon State University, 1966, p. 9; "Canyon of the Rogue ... on Foot, by Boat, by Car: Rogue River Canyon, Oregon," *Sunset* 131 (September 1963); Holm, "The Rogue," p. 4; "Regulating the Wild and Scenic Rogue," *Earthwatch Oregon*, August–September 1976.

13. Wessel Smitter, "We Ride the Fighting Rogue," *Saturday Evening Post* 221 (October 23, 1948): 36; Jeffrey M. LaLande, *Prehistory and History of the Rogue River National Forest: A Cultural Resource Overview* (Medford, OR: Rogue River National Forest, 1980), p. 161; Oregon State Game Commission, *Fish and Wildlife Plan, Rogue River: Oregon Scenic Waterways and National Wild and Scenic River Section* (Salem, 1971); National Park Service, *Recreation Resources of the Rogue River Basin, Oregon* (NPS Region Four, 1954), pp. 5, 20; Robert Dolan, "Rogue River Country," *American Forests* 64 (September 1958): 14–15, 45–48; Leo A. Borah, "Oregon Finds New Riches," *National Geographic* 90 (December 1946): 709; Richard J. Hebda and Cathy Whitlock, "Environmental History," *The Rain Forests of Home: Profile of a North American Bioregion*, edited by Peter K. Schoonmaker, Bettina von Hagen, and Edward C. Wolf (Washington, DC: Island Press, 1997), p. 228, 248.

14. Jesse Lee Gilmore, "A History of the Rogue River Valley: Pioneering Period, 1850–1862," PhD dissertation, University of California, 1952, pp. 5, 93, 239–40; *Oregonian*, January 12, 1967; LaLande, *Prehistory and History*, pp. 33, 177, 182; Stephen Dow Beckham, *Cultural Resource Overview of the Siskiyou National Forest* (Grants Pass, OR: Siskiyou National Forest, 1978), pp. 92–101, 122–23, 128.

15. Beckham, *Cultural Resource Overview*; LaLande, *Prehistory and History* pp. 40, 179, 189; *Oregonian*, May 17, 1948, August 19, 1965, January 12, 1967; T. Trueblood, "A River in His Blood," *Field and Stream* 77 (June 1972): 56–57; Davis, "Lower Rogue," pp. 36–37; "That Fish Bill," *Oregon Voter* 9 (April 28, 1917): 11–12; "Rogue River" *Oregon Voter* 71 (October 1932): 40–41.

16. First quote: *Oregonian*, January 8, 1967; second quote: Don Holm in *Medford Mail-Tribune*, January 17, May 10, June 23, 28, 1967. See *Oregonian*, January 8, 10, 11, 13, 15, 1967; *Capital Journal*, November 10, 1969, *Oregon Statesman*, November 12, 1969; Beckham, *Cultural Resource Overview*, p. 122; R.F. Rousseau, "Water Laws and Fish: Are They in Conflict?" *Oregon Wildlife* 30 (January 1975): 3; Melvin L. Saltmarsh to Kessler Cannon, April 15, 1974, Wild and Scenic Rivers #2, 75A–12, Governor McCall's Records, Box 55, RG G4, OSA; Minutes of Public Hearings Held in Regard to the Rogue

River ... (April 25, 26, 28, 1972), Scenic Waterways: Rogue PAR 12, 1968–1971, 2, especially R.H. Hamilton to Marine Board, April 28, 1972, Department of Transportation Library, Salem; Davis, "Lower Rogue," pp. 6, 36–37.

17. Richard L. Neuberger, *Our Promised Land* (New York: Macmillan, 1938), pp. 95, 100, 104–05; Ian Tyrrell, *True Gardens of the Gods: Californian–Australian Environmental Reform, 1860–1930* (Berkeley: University of California Press, 1999), pp. 39, 103; Mark Fiege, *Irrigated Eden: The Making of an Agricultural Landscape in the American West* (Seattle: University of Washington Press, 1999), p. 140.

18. Robbins, *Colony and Empire*, pp. 17, 62, 166; Earl Pomeroy, *The Pacific Slope: A History of California, Oregon, Washington, Idaho, Utah, and Nevada* (New York: Alfred A. Knopf, 1965), pp. 160–61, 314; Pisani, "Federal Water Policy," p. 128; Paula M. Nelson, "Rural Life and Social Change in the Modern West," in Hurt, ed., *The Rural West Since World War II*, pp. 38–39.

19. *Oregonian*, May 23, 1948. See ibid, July 5, 1970; W.I. Palmer in *Agate Dam and Reservoir: An Additional Feature of the Talent Division of the Rogue River Basin Reclamation Project*, 87th Cong., 1st Sess., S. 1023 (Washington, DC: Government Printing Office, 1961), p. 13; Leonard Freeman in ibid, p. 22; H.J. Rayner, "The Salmon Problem of the Rogue River Basin," *Oregon State Game Commission Bulletin* 6 (April 1951): 1, 4–5; *Oregonian*, May 9, 23, 28, June 6, 1948; Kenneth G. Denman, Charles Hoover, F.J. Mitchell in U.S. Bureau of Reclamation, *Transcript of Public Hearing Held at Medford, Oregon ... on Alternative Plans for Development of the Water Resources of the Rogue River Basin, Oregon* (Boise, Idaho, 1948), pp. 76, 149, 153, 314; J.M. Spencer in *Oregonian*, May 14, 1948.

20. John Niedermeyer in *Transcript of Public Hearing* (1948), pp. 291–92. See Kenneth G. Denman, Norman S. Sterry, Isaac Henry Cory in ibid, pp. 149, 153 158, 361; *Oregonian*, May 9, 23, 28, 1948.

21. C.M. Loring in *Oregonian*, June 16, 1948. See *Oregonian*, May 9, 17, June 6, 1948; Syl MacDowell in *Oregonian*, May 14, 1948; R.H. (Bob) Ames in *Oregonian*, June 9, 1948; George W. Olcott in *Oregonian*, June 9, 1948; Howard E. Nelson in *Oregonian*, January 14, 1956; Charles F. Grant in *Oregonian*, May 22, 1948; Ralph P. Cowgill in *Oregonian*, May 14, 1948; Victor Boehl, A.E. Brockway, Leo Gates, Clyde Willey, John Niedermeyer, William Corcoran in *Transcript of Public Hearing* (1948), pp. 24–26; 126, 236, 272, 291–92, 403; Beckham, *Cultural Resource Overview*, p. 122.

22. Richard Reinhardt, "The Short, Happy History of the State of Jefferson," *American West* 9 (May 1972): 36–41.

23. First quote: Walter F. Brittan to Douglas McKay, March 30, 1950, Rogue River Dam and Project, 1950–1952, Administrative Correspondence, Records of Governor Douglas McKay, Box 24, RG–57–98, OSA; second quote: L.J. Bailey to McKay, February 12, 1950, in ibid. See Carolyn H. Kelsey to McKay, February 4, 1950; D.E. Crabb to McKay, April 7, 1950; Stanley V. Wright to McKay, March 28, 1950; D.H. Barber to McKay, January 31, 1950; A.B. Bangs to McKay, February 11, 1950; Paul and Helen Olsen to McKay, n.d., ca. February 1950; S.D. Newhome to McKay, February 13, 1950; Rex Cummings to McKay, February 13, 1950, in ibid.

24. V.D. Bert Miller in *Oregonian*, July 2, 1948. See Elwood Hussey in ibid, May 16, 1948; A.L. Unger, Richard H. Woodson, John Niedermeyer, B.P. Lilienthal in *Transcript of Public Hearing* (1948), p. 73, 98–99, 291–92, 366.

25. On the fight against the Pelton Dam on the Deschutes River, see *Oregonian*, January 2, February 11, 1956. See also Robert W. Sawyer to *Oregonian*, January 8, 1956; William G. Robbins, *The Oregon Environment: Development vs. Preservation, 1905–1950* (Corvallis: Oregon State University, 1975), p. 38.

26. First quote: Isaac Henry Cory in *Transcript of Public Hearing* (1948), p. 158; second quote: Mrs. T.V. Scott to Douglas McKay, February 11, 1950, Rogue River Dam and Project, Jan. 1950–Feb. 14, 1950, Administrative Correspondence, Records of Governor Douglas McKay, Box 24, RG–57–98, OSA.

27. First quote: H.L. Kelly in *Oregonian*, May 30, 1948; second quote: Cole Rivers in *Oregonian*, May 27, 1948; third quote: C.J. Campbell, "Fish, Wildlife and Dams," *Oregon State Game Commission Bulletin* 8 (March 1953): 3. See Mrs. T.V. Scott to Douglas McKay, February 11, 1950, Rogue River Dam and Project, Jan. 1950–Feb. 14, 1950, Administrative Correspondence, Records of Governor Douglas McKay, box 24, RG-57-98, OSA; Smitter, "We Ride the Fighting Rogue," p. 36; Beckham, *Cultural Resource Overview*, p. 91; *Oregonian*, May 23, 1948; Anthony Netboy, "This River Is Under Arrest! Army COE vs. the Rogue," *Oregon Times* 6 (February-March 1976): 29-31; Cole M. Rivers, "Rogue and Reclamation," *Oregon State Game Commission Bulletin* 5 (March 1950): 4; National Park Service, *Recreation Resources*, p. 17; Robbins, *Oregon Environment:*, pp. 39–40; Ira N. Gabrielson, "Reclamation versus Conservation," *Oregon State Game Commission Bulletin* 3 (October 1948), p. 6; Howard E. Nelson in *Oregonian*, January 14, 1956.

28. Rivers, "Rogue and Reclamation," p. 4. See Harris Ellsworth to Douglas McKay, February 12, 1951, Rogue River Dam and Project, 1950-1952, Administrative Correspondence, Records of Governor Douglas McKay, Box 24, RG–57–98, OSA; "Resolution, Rogue River Valley Irrigation Association," ibid; telegram from Douglas McKay to Guy Cordon, February 6, 1951 in ibid; D.H. Barber to McCay, February 28, 1950, in ibid; Louis F. Schmitt to McKay, February 13, 1950, in ibid

29. John Clark Hunt, "Lucky Rivers," *Oregonian Northwest Magazine*, June 28, 1970, p. 7; *Oregonian*, January 3, 1956; National Park Service, *Recreation Resources*, p. 17; Charles Konopa, "The Rogue River Trail," *Oregonian Northwest Magazine*, June 25, 1967.

30. Gabrielson, "Reclamation versus Conservation," pp. 1, 6.

31. Starnes, "Shall Our Wild Rivers Be Eternally Dammed?" p. 14. See Charles H. Callison, "A New Kind of National Park Area," *Audubon* 66 (May–June 1964): 178; Callison, "House Committee Must Choose between Four Wild River Bills," *Audubon* 69 (November–December 1967): 70-71; Callison, "Saylor Bill Offers Real Protection for America's Remaining Wilderness Rivers," *Audubon* 68 (July–August 1966): 226-27; Callison, "Wild Rivers Bill Lacks Teeth," *Audubon* 68 (March 1966): 92; "Rivers, Recreation, and You: Interview [Sen. Frank Church]," *Field and Stream* 71 (July 1966): 55; "Of Time and the Rivers," *Newsweek* 69 (June 19, 1967): 86.

32. Governor Tom McCall in *Wild and Scenic Rivers: Hearings before the Committee on Interior and Insular Affairs*, 90th Cong, 1st Sess., S. 119 and S. 1092 (Washington, DC: Government Printing Office, 1967), p. 157; Charles S. Collins in *Wild Rivers System— St. Croix Waterway: Hearings before the Committee on Interior and Insular Affairs*, 89th Cong., 1st Sess., S. 1446 and S. 897 (Washington, DC: GPO, 1965), pp. 316, 334-35.

33. Hunt, "Lucky Rivers,"p. 7. See *Oregonian*, January 8, 10, 12, 13, 16,1967; Netboy, "River Is Under Arrest," 29-31; Davis, "Lower Rogue," p. 35; *Capital Journal*, November 10, 1969; *Oregon Statesman*, July 14, 1969, November 12, 1969; *Medford Mail-Tribune*, June 23, 1967; Oregon State Game Commission, *Fish and Wildlife Plan*, pp. 2–3.

34. G. Douglas Hofe, Jr., "Wild Rivers," *Parks and Recreation* 5 (February 1970): 23. See Robert E. Pfister and Robert E. Frenkel, *Rogue River Study, Report 1: Field Investigations of River Use within the Wild River Area of the Rogue River, Oregon* (Corvallis: Oregon State University, 1976), p. 4; John J. Craighead, Orville Freeman, Sigurd F. Olson in *Wild Rivers System* (1965), pp. 24–25, 72, 238; Stewart L. Udall in *National Scenic Rivers System: Hearings before the Subcommittee on National Parks and Recreation of the Committee*

on Interior and Insular Affairs, 90th Cong., 2nd Sess., H.R. (Washington, DC: 1968), pp. 20, 24; Hunt, "Lucky Rivers," p. 7; *Oregon Statesman* September 9, 1965; Gaylord Nelson in *Wild and Scenic Rivers* (1967), p.19.

35. John P. Saylor, "Once Along a Scenic River," *Parks and Recreation* 3 (August 1968): 21; second quote: W.P. (Duke) Simons in *Wild Rivers System* (1965), p. 205; Glenn Thompson in ibid, p. 209. See Martin Hanson, Anthony Wise, Bill Isaacs, Mr. and Mrs. W.A. Burzlander in ibid, pp. 247, 255, 339, 339–40; Walter F. Mondale in *Wild and Scenic Rivers* (1967), p. 30.

36. John A. Wethington, Jr., in *Wild and Scenic Rivers* (1967), p. 179.

37. "Rivers, Recreation, and You: Interview [Sen. Frank Church]," *Field and Stream* 71 (July 1966): 11–12 (our emphasis). See William Cronon, "The Trouble with Wilderness: Or, Getting Back to the Wrong Nature," in *Out of the Woods: Essays in Environmental History*, edited by Char Miller and Hal Rothman (Pittsburgh, PA: University of Pittsburgh Press, 1997), pp. 28–50.

38. Edward C. Crafts in *National Scenic Rivers System* (1968), pp. 152, 156–57. See *Capital Journal*, January 8, 1970.

39. *Oregon Statesman*, October 18, 1965. See "The BLM and the Rogue River, Draft of Proposals, November 16, 1969," Rivers, Rogue, Articles, Reports, Etc., 1969–73, Oregon Environmental Council Papers (hereafter OECP), Box 27, Ms. 2386, OHS; Stewart L. Udall, *Wild Rivers System* (1965), p. 66; Oregon State Game Commission, *Fish and Wildlife Plan*, p. iv; appendix 1, p. 26; *Oregon Environmental Council Newsletter*, November 1969, p. 4.

40. Harvey, *Symbol of Wilderness*, p. 59; Courtland Smith, *Public Participation in Willamette Valley Environmental Decisions* (Corvallis: Oregon State University, 1973), p. 62; Douglas R. Reed in "March 30–31, 1962," Minutes Interim Committee on Natural Resources, 1961–1962, Box 47, OSA.

41. State Game Commission, "Rogue Wild and Scenic River: Fish and Wildlife Plan" (March 15, 1971), in *Scenic Waterways: Rogue*, PAR 12, 1968–1971, #1, Department of Transportation Library; James A. Pritchard, *Preserving Yellowstone's Natural Conditions: Science and the Perception of Nature* (Lincoln: University of Nebraska Press, 1999), p. 203.

42. Lawrence F. Williams, "Statement by the Sierra Club and the Federation of Western Outdoor Clubs before the BLM and Forest Service Advisory Committee ...," June 25, 1969," in Rivers, Rogue, Articles, Reports, Etc., 1969–73, Box 27, Ms. 2386, OECP; Robert W. Straub in *Oregonian*, November 9, 17, 1969; in *Capital Journal*, October 29, November 10, 12, 1969; and in *Oregon Statesman*, November 15, 1969. See "Comments of Sierra Club on Siskiyou National Forest preliminary Plan for the Rogue River", n.d.; Ernest M. Dickerman to Lawrence F. Williams, December 22, 1969; Anthony N. Cary to H. Don Harris, Jr., August 12, 1969; Margaret Wood to Lawrence Williams, January 28, 1970, Rivers, Rogue, Articles, Reports, Etc., 1969–73, Box 27, Ms. 2386, OECP; Oregon Environmental Council *Newsletter*, November 1969; *Oregon Statesman*, December 19, 1969; *Capital Journal*, November 8, 12, 13, 17, 1969; *Oregon Statesman*, November 12, 13, 1969.

43. *Medford Mail-Tribune*, October 19, 1969.

44. Pisani, "Federal Water Policy," pp. 138–39; Charles H. Callison, "Dams Threaten the Grand Canyon," *Audubon* 67 (March–April, 1965): 114.

45. *High Country News*, June 18, 1975. See ibid, April 26, 1974; "'A Dam Is Not Difficult to Build Unless It Is in the Wrong Place." *Audubon* 78 (November 1976): 132–33.

46. Barry Goldwater in *High Country News*, May 10, 1974. On the Platte, see ibid, July 5, 1974; on the Ohio, see *Waterlog* (Kentucky Rivers Coalition), March 26, 1979, in State

Files, Kentucky, Alabama, Samuel P. and Barbara Hays Environmental Archives, University of Pittsburgh.

47. Pisani, "Federal Water Policy"; Fiege, *Irrigated Eden*, p. 140.

48. Robert M. Collins, "Growth Liberalism in the Sixties: Great Societies at Home and Grand Designs Abroad," in *The Sixties*, edited by Farber, pp. 11–44; Sherow, "Environmentalism and Agriculture," pp. 65–66.

49. W.I. Palmer in *Agate Dam and Reservoir* (1961), pp. 9, 11; *Oregonian*, January 13, 1967; Holm, "The Rogue," p. 4; Netboy, "River Is Under Arrest," pp. 29–31; *Medford Mail-Tribune*, February 13, 1967; *Capital Journal*, July 24, 1969; *Oregonian*, January 13, 1967; Frederic L. Fleetwood to Oregon State Game Commission, December 18, 1972, Rivers: Rogue, Correspondence, 1972, Box 27, Ms. 2386, OECP, Oregon Environmental Council *Newsletter*, April 1972, p. 4; Al Smith, "Evaluation of the Lost Creek Dam" *Oregon Wildlife* 31 (July 1976): 10–20.

50. *Medford Mail-Tribune*, October 19, 1969. See *Salmon Trout Steelheader* magazine, October–November 1972; Robert Bunting, *The Pacific Raincoast: Environment and Culture in an American Eden, 1778–1900* (Lawrence: University of Kansas Press, 1997), p. 161; Smith, "Evalution of the Lost Creek Dam"; Oregon State Game Commission, *Fish and Wildlife Plan*, pp. 2–3, 5; Oregon Environmental Council *Newsletter*, April 1972; *Medford Mail-Tribune*, February 13, October 6, 1967; Netboy, "River Is Under Arrest," pp. 29–31.

51. Don Holm, "Wild Steelhead Survive Longer," *Oregonian*, March 4, 1975. See Martin K. Bovey in *Wild and Scenic Rivers* (1967), p. 172; Leon A. Verhoeven in ibid, p. 102.

52. *Medford Mail-Tribune*, October 6, 1967; *Oregonian*, February 16, 25, March 11, 28, April 19, 22, 24, 26, May 10, June 10, July 29, 1975; Netboy, "River Is Under Arrest," pp. 29–31; Anthony J. Golden in *Oregonian*, April 30, 1975; Citizens League for Emergency Action on the Rogue (CLEAR), "Rogue River Basin," April 28, 1975, and Bill Meyer, "Opposition to Elk Creek Dam," in Rivers–Rogue–Elk Creek, Articles, Preliminary Plans, Etc., Box 27, Ms. 2386, OECP.

53. Netboy, "River Is Under Arrest," pp. 29–31; George McKinley and Doug Frank, *Stories on the Land: An Environmental History of the Applegate and Upper Illinois Valleys* (Medford, OR: Bureau of Land Management, Medford District, 1995), pp. 200, 202–203; *Oregonian*, Feburary 25, April 19, 22, 29, September 4, 1975; Andy Kerr, William Meyer, Alice D. Blair, Kenneth Lund in Department of the Army Portland District Corps of Engineers, *Applegate Lake, Rogue River Basin, Oregon: Transcript of Public Meeting for the Presidential Review of Water Resource Projects* (Grants Pass, Oregon: Portland District Corps of Engineers, 1977), pp. 32, 157, 209, exhibit 83; Anthony Netboy to Mike Frome, December 17, 1972; Frome to Netboy, December 29, 1972, Rivers: Rogue, Correspondence, 1972, Box 27, Ms. 2386, OECP; Gladwin Hill, "Oregon's Rogue: Lonely Example of Scenic Rivers Progress," *New York Times* News Service, and Lee Johnson to Robert Straub, December 26, 1974, 907.1 Rogue River, Administrative Correspondence, 1975–1978, Box 85, RG–90–A, Office of Governor Robert W. Straub Records, 1974–1979, OSA.

54. *Oregonian*, April 16, 19, 29, 1977; Jeffrey K. Stine, "Environmental Policy during the Carter Presidency," in *The Carter Presidency: Policy Choices in the Post–New Deal Era*, edited by Gary M. Fink and Hugh Davis Graham (University of Kansas Press, 1998), p. 185.

55. First quote: James W. Bayliss; second quote: Myrtle I. Krouse, Portland District Corps of Engineers, *Applegate Lake* (1977), exhibit 17 and p. 236. See Ed Zajonc (for Sen. Robert Packwood), Pete Loughridge, Ms. Ned Dunnkrack, Greydon Gilmer, Cathy J. Ross, L. Jean Soltis, Jeanette R. Head, directors of the Josephine Soil and Water Conservation District, George W. Pace and Josephine M. Pace, William L. Jess, in ibid, pp. 15, 50, 79,

93, 253, 257, Exhibit 62, Exhibit 77, Exhibit 103, Exhibit 115; *Medford Mail–Tribune*, October 25, 1976.

56. First quote: *Medford Mail–Tribune*, October 25, 1976; second quote: Jerome G. Smith, Portland District Corps of Engineers, *Applegate Lake* (1977), p. 256. See Kathering Ging, Leslie Dunnkrack, George Calvert, Eleanor Bradley, Sara Polenick, Karen J. Winningham, Steven J. James, Neal A. Pondelick, Cathy J. Ross, Clyde and Diana Bergman, Tom Drew, Russell Mitchell, Portland District Corps of Engineers, *Applegate Lake* (1977), pp. 24, 27, 30, 53, 68, 70, 164, 183–84, 229, 250, 253, Exhibit 19, Exhibit 36, Exhibit 43, in Portland District Corps of Engineers, *Applegate Lake* (1977); *Oregonian*, April 19, 1977.

57. Quotes: Kathering Ging in Portland District Corps of Engineers, *Applegate Lake* (1977), p. 27. See Pete Loughridge, Royal B. DeLand, William Meyer, Sara Polenick, Cathy J. Ross, L. Jean Soltis, Eleanor Ramsay in ibid, pp. 50, 117, 119, 158, 160, 164, 253, 257, Exhibit 113; *Medford Mail–Tribune*, October 25, 1976.

58. McKinley and Frank, *Stories on the Land*, pp. 200–201.

59. Ms. Ned Dunnkrack. See Edwin B. Abbott, Portland District Corps of Engineers, *Applegate Lake* (1977), Exhibit 12, p. 79.

60. Davis, "Lower Rogue," p. 20.

61. William Cronon, "The Trouble with Wilderness: Or, Getting Back to the Wrong Nature," in *Uncommon Ground: Toward Reinventing Nature*, edited by Cronin (New York: W.W. Norton, 1995), p. 80.

62. William O. Douglas, "Allagash," in *My Wilderness: East to Katahdin* (Garden City, NY: Doubleday, 1961), pp. 260, 262–63; Douglas, "Why We Must Save the Allagash," *Field and Stream* 68 (July 1963): 25.

63. Maine State Legislature *Legislative Record* (Augusta 1963), p. 2662. See ibid, pp. 2200, 2615; Richard Saltonstall, Jr., *Maine Pilgrimage: The Search for an American Way of Life* (Boston: Little, Brown, 1974), p. 225; *Maine Sunday Telegram*, February 24, 1963, December 3, 1972; *Portland Press Herald*, April 30, 1965; E. Melanie DuPuis, "In the Name of Nature: Ecology, Marginality, and Rural Land Use Planning During the New Deal," in *Creating the Countryside: The Politics of Rural and Environmental Discourse*, edited by DuPuis and Peter Vandergeest (Philadelphia: Temple University Press), 1996, pp. 100–01.

64. Douglas, "Why We Must Save the Allagash," p. 28.

65. John Shanklin, *Multiple Use of Land and Water Areas: Report to the Outdoor Recreation Resources Review Commission*, Study Report No. 17 (Washington, DC, 1962), p. 28; Mike Frome, "Questioning the Future of Woods in a Rapidly Changing Land," *American Forests* 73 (May 1967): 5.

66. J.W. Penfold, "Wilderness East: A Dilemma," *American Forests* 78 (April 1972): 24; and Fred C. Simmons, "Wilderness East? No," *American Forests* 78 (July 1972): 3, 44–45; Edwin H. Ketchledge, "Born-Again Forest," *Natural History*, May, 1992, pp. 34–39; Henry Clepper, "Waters of the Allagash," *American Forests* 70 (November 1964): 46; Christopher McGory Klyza, "An Eastern Turn for Wilderness," in *Wilderness Comes Home: Rewilding the Northeast*, edited by Klyza (Hanover, NH: Middlebury College Press, 2001), pp. 3–26.

67. Richard W. Judd, *Aroostook: A Century of Logging in Northern Maine* (Orono: University of Maine Press, 1989); Saltonstall, *Maine Pilgrimage*, p. 121; A.L. Leighton to Lawrence Stuart, December 27, 1966, Seven Islands (Ownerships – Cutting Plans); Jerome Matus, "Taxation, Allagash," MSA; *Portland Telegram*, July 9, 1961; Arch Soutar, "Should the

Allagash Be Turned into National Recreation Park Area?," *Lewiston Evening Journal Magazine*, May 5, 1962; *Bangor News*, August 3, 1938; *Lewiston Evening Journal*, August 12, 1944; *Kennebec Journal*, October 2, 1963; *Portland Sunday Telegram*, February 24, 1963; "Report on the Allagash," *Appalachia* 34 (June 1962): 153; Maine Legislature, *Legislative Record* (Augusta, 1959), p. 1237.

68. First quote: Maine Legislature, *Legislative Record* (Augusta 1965), p. 2929; Second quote: *Kennebec Journal*, January 7, 19, 1955. See ibid, May 2, 1961; *Lewiston Evening Journal*, January 17, 1955.

69. The idea of an Allagash national park was broached in the *New York Times* as early as 1905. See *Bangor Daily News*, June 30, 1961; Maine Legislature, *Legislative Record* (Augusta, 1933), pp. 144, 147, 161–62, 167, 169–70, 208–10, 236, 262 (1935), pp. 170, 840, 844–46, 1023–26, 1032, 2935; *The Enterprise* (Lisbon Falls), August 20, 1949; "Report on the Allagash," p. 153.

70. *Bangor Daily News*, August 26, 1958; *Portland Press Herald*, April 5, 1963; Maine Legislature, *Legislative Record* (Augusta 1933), pp. 170, 237, 263; (1935), pp. 1023; (1951), pp. 468–69, 470–71, 1027; (1963), p. 2151; *Lewiston Evening Journal*, August 5, 1959; *Maine Times*, June 21, 1974; *Kennebec Journal*, April 24, 1959.

71. *Kennebec Journal*, June 28, 1961; *Maine Sunday Telegram*, June 10, 1984. See Clarence Roth in Allagash Study Committee, 1965-1966 Committee Minutes and Public Hearing Notes, August 17, 1965, to December 1, 1965, Legislative Committees, Box 1, MSA.

72. James Carr in *Lewiston Daily Sun*, December 30, 1959. See John Hakola, *Legacy of a Lifetime: The Story of Baxter State Park* (Woolwich, ME: TBW Books, 1981); Botkin, *No Man's Garden*, p. 151.

73. *Lewiston Daily Sun*, April 28, 1959. See National Park Service, *Proposed Allagash National Recreation Area* (Washington, DC, n.d., ca. 1961); Clepper, "Waters of the Allagash," p. 49; *Kennebec Journal*, June 28, 1961; *Portland Press Herald*, January 3, 1962; *Lewiston Evening Journal*, August 5, 1959; *Portland Telegram* July 9, 1961; Soutar, "Should the Allagash Be Turned?"

74. Gene L. Letourneau in *Portland Sunday Telegram*, May 15, 1960. See the *Enterprise*, November 9, 1960; *Lewiston Evening Journal*, May 1, August 5, 1959; Soutar, "Should the Allagash Be Turned?; *Kennebec Journal*, March 23, 1960, May 6, 1961, December 28, 1962; *Portland Telegram* July 9, 1961; *Bangor Daily News*, April 21, 1960.

75. James B. Craig, "Maine Asks for a Chance on the Allagash," *American Forests* 69 (October 1963): 11. See *Lewiston Evening Journal*, September 6, 1962; "New England's Last Frontier," *New England Business Review*, September 1964, pp. 3, 8.

76. First quote: Michael Huston in *Maine Times* January 13, 1978; second quote: Austin Wilkins in Craig, "Maine Asks for a Chance," p. 11. See *Kennebec Journal*, November 7, 1976; *Bangor Daily News*, September 2, 1977; *Waterville Sentinel*, December 7, 1976; Frome, "Questioning," p. 48.

77. Clepper, "Waters of the Allagash," pp. 48–49; Soutar, "Should the Allagash Be Turned?" See Maine Legislature, *Legislative Record* (Augusta, 1963), pp. 2623–24; (1961), pp. 3052–53; "Report on the Allagash," p. 155; *Kennebec Journal*, July 29, 1961; *Bangor Daily News*, June 26, 1960; *The Enterprise*, September 13, 1962; *Portland Telegram*, July 9, 1961.

78. Brad Wellman, October 26, 1965, Allagash Study Committee, 1965–1966, Committee Minutes and Public Hearing Notes, August 17, 1965, to December 1, 1965, Legislative Committees, Box 1, Maine State Archives; *The Enterprise*, September 13, 1962. See Clepper, "Waters of the Allagash," p. 48–49; Frome, "Questioning," p. 48; James B.

Craig, "The Allagash: A Pattern Emerges," *American Forests* (August 1966): 27; *Portland Press Herald*, July 10, 1962, February 12, 1963, April 25, 1964; *Portland Evening Express*, October 26, 1965; *Lewiston Daily Sun*, August 3, 1964; *Lewiston Evening Journal*, August 5, 1959; *Bangor Daily News*, January 12, 1962, July 30, 1965; *Kennebec Journal*, February 20, 1962.

79. Maine Legislature, *Legislative Record* (Augusta 1965), p. 2929.

80. *Kennebec Journal*, July 12, 1962. See Maine Legislature, *Legislative Record* (Augusta, 1961), pp. 3052–53; *Kennebec Journal*, June 28, 1961; *Bangor Daily News*, September 5, 1962; *Portland Evening Express*, September 5, 1962; *Portland Press Herald*, March 29, July 10, September 17, 1962: *Lewiston Evening Journal*, September 6, October 1, 1962; *The Enterprise*, September 13, 1962.

81. *Portland Sunday Telegram*, February 28, 1965. See *Bangor Daily News*, August 26, 1958, December 28, 1962; *Portland Press Herald*, January 4, February 12, May 16, 1963; *Kennebec Journal*, January 4, 9, April 4, 1963; *Portland Sunday Telegram*, December 15, 1963, March 21, 1965; *Portland Press Herald*, April 5, 1963; *Houlton Pioneer Times*, January 17, 1963; "New Threat to the Allagash," *Audubon* 65 (July–August 1963): 211.

82. Clepper, "Waters of the Allagash," p. 49; Edward Cyr in Maine Legislature, *Legislative Record* (Augusta, 1963), p. 2150–51. See ibid (1963), pp. 2150–54, 2201–03, 2610–13, 2986, 3060; *Kennebec Journal*, March 21, 1963.

83. Maine Legislature, *Legislative Record* (Augusta, 1963), pp. 2155–56, 2614, 2617–20, 2986, 2989, 3061–62, 3155; *Kennebec Journal*, March 21, June 12, 1963.

84. First quote: Maine Legislature, *Legislative Record* (Augusta, 1963), p. 2155; second quote: ibid, p. 3063. See ibid, 2623, 2971; *Bangor Daily News*, May 23, 1963; *Lewiston Evening Journal*, May 24, June 7, 1963.

85. *The Enterprise*, September 13, 1962. See Craig, "Allagash"; William A. King, "Report on the Allagash," pp. 802–3; Maine Legislature *Legislative Record* (Augusta, 1963), p. 2989; *Kennebec Journal*, October 2, 1963.

86. Craig, "Maine Asks for a Chance," p. 10. See *Kennebec Journal*, June 12, 15, 1963; *Portland Press Herald*, April 29, 1965; Maine Legislature, *Legislative Record* (Augusta, 1965), p. 2929; (1966), p. 2149.

87. *Maine Times*, June 21, 1974. See "New England's Last Frontier," pp. 4–6; *Portland Press Herald*, December 31, 1964, February 28, 1965; *Houlton Pioneer Times*, January 17, 1963, January 28, 1965; *Portland Sunday Telegram*, December 15, 1963; *Bangor Daily News*, February 5, 1965; *Portland Sunday Telegram*, December 15, 1963.

88. *Bangor Daily News*, February 5, 1965.

89. Maine Legislature, *Legislative Record* (Augusta, 1965), pp. 2982–85, 3054. See ibid, p. 2986; *Portland Press Herald*, February 12, April 1, 30, June 4, 5, August 19, 20, 1965; *Lewiston Evening Journal*, August 19, 1963, April 30, 1965; *Kennebec Journal*, March 22, April 4, 26, 1963, February 10, 1965; *Sanford Tribune*, January 28, 1965; *Portland Sunday Telegram*, February 28, April 25, 1965; *Portland Evening Express*, March 29, 1960; *Boston Globe*, April 30, 1965; Robert Haskell in Allagash Study Committeee, 1965–1966 Committee Minutes and Public Hearing Notes, August 17, 1965 to December 1, 1965, Legislative Committees, Box 1, MSA.

90. First quote: Elmer Violette in Allagash Study Committee, 1965–1966, Committee Minutes and Public Hearing Notes, August 17, 1965, to December 1, 1965, Legislative Committees, Box 1, MSA; second quote: Willard Jalbert in ibid; third quote: Robert Patterson in ibid. See George Sawyer, Helen Beady, Ted Grant, Arthur Bennett in ibid; *Kennebec Journal*, March 27, 1964; *Bangor Daily News*, April 22, 1964, October 15, 1965.

91. Maine Legislature, *Legislative Record* (Augusta, 1965), p. 2993. See ibid, pp. 2929, 2985, 2988, 2990; ibid (1966), p. 328; *Portland Press Herald*, July 22, 1965; *Bangor Daily News*, September 22, October 21, December 25–26, 1965, August 11, 1966; *Lewiston Evening Journal*, July 10, 1965, May 6, September 23, 1966; *Portland Evening Express*, June 4, 1965.

92. Edward C. Crafts, Lawrence Stuart, Elmer Violette in Allagash Study Committee, 1965–1966, Committee Minutes and Public Hearing Notes, August 17, 1965 to December 1, 1965, Legislative Committees, Box 1, MSA; *Portland Press Herald*, November 27, 1965, January 17, 27, 1966; *Kennebec Journal*, January 14, 20, 1966; *Bangor Daily News*, October 27, 1965; Maine Legislature, *Legislative Record* (Augusta, 1965), p. 2990; Craig, "Allagash"; Allagash River Authority, *Allagash Wilderness Waterway: A Proposal for State Control* (Augusta, 1965); Maine Legislature, Interim Joint Committee, *Findings and Recommendations for Preservation of the Allagash Wilderness Waterway* (Augusta, 1966), p. 6, 8; "Allagash Victory Appears Near," *Audubon* 68 (July–August 1966): 213. *Bangor Daily News*, May 15, 1968; *Maine Sunday Telegram*, August 25, 1968, July 12, 19, 1970; *Kennebec Journal*, July 20, 1970; *Portland Press-Herald*, July 11, 1970; Lawrence Stuart, "The Allagash: Three Years Later," *Parks and Recreation* 5 (April 1970): 22–24.

93. *Maine Sunday Telegram*, September 1, 1968, October 12, 1969. See *Kennebec Journal*, July 27, 1967; *Bangor Daily News*, July 10, August 6, 1965, March 2, 1966, June 13, 21, 25, July 17, 1968, June 16, November 13, 1969, April 3, June 25, 1970; *Lewiston Evening Journal*, July 10, August 10, 1965; *Waterville Sentinel*, September 9, 1977. On the floods, see *Maine Times*, June 14, June 21, 1974; on the utilities companies, see *Maine Sunday Telegram*, April 7, 28, 1974; *Bangor Daily News*, April 25, 1974.

94. *Maine Times*, June 14, 1974. See ibid, November 19, 1976, February 25, March 18, 25, April 29, September 2, 8, 1977; *Maine Sunday Telegram*, February 20, March 20, September 11, 1977; *Waterville Sentinel*, November 10, December 6, 8, 1976, February 9, April 22, August 26, 27, 1977; *Portland Press Herald*, July 25, 1977; *Bangor Daily News*, April 1, 1977; *Kennebec Journal*, February 21, 1977.

95. *Lewiston Evening Journal*, August 15, 1968. See *Maine Sunday Telegram*, October 2, 1977; *Maine Times*, June 21, 1974; *Bangor Daily News*, May 29, 1974.

96. *Bangor Daily News*, May 29, 1974.

97. *Bangor Daily News*, May 29, 1974. See ibid, May 29, 1974; *Maine Times* June 21, 1974.

98. Robert Jalbert in *Maine Times*, June 21, 1974. See Leonard McBreairty in *Maine Sunday Telegram*, October 2, 1977.

99. *Maine Times*, November 26, 1976. See ibid, October 28, 1977; *Maine Sunday Telegram*, October 2, 16, 1977; *Bangor Daily News*, October 8–9, 1977.

100. *Maine Times*, November 26, 1976. See ibid, October 28, 1977; Saltonstall, *Maine Pilgrimage*, p. 173.

101. *Kennebec Journal*, November 7, 1976. See *Maine Times*, July 23, 1976, May 20, October 28, December 16, 1977, January 13, 1978; *Maine Sunday Telegram*, September 19, 1976, October 23, 1977; April 2, 1977; *Portland Press Herald*, August 5, 1971, September 6, 1977; *Waterville Sentinel*, April 5, 1977; *Bangor Daily News*, February 11, 1976, September 10–11, 1977, September 9, 1981; *Waterville Sentinel*, September 9, 1977.

102. D.W. Scott, "Is Purity a Virtue?" *Living Wilderness* 36 (Winter 1972): 2. See Dennis M. Roth, *The Wilderness Movement and the National Forests* (College Station, TX: Intaglio Press, 1988), pp. 39–45.

CHAPTER 4: SAVING NATURE'S ICONS

1. Maine State Legislature, *Legislative Record* (Augusta, 1970), pp. 682–83; McCall in *Capital Journal*, November 13, 1969; Charles H. Callison, "'Environmental Crisis' Is the New 'In' Phrase," *Audubon* 71 (July 1969): 104; "The Audubon Viewpoint," *Audubon* 72 (May 1970): 109.

2. "Fellow Americans: Keep Out!" *Forbes* 107 (June 15, 1971): 22–30; Russell W. Peterson in Richard Saltonstall, Jr., *Maine Pilgrimage: The Search for an American Way of Life* (Boston: Little, Brown, 1974), p. 65; *What Kind of Coast Will You Leave for Your Children?* (Maine Regional Group of the Sierra Club, n.d., ca. 1967), "Machiasport," vertical file, Mantor Library, University of Maine at Farmington (hereafter ML–UMF).

3. First quote: Lawrence Halprin & Associates, *The Willamette Valley: Choices for the Future* (Willamette Valley Environmental Protection and Development Planning Council, 1972), p. 7; second quote: ibid, p. 3; third quote: Robert Straub in *Oregon Statesman*, December 9, 1969. See ibid, December 18, 21, 1969; *Oregonian*, March 14, 1970.

4. Ian McHarg, "Blight or a Noble City?" *Audubon* 68 (January–February 1966): 47; Anthony Netboy, "Cleaning Up the Willamette," *American Forests* 78 (May 1972): 15; Netboy, "French Pete for People," *American Forests* 76 (May 1970): 59; Keith Montgomery Carr, "Changing Environmental Perceptions, Attitudes, and Values in Oregon's Willamette Valley, 1800 to 1978," master's thesis, University of Oregon, 1978.

5. Samuel P. Hays, "From Conservation to Environment: Environmental Politics in the United States since World War II," in *Out of the Woods: Essays in Environmental History*, edited by Char Miller and Hal Rothman (Pittsburgh: University of Pittsburgh Press, 1997), pp. 115–16; Hays, *Beauty, Health, and Permanence: Environmental Politics in the United States, 1955–1985* (New York: Cambridge University Press, 1987), p. 52; Frank J. Popper, "Rural Land Use Policies and Rural Poverty," *Journal of the American Planning Association* 3 (Summer 1984): 329.

6. Lynton K. Caldwell, Lynton R. Hayes, Isabel M. MacWhirter, *Citizens and the Environment: Case Studies in Popular Action* (Bloomington: Indiana University Press, 1976), pp. 7–8, 64–66, 17–22, 154–170; *Oregonian*, February 18, 1973; *Medford Mail–Tribune*, October 26, 1970; "Fellow Americans: Keep Out,"pp. 22–23; Pacific Power and Light official in Carr, "Changing Environmental Perceptions," p. 137; Stephanie S. Pincetl, *Transforming California: A Political History of Land Use and Development* (Baltimore: Johns Hopkins University Press, 1999), pp. 149, 159, 194; Lyndon B. Johnson, in U.S. Department of the Interior, *Quest for Quality: U.S. Department of the Interior Conservation Yearbook* (Washington, DC, 1965), p. 15; William G. Robbins, *The Oregon Environment: Development vs. Preservation, 1905–1950* (Corvallis: Oregon State University, 1975), pp. 4–5.

7. Rachel Carson, *Edge of the Sea*, in Daniel B. Botkin, *No Man's Garden: Thoreau and a New Vision for Civilization and Nature* (Washington, DC: Island Press/Shearwater Books, 2001), p. 174.

8. L. Morrill Burke, "The Room We Live In," *Maine Sunday Telegram*, July 13, 1969.

9. Walt and Rosaline Morey, *Land Use Oregon* 1 (June 15, 1974): 3.

10. Norman Sanders, "The New Tide of Coastal Legislation," *Sierra Club Bulletin* 58 (February 1973): 10.

11. Kenneth M. Curtis in Saltonstall, *Maine Pilgrimage*, pp. 49–50. See Richard Pardo, "Something Up, Down East," *American Forests* 75 (May 1969): 45; *Maine Times*, February 21, 1969, March 12, 1975; *Maine Sunday Telegram*, February 20, November 5, 1972.

12. First quote: *Maine Sunday Telegram*, November 5, 1972; second quote: *Kennebec Journal*, June 10, 1970. See Neal R. Peirce, *The New England States: People, Politics, and Power in the Six New England States* (New York: W.W. Norton, 1976), pp. 412–13; *Maine Times*, February 21, 1969, March 12, 1975; *Kennebec Journal*, June 10, 1970.

13. Maine Legislature, Legislative Research Committee, *Report on Coastal Conveyance of Petroleum Products to First Special Session of the 104th Legislature* (Augusta, 1970), p. 1.

14. A. Thomas Easley, *An Atlantic World Port and Oil Refinery for New England* (Boston: New England Council for Economic Development, 1968), "Machiasport," vertical file, ML–UMF.

15. Keith Roberts, *Machiasport: Oil and the Maine Coast* (Boston: Eastern New England Groups of the Sierra Club, August 1969), pp. ii, iii, 26–28, "Machiasport," vertical file, ML–UMF.

16. Roberts, *Machiasport*, pp. i, v, appendix B, p. 30.

17. James Storer to Governor Kenneth Curtis, September 17, 1968, Box 17, Machiasport Project, Governor Kenneth Curtis Papers (hereafter KCP), Maine State Archives (hereafter MSA); list of attendance, minutes of October 1, 1968 meeting of the Conservation and Planning Committee for the Machiasport Project, KCP.

18. Maine Legislature, *Legislative Record* (Augusta, 1969), p. 4569; Peter A. Bradford, *Fragile Structures: A Story of Oil Refineries, National Security, and the Coast of Maine* (New York: Harper's Magazine Press, 1975), pp. 42, 145.

19. Stephanie S. Pincetl, *Transforming California*, pp. 157, 182; Bradford, *Fragile Structures*, pp. 148, 150, 242, 243, 250, 251; Peirce, *New England States*, p. 400.

20. Frank Graham, *Oil and the Maine Coast: Is It Worth It?* (Augusta: Natural Resources Council of Maine, 1970), pp. 16, 19.

21. Graham, *Oil and the Maine Coast*, pp. 21, 27.

22. *Maine Times*, December 10, 1976; Saltonstall, *Maine Pilgrimage*, pp. 7–9.

23. *Maine Times*, December 10, 1976; Kenneth P. Hayes, *As Envisioned by Maine People: A Report to the Commission on Maine's Future* (Orono: University of Maine, 1976), p. 87.

24. Graham, *Oil and the Maine Coast*, pp. 2, 16, 35–36.

25. *Newsweek*, February 16, 1970; Maine State Legislature, *Legislative Record* (Augusta, 1969), p. 4569. See Russell D. Butcher, "Maine to Regulate Commercial Development," *Living Wilderness* 34 (Summer 1970): 54–5; *Public Laws 1969*, Ch. 572, "An Act Relating to Coastal Conveyance of Petroleum"; Legislative Research Committee, Maine State Legislature, *Report on Coastal Conveyance of Petroleum Products* (Augusta, 1970), p. 1.

26. Maine State Legislature, *Legislative Record* (Augusta, 1970), p. 679.

27. Maine Legislature, *Legislative Record* (Augusta, 1970), pp. 681–82.

28. Peirce, *New England States*, pp. 399–402. See Bradford, *Fragile Structures*, p. 134.

29. *New York Times*, February 12, 1970.

30. *Maine Public Laws*, 1969, Ch. 571, 1970 (Special Session); Bradford, *Fragile Structures*, pp. 148, 150; Bennett D. Katz in Maine Legislature, *Legislative Record* (Augusta, 1970), p. 734. See ibid, pp. 676–678, 729–734.

31. *Portland Press Herald*, May 20, 1970.

32. *Portland Press Herald*, May 23, 1970. See Environmental Improvement Commission, *Application of King Resources Company*, vol. 1 (Augusta, 1970), pp. 4–7, 49–51, 67–69, 79, 85, 110, 117, 128, 153–156, 176.

33. Environmental Improvement Commission, *Application of King Resources Company*, 1: 9–15, 18–22, 26–31.

34. Environmental Improvement Commission, *Application of King Resources Company*, 1: 117, 220. See ibid, pp. 88–104.

35. Environmental Improvement Commission, *Application of King Resources Company*, 1: 84, 100, 154–157, 202–211.

36. Environmental Improvement Commission, *Application of King Resources Company*, 1: 190; letters in volume 5. See Mary Norton, ibid, 1: 200–02; ibid, pp. 153–76.

37. Environmental Improvement Commission, *Application of King Resources Company*, 1: 124, 135, 195–199.

38. Bradford, *Fragile Structures*, pp. 273–78; *Maine Sunday Telegram*, May 24, 1970; *Kennebec Journal*, July 9, 1970; *King Resources versus EIC*, ME. 270 A2d 863, November 19, 1970.

39. U.S. Department of Interior, Bureau of Outdoor Recreation Northeast Region Office, *Island Study Phase III Report on Casco Bay, Maine* (Washington, DC, 1967).

40. *Bangor Daily News*, February 8, 1971; *Waterville Sentinel*, July 20, 1970; *Maine Sunday Telegram*, September 27, 1970; Bradford, *Fragile Structures*, pp. 302–03.

41. *Lewiston Journal*, April 11, 1964; *Bangor Daily News*, March 14, 1970; *Waterville Sentinel*, February 11, March 3, 1971; Bradford, *Fragile Structures*, p. 303.

42. Bradford, *Fragile Structures*, pp. 304, 310; *Waterville Sentinel*, February 4, 6, 1971; *Bangor Daily News*, February 19, 1971.

43. "Maine Clean Fuels Application," "Searsport Transcript," vol. 1, pp. 1–603, 500, 512–66, 567, Box 16, 10–12; "Findings of Fact and Order," July 23, 1971; "Record of Intent," Box 17, pp. 8, 18, Department of Environmental Protection Collection, MSA; *Bangor Daily News*, March 23, 1971; *Waterville Sentinel*, March 24, 1971; *Kennebec Journal*, March 23, April 13, 1971.

44. *Maine Sunday Telegram*, March 28, 1971.

45. Maine Clean Fuels, "Searsport Transcript," Box 18, vol. 1, pp. 23, 27, 30, 35, 38; 519–530, Department of Environmental Protection Collection, MSA; Donald Hansen, *Waterville Sentinel*, February 11, March 3, 1971.

46. Maine Clean Fuels, "Searsport Transcript," Box 18, pp. 613–1223, Department of Environmental Protection Collection, MSA. Proponents offered about 3,000 signatures. See ibid, Box 16.

47. Ossie Beal, Maine Clean Fuels, "Searsport Transcript," Box 18, pp. 756–66, Department of Environmental Protection Collection, MSA. See Beal in Saltonstall, *Maine Pilgrimage*, p. 43.

48. Maine Clean Fuels, "Searsport Transcript," Box 18, pp. 1243–1251, 1227–1290, Department of Environmental Protection Collection, MSA. See *Maine Times*, July 28, 1978.

49. *Bangor Daily News*, May 21, 1971.

50. Bradford, *Fragile Structures*, p. 337; *Portland Press Herald*, July 31, 1971. See *In re Maine Clean Fuels, Inc.* Me. 310 A2d 736, Oct. 17, 1973, pp. 742, 747.

51. Maine Governor's Task Force on Energy, Heavy Industry, and the Maine Coast, *Final Report* (Augusta, 1972), pp. ix–x.

52. Maine Governor's Task Force, *Final Report*, pp. 2, 6–7, 15–18.

53. *Waterville Sentinel*, March 3, 1971; *Kennebec Journal*, June 10, 1970. See *Maine Sunday Telegram*, December 19, 1971.

54. John Cole in Saltonstall, *Maine Pilgrimage*, p. 66; *Maine Sunday Telegram*, March 5, 1972; *Kennebec Journal*, May 16, 1972; *Portland Press Herald*, May 24, 1972; March 15, May 10, 1973; *Maine Times*, January 13, 1978; Maine State Legislature, *Legislative Record* (Augusta, 1973), p. 3681.

55. *Maine Sunday Telegram*, October 21, December 2, 1973, October 6, 1974; *Waterville Sentinel*, July 12, August 8, October 23, 1973, November 27, 1974; *Kennebec Journal*, June 5, November 14, 1973; *Portland Press Herald*, December 21, 1973, February 14, April 5, August 30, September 28, 1974, January 15, June 10, 1975; Bradford, *Fragile Structures*, pp. 350, 363, 366; *Maine Sunday Times*, October 14, 1973.

56. *Maine Times*, March 7, 1975; Frank Graham Jr., "Decision at Eastport: The Oilmen Cometh," *Audubon* 76 (March 1974): 97.

57. *Maine Times*, February 25, 1977, July 28, 1978, August 25, 1978, September 29, 1978.

58. *Portland Press Herald*, September 30, 1971; *Waterville Sentinel*, March 29, 1973; Saltonstall, *Maine Pilgrimage*, p. 53.

59. *Maine Times*, December 12, 1975, December 10, 1976, July 28, 1978.

60. Saltonstall, *Maine Pilgrimage*, pp. 45–46. See *Maine Times*, June 4, 1976.

61. John N. Cole, "Letter from Maine: Oil and Water," *Harper's* 243 (November 1971): 50.

62. *Portland Press Herald*, September 8, 1971. See *Maine Sunday Telegram*, August 3, 1969.

63. David Talbot, "Statement of State Highway Department in support of HB 1601, March 7, 1967," in 1967 House Highway Committee Minutes and Exhibits, H.B. 1601, Committee Minutes, Oregon State Archives, Salem (hereafter OSA).

64. *Oregonian*, November 29, 30, 1967.

65. *Oregonian*, November 29, 1967.

66. *Oregonian*, November 29, 1967; Catherine Williams, "The People's Fight to Save the Oregon Beaches, 1965–72" (typescript), in Beaches, Articles, Reports, 1970–73, File 16, Box 2, Ms. 2386, Oregon Environmental Council Papers (hereafter OEC), Oregon Historical Society, Portland (hereafter OHS).

67. Williams, "The People's Fight to Save the Oregon Beaches."

68. J. Richard Byrne to Glenn Jackson, July 14, 1966; Jackson to Byrne, July 22, 1966, Committee Minutes, Exhibits for March 23, 1967 (H.B. 1601), House Highway Committee Minutes and Exhibits, 1967, OSA.

69. William Noyes to G.E. Rhode, August 15, 1966, March 7, 1967; Lawrence Bitte to Governor McCall, August 25, 1966, Highway Committee Minutes, Exhibits for March 7, 1967 (H.B. 1601), House Highway Committee Minutes and Exhibits, 1967, OSA.

70. Loran L. Stewart in Highway Committee Minutes and Exhibits for March 7, 1967 (H.B. 1601), House Highway Committee Minutes and Exhibits, 1967, OSA; *Daily Astorian*, May 10, 1967.

71. House Highway Committee Minutes and Exhibits for March 7, 1967 (H.B. 1601), House Highway Committee Minutes, OSA.

72. *Daily Astorian*, May 12, 24, 1967, January 16, 1968. See Dann DeBernardi in House Highway Committee Minutes and Exhibits for March 23, 1967 (H.B. 1601), House Highway Committee Minutes and Exhibits, 1967, OSA.

73. Notes for a speech to House (probably by Sidney Bazett), Beach Bill 1968–69, Sidney Bazett Papers, OHS; Lawrence Bitte to Bazett, April 27, 1967, Minutes, May 2, Public Hearing on HB1601, in House Highway Minutes, April 27, OSA; *Daily Astorian*, May 15, 1967.

74. "Recommendation by Gov. McCall for Preservation of Oregon's Beaches," May 11, 1967 (submitted to committee); "Statement of State Treasurer Robert Straub," May 11, 1967, Minutes, House Highway Committee, 1967, Minutes and Exhibits on H.B. 1601, OSA.

75. "Statement of House Committee on Highways concerning H.B. 1601," Minutes and Exhibits for May 18, 1967, House Highway Committee, Minutes and Exhibits on H.B. 1601, OSA. See Joe Richards, "HB1601: The Beach Bill," n.d., 1601 Beach Bill (1967), Box 1, Sidney Bazett Papers, OHS; *Daily Astorian*, May 15, 1967.

76. Tom McCall in *Daily Astorian*, July 6, 1967. See ibid, May 15, 22, July 6, 1967; Walth, *Fire at Eden's Gate*, p. 191; Joe Richards, "HB1601: The Beach Bill," n.d.; KGW TV–8 editorial (copy), June 16, 1967, 1601 Beach Bill (1967), Box 1, Sidney Bazett Papers, OHS.

77. Walth, *Fire at Eden's Gate*, p. 194.

78. Williams, "The People's Fight to Save the Oregon Beaches." See Walth, *Fire at Eden's Gate*, pp. 195–96; *Daily Astorian*, September 20, 1967.

79. Paul Hill, Nancy Shoemaker, Vera Springer, Janet McLennan (exhibits, petitions), Howard Glazer, Ted Howe, Howard Hutchins, Truman Robbins, Walter Otzen, Gordon Guild, "Nestucca Bay Hearing Transcript," pp. 38, 41, 49, 55, 57, 61–62, 69, 72–74, file 25, Box 2, OEC.

80. E.M. Potter in *Oregonian*, December 2, 1967. See ibid, November 29, 1967; *Daily Astorian*, November 28, 30, 1967.

81. Williams, "The People's Fight to Save the Oregon Beaches."

82. *Oregonian*, November 29, 30, 1967; Liz Millar and Jan Byerlee in "Nestucca Bay Hearing Transcript," pp. 44–47.

83. *Oregonian*, November 30, 1967; *Daily Astorian*, November 30, 1967; Robert W. Straub in "Nestucca Bay Hearing Transcript," pp. 12–16.

84. *Oregonian*, December 6, 1967. According to McCall biographer Brent Walth, McCall had already ordered termination of the highway project before the hearing. See *Fire at Eden's Gate*, pp. 197–98. See *Daily Astorian*, December 27, 1968.

85. *Daily Astorian*, November 9, 15, 28, 1967, February 8, 1968.

86. *Daily Astorian*, March 27, 1968.

87. *Daily Astorian*, October 31, 1968; *Oregonian*, November 8, 1968; Williams, "The People's Fight to Save the Oregon Beaches."

88. Williams, "The People's Fight to Save the Oregon Beaches"; Robert W. Straub to Lawrence Williams, September 24, 1968, and Keith Drury to Lawrence Williams, November 20, 1968, "Nestucca Bay Correspondence, 1968–70," File 24, Box 2, Ms. 2386, OEC; *Oregon Journal*, October 23, 1968.

89. Williams, "The People's Fight to Save the Oregon Beaches"; R.L. Bacon to "Dear Citizen," March 25, 1968; Robert W. Straub in mimeo by Bacon, April 20, 1968; Straub, "Constitutional Amendment" (copy); Committee to Save Oregon Beaches amendment (copy), n.d., file 16, Box 2, OEC.

90. *Daily Astorian*, November 8, 1968; *Oregonian*, November 1, 8, 1968.

91. Sidney Bazett in "Minutes, February 24, 1969," House Judiciary Committee, Minutes, February 5, 1969, OSA; "Exhibits in Connection with HB1045," in ibid; "Ballot title: State Acquisition of Ocean Shore"; Robert Straub to Don Wilson, March 31, 1969; Robert Bacon, Laurence Bitte, and Jefferson Conor, "Concerning Public and Private Rights on Oregon Beaches," 1969 Beach Bill (H.B. 1045), April 30, 1969, in Office of the Majority Leader, House of Representatives, House Judiciary Committee, OSA;

Kessler Cannon to Livable Oregon Steering Committee, November 7, 1968; "Statement by Gov. McCall on 5-Point Beach Program," November 7, 1968; "Proposal by Governor Tom McCall on Oregon Beaches and Willamette River Park System," December 27, 1968," in "Beach Legislation 1967," Sidney Bazett papers, OHS; *Salem Journal*, January 21, 1969.

92. Robert W. Straub in *Oregonian*, October 17, November 1, 1968; McCall in *Daily Astorian*, December 27, 1968; Williams, "The People's Fight to Save the Oregon Beaches."

93. Anne W. Simon, *The Thin Edge: Coast and Man in Crisis* (New York: Harper & Row, 1978), pp. 130, 132; U.S. Congress, *Legislative History of the Coastal Zone Management Act of 1972*, 94th Cong., 2nd sess. (Washington, DC: Government Printing Office 1976), pp. 1–7, 13–15, 24–25, 27.

CHAPTER 5: NAVIGATING THE NATURAL STATE

1. E. Nadine Harrang to Robert Straub, May 13, 1975, 1150.3.1, Willamette Greenway, Box 96, RG 90A-30, Office of Governor Robert W. Straub Records (hereafter RWSR), 1974–1979, Oregon State Archives, Salem (hereafter OSA).

2. First quote: Helmut Koenig and Gea Koenig, "Thoreau's Maine," *Travel* 142 (October 1974): 44, 67; second quote: Eliot Porter, "Summers in Penobscot Country," *Natural History* 75 (August 1966): 34.

3. James E. Sherow, "Environmentalism and Agriculture in the American West," in *The Rural West Since World War II*, edited by R. Douglas Hurt (Lawrence: University of Kansas Press, 1998), p. 59; Philip G. Terrie in E. Melanie DuPuis, "In the Name of Nature: Ecology, Marginality, and Rural Land Use Planning During the New Deal," in *Creating the Countryside: The Politics of Rural and Environmental Discourse*, edited by DuPuis and Peter Vandergeest (Philadelphia: Temple University Press, 1996), p. 103.

4. Lloyd C. Irland, *The Northeast's Changing Forest* (Petersham, MA: Harvard Forest, 1999), p. 130.

5. E.B. White, "Letter from the East," *New Yorker* 36 (December 3, 1960): 238. See Richard Wescott and David Vail, "The Transformation of Farming in Maine, 1940–1985," *Maine Historical Society Quarterly* 28 (Fall 1988): 66–84.

6. David Burner, *Making Peace with the 60s* (Princeton, NJ: Princeton University Press, 1996), p. 221; Warren Susman in Andrew Jamison and Ron Eyerman, *Seeds of the Sixties* (Berkeley: University of California Press, 1994), p. 8; Melanie DuPuis and Peter Vandergeest, "Introduction," in DuPuis and Vandergeest, *Creating the Countryside*, p. 3; Sewall Pettingill, "Maine," *Audubon* 61 (May–June 1959): 128; Paul Lyons, *New Left, New Right, and the Legacy of the Sixties* (Philadelphia: Temple University Press, 1996), p. 39.

7. See U.S. Department of the Interior, *Quest for Quality: U.S. Department of the Interior Conservation Yearbook* (Washington, DC, 1965), p. 13; Virgil H. Freed to Kessler Cannon, August 14, 1968, "Correspondence, August–October, 1968," Governor's Committee for a Livable Oregon, Ms. 2537, Oregon Historical Society, Portland (hereafter OHS); Stuart U. Rich, *Ecology, Environmentalism, and Future Timber Supply: Proceedings of a Current Issues Conference, March 18, 1975* (Eugene: University of Oregon, 1975), pp. 36–37.

8. Owen Ulph, "On the Limits to Growth," *Sierra Club Bulletin* 58 (April 1973): 12. See Jamison and Eyerman, *Seeds of the Sixties*, p. 93; *Oregonian*, November 18, 1967; Keith

Montgomery Carr, "Changing Environmental Perceptions, Attitudes, and Values in Oregon's Willamette Valley, 1800 to 1978," master's thesis, University of Oregon, 1978, p. 116; "The Malthusian Dilemma Updated," in Robert C. Paehlke, *Environmentalism and the Future of Progressive Politics* (New Haven, CT: Yale University Press, 1989), pp. 41–75.

9.   William Cronon, "The Trouble with Wilderness: Or, Getting Back to the Wrong Nature," in *Out of the Woods: Essays in Environmental History*, edited by Char Miller and Hal Rothman (Pittsburgh, PA: University of Pittsburgh Press, 1997), p. 39; Robert G. Healy, "Forests in an Urban Civilization: Land Use, Land Markets, Ownership, and Recent Trends," in *Land Use and Forest Resources in a Changing Environment: The Urban/Forest Interface*, edited by Gordon A. Bradley (Seattle: University of Washington Press, 1984), pp. 17–18.

10. Anthony Netboy, "French Pete for People," *American Forests* 76 (May 1970): 59. See C. Clare Hinrichs, "Consuming Images: Making and Marketing Vermont as Distinctive Rural Place," in DuPuis and Vandergeest, *Creating the Countryside*, p. 260.

11. "Oregon Environmental Council Newsletter," July 1970, p. 3; "Wilderness for Oregon," *Earthwatch Oregon: News Report of the Oregon Environmental Council*, March–April 1978, p. 8.

12. *Kennebec Journal*, September 2, 1982, July 1, 1985, April 5, 1986; *Waterville Sentinel*, November 2, 1973; Douglas Hofe, Jr., "Wild Rivers," *Parks and Recreation* 5 (February 1970): 53–55; Mark Dowie, *Losing Ground: American Environmentalism at the Close of the Twentieth Century* (Cambridge, MA: MIT Press, 1995), pp. 28, 32.

13. Hofe, "Wild Rivers," p. 23; Frank J. Popper, "Rural Land Use Policies and Rural Poverty," *Journal of the American Planning Association* 3 (Summer 1984): 326–34.

14. First quote: Dorotheen Wilson to Robert W. Straub, March 8, 1975, 907 Waterways, Administrative Correspondence, 1975–1978, Box 85, RG–90–A, RWSR; second quote: Glen and Rhoda Love to Tom McCall, June 14, 1974, Water Resources: Wild and Scenic Rivers, File 1, 75A–12, Box 55, RG–G4, Governor Tom McCall's Records (hereafter TMR), OSA; third quote: *Kennebec Journal* in Maine Legislature, *Legislative Record* (Augusta, l965), p. 2929. See Sam Ball and Eleanor C. Ball to Glenn Jackson, April 19, 1974, Wild and Scenic Rivers 2, 75A–12, Box 55, RG–G4, TMR.

15. Doug Adams to Governor Straub, November 17, 1975, PAR 14, Willamette River Parks System, General, 1975, Department of Transportation Library, Salem, OR (hereafter DOTL. See Ethel A. Starbird, "A River Restored: Oregon's Willamette," *National Geographic* 141 (June 1972): 828; Mrs. Kenneth Lorber in *Oregonian*, November 8, 1967; George W. Gleeson, *The Return of a River: The Willamette River, Oregon* (Corvallis: Oregon State University, 1972); George W. Churchill, "The Story of a Great River," *Parks and Recreation* 7 (January 1972): 83, 113, 103; Robert N. Royston, et al., *Preliminary Williamette River Greenway* (Salem: Oregon Department of Transportation, 1974), p. 54; *Oregonian*, March 23, 1975; "Willamette: Oregon's River Parkway," *Sunset* 149 (July 1972): 55; *Medford Mail-Tribune*, August 8, November 15,1967; *Earthwatch Oregon*, April 1975, p. 6.

16. Loring LaB. Schwarz, Charles A. Flink, Robert M. Searns, *Greenways: A Guide to Planning, Design, and Development* (Washington, DC: Island Press, 1993), p. xii; Rutherford H. Platt, Rowan A. Rowntree, and Pamely C. Muick, editors, *The Ecological City: Preserving and Restoring Urban Biodiversity* (Amherst: University of Massachusetts Press, 1994), p. 32.

17. *Earthwatch Oregon*, June–July 1977, p. 2; David G. Talbot to Keith Burns, February 18, 1975, 1150.3.1, Willamette Greenway, Box 96, RG90A–30, RWSR. See Anthony Netboy, "Willamette Greenway Act," *Oregonian Northwest*, November 5, 1967, p. 14; *Oregonian*, February 26, 1967, March 5, 1975; Harry Bodine, "Roller Coaster Politics Put Greenway Back on Shelf" *Oregonian Forum*, June 1,1975, p. 1.

18. McCall to Johnson, March 4, 1974, Willamette River Parkway System, 75A–12, Item 1, Box 57, RG–G4, TMR; Churchill, "Story of a Great River," p. 104. See David G. Talbot to Keith Burns, February 18, 1975, 1150.3.1, Willamette Greenway, Box 96, RG–90A–30, RWSR; *Oregonian*, June 1, 1967; *Earthwatch Oregon*, June–July 1977, p. 2; "Greenway Gripes," *Oregon Voter* 130 (September 2, 1967): 11–12; *Medford Mail–Tribune*, August 8, 1967.

19. Stewart Udall in Churchill, "Story of a Great River," p. 113. See William D. Honey, Jr., *The Willamette River Greenway: Cultural and Environmental Interplay* (Corvallis: Oregon State University, 1975), p. 3; Robert N. Royston, et al., *Willamette River Greenway* (Salem: Oregon Department of Transportation, n.d.), p. 5; *Oregonian*, March 2, 1967, April 12, 1975.

20. Churchill, "Story of a Great River," p. 105.

21. Rep. Dick Magruder in Bodine, "Roller Coaster Politics," p. 1; *Oregonian*, June 7, 1967. See Governor Tom McCall to Governor-Elect Robert Straub, January 8, 1975, 1150.3.1, Willamette Greenway, Box 96, RG 90A–30, RWSR; *Earthwatch Oregon*, June–July 1977, p. 2; Lily Kaup to Robert W. Straub, April 19, 1975, 1150.3.1, Willamette Greenway, Box 96, RG 90A–30, RWSR; Keep Oregon Livable, *Editorial Digest*, May 9, 1972; "Willamette Greenway Bulletin," Correspondence, January–March, 1968, Governor's Committee for a Livable Oregon, Ms. 2537, OHS.

22. Greg Morley, May 22, 1970, and George M. Baldwin, August 1, 1972, to Oregon State Highway Commission, PAR 14–2, Willamette River Park System, K–M (Milwaukie), DOTL; Schwarz, Flink, and Searns, *Greenways*, pp. 7–8. See Honey, *Willamette River Greenway*, p. 56; "Willamette: Oregon's River Parkway," p. 55; H.C. Johnson to Tom McCall, February 8, 1974, Willamette River Parkway System, 75A–12, Item 1, Box 57, RG G4, TMR.

23. *Earthwatch Oregon*, June–July 1977, p. 6; Ken Johnson to Libby Loftin, May 4, 1977, 1150.3.1, Willamette Greenway, Box 96, RG 90A–30, RWSR; Honey, *Willamette River Greenway*, pp. 56–57; *Oregonian*, April 6, 1975; Governor Tom McCall to Governor-Elect Robert Straub, January 8, 1975, 1150.3.1, Willamette Greenway, Box 96, RG 90A–30, RWSR.

24. On the Corvallis farmer, see *Oregonian*, April 12, 1975. See also *Oregonian*, April 6, 9, 1975; Robert E. Frenkel, Eric F. Heinitz, S. Nimal Wickramaratne, *Vegetational Changes in the Willamette River Greenway in Benton and Linn Counties, 1972–1981*, WRRI–79 (Corvallis: Oregon State University, 1983), p. 4.

25. Governor Tom McCall to Governor-Elect Robert Straub, January 8, 1975, Straub to McCall, January 9, 1975, David G. Talbot to Keith Burns, February 18, 1975, 1150.3.1, Willamette Greenway, Box 96, RG 90A–30, RWSR. See *Oregonian*, March 5, 1975.

26. Rebecca Jones to Robert W. Straub, March 6, 1975, 1150.3.1, Willamette Greenway, Box 96, RG 90A–30, RWSR. See *Oregonian*, March 23, 1975; Ned Duhnkrack, "Can Scenic Waterways Grow Up?" *Earthwatch Oregon*, April–May 1979, pp. 8–9; *Earthwatch Oregon*, April 1975: p. 6, ibid, March April 1978, p. 8; *Oregonian*, March 7, 1975; Martin Gates (n.d.), Julie Malcom (n.d., ca. November 3, 1975), and Mike Swan (May 20, 1976) to Robert W. Straub, 1120, Pollution, Administrative Correspondence, 1975–1978, Box 85, RG 90–A, RWSR; memo from Bob Potter, April 2, 1975, PAR 14, Willamette River Parks System, General, 1975, DOTL; Robert M. Johnson and Lynne C. Johnson to Straub, March 28, 1975, and Keith Martin to Straub, March 6, 1975, 1150.3.1, Willamette Greenway, Box 96, RG 90A–30, RWSR.

27. Robert W. Straub to Frank A. Elliott, April 15, 1976, 1150.3.1, Willamette Greenway, Box 96, RG 90A–30, RWSR. See Ken Johnson to Libby Loftin, May 4, 1977, and Straub to

Mrs. Jack Pyburn, May 12, 1975, in ibid; "Governor Straub's Greenway Legislation (H.B. 3225), March 1975," in PAR 14, Willamette River Parks System, General, 1975, DOTL; *Oregonian*, March 5, 8, 23, April 6, May 7, 1975; *Earthwatch Oregon*, April 1975.

28. Bodine, "Roller Coaster Politics"; *Oregonian*, April 6, May 13, 1975; *Earthwatch Oregon*, June–July 1977, p. 6; Carleton Whitehead to Robert W. Straub, June 4, 1975, and Frank A. Elliott to Straub, April 5, 1976, 150.3.1, Willamette Greenway, Box 96, RG 90A–30, RWSR.

29. First quote: Bodine, "Roller Coaster Politics," p. 1; *Oregonian* May 7, 1975; Johnson in Ken Johnson to Robert W. Straub, April 4, 1975, 1150.3.1, Willamette Greenway, Box 96, RWSR. See Johnson to Libby Loftin, May 4, 1977, and Straub to Frank A. Elliott, April 15, 1976, in ibid.

30. Mrs. Melvin F. Wood to Robert W. Straub, April 8, 1975, 1150.3.1, Willamette Greenway, Box 96, RG 90A–30, RWSR. See Harold E. Wilde to Straub, n.d., ca. March 20, 1975, Mrs. Laird Kaup to Straub, June 3, 1975, and Bruce M. Hall to Straub, October 14, 1975, in ibid.

31. Neva Alford to Straub, April 28, 1975, 1150.3.1, Willamette Greenway, Box 96, RG 90A–30, RWSR; second quote: Mable Hupp to Straub, May 14, 1975, in ibid. See Irene M. Fraser to Straub, June 11, 1975, in ibid.

32. Rodney Hammagren to Robert W. Straub, May 5, 1975, 1150.3.1, Willamette Greenway, Box 96, RG 90A–30, RWSR.

33. Robert E. Farrell to Straub, April 28, 1975, 1150.3.1, Willamette Greenway, Box 96, RG 90A–30, RWSR. See Straub to Farrell, May 12, 1975, in ibid.

34. First quote: Thomas H. Reink to Robert W. Straub, October 11, 1975, 1150.3.1, Willamette Greenway, Box 96, RG 90A–30, RWSR; second quote: Straub to Reink, October 15, 1975, in ibid. See Arthur C. Johnson to Straub, April 1, 1976, in ibid; Frank Hammel to Straub, n.d., Straub to John W. Nielson, April 9, 1976, and Straub to Lily Kaup, April 23, 1975, in ibid.

35. Lily Kaup to Robert W. Straub, April 19, 1975, 1150.3.1, Willamette Greenway, Box 96, RG 90A–30, RWSR. See Rodney Hammagren to Straub, May 5, 1975, Mr. and Mrs. W.R. Robnett to Straub, April 28, 1975, Patricia Croman to Straub, April 17, 1975, in ibid.

36. Mr. and Mrs. H. Tipton to Straub, May 1, 1975, 1150.3.1, Willamette Greenway, Box 96, RG 90A–30, RWSR. See Bruce M. Hall to Straub, October 14, 1975, and Mrs. Jack Pyburn to Straub, May 12, 1975, in ibid.

37. Robert W. Straub to William G. Pearcy, April 22, 1975, 1150.3.1, Willamette Greenway, Box 96, RG 90A–30, RWSR; Mike Frome, "Questioning the Future of the Woods in a Rapidly Changing Land," *American Forests* 73 (May 1967): 48.

38. William G. Pearcy to Robert W. Straub, March 6, 1975, 1150.3.1, Willamette Greenway, Box 96, RG 90A–30, RWSR.

39. Erma Jean Wood to Robert W. Straub, April 9, 1975, 1150.3.1, Willamette Greenway, Box 96, RG 90A–30, RWSR. See Charles S. Collins to Straub November 21, 1975, in ibid.

40. Roger E. Martin to Robert W. Straub, April 10, 1975, 1150.3.1, Willamette Greenway, Box 96, RG 90A–30, RWSR.

41. First quote: Archie Weinstein to Department of Transportation, July 21, 1977, PAR 14, Willamette River Greenway, 1978–1989, DOTL; second quote: Robert W. Straub to William G. Pearcy, April 22, 1975, 1150.3.1, Willamette Greenway, Box 96, RG 90A–30, RWSR. See Straub to Mable Hupp, May 21, 1975, in ibid; Straub to Frank A. Elliott, April 15, 1976, in ibid; Honey, *Willamette River Greenway*, pp. 59–61.

42. First quote: Bruce M. Hall to Robert W. Straub, October 23, 1975, 1150.3.1, Willamette Greenway, Box 96, RG 90A–30, RWSR; second quote: Royston, et al., *Willamette River Greenway*, p. 5.

43. Honey, *Willamette River Greenway*, p. 116. See Stan Bunn in *Oregonian*, May 7, 1975; *Earthwatch Oregon*, June–July, 1977, p. 6; Willamette River Greenway Committee meeting minutes July 7, 1978, PAR 14, Willamette River Greenway, 1978–1989, DOTL.

44. David G. Talbot to Pat Amedeo, February 28, 1979, PAR 14, Willamette River Greenway, 1978–1989, DOTL. See Talbot, "Memo to State Parks and Recreation Advisory Committee," October 30, 1978, and Forrest Cooper, "Marshall Island," August 15, 1969, in PAR 14–2, Willamette River Park System, K–M, in ibid.

45. Ernest Drapela to Willamette River Greenway Commission, September 8, 1976, PAR 14, Willamette River Greenway, 1976, DOTL. See *Earthwatch Oregon*, June–July 1977, p. 2.

46. *Bangor Daily News*, September 9–10, 1995; *Waterville Sentinel*, February 27, April 6, 1970, April 9, 1971; Aimé Gauvin, "The River the Loggers Stole," *Audubon* 73 (July 1971): 76.

47. Gauvin, "River the Loggers Stole," p. 76. See *Bangor Daily News*, September 9–10, 1995; *Waterville Sentinel*, January 20, April 9, 1971, November 10, 1979; *Maine Sunday Telegram*, June 13, 1971, August 19, 1979; Neal R. Peirce, *The New England States: People, Politics, and Power in the Six New England States* (New York: W.W. Norton, 1976), pp. 408–09; Richard Saltonstall, Jr., *Maine Pilgrimage: The Search for an American Way of Life* (Boston: Little, Brown, 1974), pp. 149, 293.

48. *Bangor Daily News*, December 16, 1970; Gauvin, "River the Loggers Stole," pp. 70–78; *Waterville Sentinel*, March 17, April 9, 1971.

49. *Kennebec Journal*, July 14, 1976; *Maine Times*, December 13, 1974; *Maine Sunday Telegram*, October 10, 1971. For the Saco River, see *Maine Times*, July 18, 1969.

50. *Maine Sunday Telegram*, March 8, 1970. See *Kennebec Journal*, December 4, 1969; Maine State Park and Recreation Commission, *Saco River Corridor: Open Space and Recreation Potential: A Proposal for Joint Action* (Augusta, 1969), pp. 7, 10; *Maine Sunday Telegram*, March 8, 1970.

51. "Saco River Corridor: Newsletter of the Saco River Corridor Commission," June 1979, p. 3. See *Waterville Sentinel*, November 2, 1973; *Portland Press Herald*, February 19, 1973, September 2, 1977; *Maine Sunday Telegram*, March 8, July 19, 1970, July 8, 1973; *Kennebec Journal*, December 4, 1969, August 22, 1979; *Maine Times*, December 5, 1969; Park and Recreation Commission, *Saco River Corridor*, pp. 7, 16, 38.

52. Clinton B. Townsend [Kennebec River Corridor Committee], "Kennebec River Plan," July 16, 1971, Donaldson Koons papers, private possession of Koons; *Waterville Sentinel*, November 8, 1971, November 11, 1974; Gauvin, "River the Loggers Stole," p. 76; *Maine Sunday Telegram*, October 10, 1971; Appendix 5, "Management," in E. Lyle Flynn, Jr., Gerald Bernstein, and Donald Meagher (North Kennebec Regional Planning Commission), *Kennebec River Corridor Project* (Augusta, 1977), pp. 1, 7; *Waterville Sentinel*, November 8, 1971, December 12, 1974.

53. Flynn, Bernstein, and Meagher, "Management," pp. 9–11.

54. *Waterville Sentinel*, November 8, 1971; November 12, 1974. See ibid, November 16, December 12, 1974; *Portland Press Herald*, September 16, 1982; *Maine Sunday Telegram*, December 5, 1982; Townsend, "Kennebec River Plan," pp. 8–9.

55. Townsend, "Kennebec River Plan," p. 9. See *Maine Sunday Telegram*, October 10, 1971; *Waterville Sentinel*, November 16, 1974; Flynn, Bernstein, and Meagher, "Management," p. 2.

290 NATURAL STATES

56. William Gilbert in *Maine Sunday Telegram*, October 10, 1971; *Kennebec Journal*, July 1, 1985; Flynn, Bernstein, and Meagher, "Management," pp. 11–13.

57. *Kennebec Journal*, July 1, 1985; *Waterville Sentinel*, November 28, 1974. See *Kennebec Journal*, October 17, 1975; *Maine Sunday Telegram*, July 22, 1979; *Waterville Sentinel*, November 11, 28, 1974; Townsend, "Kennebec River Plan," p. 4.

58. Flynn, Bernstein, and Meagher, "Management," pp. 3–6.

59. First quote: *Portland Press Herald*, March 27, 1981; second quote: Flynn, Bernstein, and Meagher, "Management, " p. 1.

60. *Kennebec Journal*, July 14, 1976; *Waterville Sentinel*, November 11, 1974.

61. Eben Ellwell in *Kennebec Journal*, July 1, 1985. See *Portland Press Herald*, October 22, 1991.

62. *Oregon Environmental Council Newsletter*, October 1969, March 1970; *Earthwatch Oregon*, March 1977, March–April, 1978; Duhnkrack, "Can Scenic Waterways Grow Up?" pp. 8–9; *Oregonian*, June 24, 28, 1970, November 3, 1974; *Capital Journal*, February 11, 1970; J. Herbert Stone to Kessler R. Cannon, June 3, 1968, Stone to Scott Warren, June 17, 1968, Correspondence, June–July, 1968, Governor's Committee for a Livable Oregon, Ms. 2537, OHS; Scott Warren to Donald E. Jimerson, and Herbert Stone to Warren, May 14, 1968, Correspondence, April–May 1968, in ibid.

63. Lee Johnson to Robert W. Straub, December 26, 1974, 907.1, Rogue River, Administrative Correspondence, 1975–1978, Box 85, RG–90–A, RWSR. See Mrs. Sandra Hays to Straub, October 14, 1977, 1130.1 Scenic Waterways, in ibid, Box 95; John Daniel Callaghan to Glenn L. Jackson, April 23, 1974; Alan Dickman to Governor Tom McCall, July 14, 1974, Water Resources: Wild and Scenic Rivers, 1, 75A–12, Box 55, RG G4, TMR; Mrs. Fred B. Jennings to Glenn Jackson, April 25, 1974; Judy Nolte to McCall, April 25, 1974, File 2, in ibid; Forrest Cooper to Fred J. Overly, November 6, 1968 PAR 12, Scenic Waterways: Rogue, 1968–1971, no.1, DOTL.

64. First quote: Jack Steiwer in Transcripts of Scenic Waterway Hearings ... Fossil, March 30, pp. 25–26, in Rivers: Oregon Scenic Waterways, Testimony, Statements, 1970 and 1972, Box 26, Ms. 2386, Oregon Environmental Council Papers (hereafter OECP), OHS; second quote: Robert W. Straub to Maurice H. Lundy, June 21, 1976, 1130.1, Scenic Waterways, Administrative Correspondence, 1975–1978, Box 95, RG 90A–30, RWSR. See Sam Ball and Eleanor C. Ball to Glenn Jackson, April 19, 1974, Wild and Scenic Rivers no. 2, 75A–12, Box 55, RG G4, TMR; Mary Tankersley in Transcripts of Scenic Waterway Hearings ... Fossil, March 30, p. 32, in Rivers: Oregon Scenic Waterways, Testimony, Statements, 1970 and 1972, Box 26, Ms. 2386, OECP.

65. Glenn Jackson in *Oregonian*, March 7, 1973. See ibid, February 25, March 16, 1973.

66. Betty J. Rust to Tom McCall, May 6, 1974; Fred E. Parker to McCall, May 2, 1974, Water Resources: Wild and Scenic Rivers, no. 1, Box 55, 75A–12, RG G4, TMR.

67. Jim Ellett, Loretta Ellett, Vance Ellett, Vickie Ellett, April 19, 1972 (submitted statements), Transcripts of Scenic Waterway Hearings ... p. 92, in Rivers: Oregon Scenic Waterways, Testimony, Statements, 1970 and 1972, Box 26, Ms. 2386, OECP.

68. Glen and Rhoda Love to Tom McCall, June 14, 1974, Water Resources: Wild and Scenic Rivers, no. 1, Box 55, 75A–12, RG G4, TMR. See W. Stanley Knouse to McCall April 29, 1974 (with clipping), in ibid; John P. Tribe to McCall, April 23, 1974, in ibid, File 2; Joan M. Knudtson to Straub, March 14, March 28, 1975; Knudtson to Richard Bonebrake, March 14, 1975, 1130.1, Scenic Waterways, Administrative Correspondence, 1975–1978, Box 95, RG 90A–30, RWSR; Patricia Harris to Floyd Query, March 9, 1972, in Rivers: Oregon Scenic Waterways, Correspondence, 1970–1973, Box 26, Ms. 2386, OECP.

69. First quote: Russell F. Hill, March 24, 1972 (submitted statements), p. 62, Transcripts of Scenic Waterway Hearings ... in Rivers: Oregon Scenic Waterways, Testimony, Statements, 1970 and 1972, Box 26, Ms. 2386, OECP; second quote: Mr. & Mrs. John H. Collins, April 21, 1972, in ibid. See Mary Tankersley, ibid, p. 31; H.C. Wright, ibid, p. 27; Arthur L. McGreer (submitted statements), ibid; Russell F. Hill (submitted statements), ibid; Kelly McGreer (submitted statements), ibid; Mrs.Virgil (Mary) Misener (submitted statements), ibid; Sam Ball and Eleanor C. Ball to Glenn Jackson, April 19, 1974, Mrs. Vern Shrum to Jackson, April 15, 1974, Mrs. Ned Bardeen to Tom McCall, February 14, 1974, Wild and Scenic Rivers, File 2, 75A–12, Box 55, RG G4, TMR; Mrs. Corinne L. Rust to Robert W. Straub, April 12, 1975, 1130.1, Scenic Waterways, Administrative Correspondence, 1975–1978, Box 95, RG–90–A, RWSR; Dorotheen Wilson to Straub, March 8, 1975, 907, Waterways, Box 85, ibid.

70. First quote: Phil Marincic, Jr., in U.S. Congress, *Wild Rivers System: St. Croix Waterway— Hearings before the Committee on Interior and Insular Affairs*, 89th Cong., 1st Sess. (Washington, DC: Government Printing Office, 1965), p. 458; second quote: Mrs. Ward McMann in ibid, p. 450; third quote: James A. Greenwood, Jr. in ibid, p. 81; fourth quote: Mrs. John D. Story in ibid, p. 461. See Milton David and Ethel David in U.S. Congress, *Wild Rivers System*, p. 481.

71. Max Ekenberg, April 18, 1972 (submitted statements), Transcripts of Scenic Waterway Hearings ... in Rivers: Oregon Scenic Waterways, Testimony, Statements, 1970 and 1972, Box 26, Ms. 2386, OECP. See David Richey, April 7, 1972, in ibid.

72. Lee Johnson to Straub, December 26, 1974, 907.1, Rogue River, Administrative Correspondence, 1975–1978, Box 85, RG–90–A, RWSR; John R. Pearson to Tom McCall, June 17, 1974, Water Resources, Wild and Scenic Rivers, 1, 75A–12, Box 55, RG G4, TMR. See J.H. Terry to McCall, June 1, 1974, in ibid; Jack Steiwer (p. 25) and Mary Tankersley (p. 32) in Transcripts of Scenic Waterway Hearings ... Fossil, March 30, in Rivers: Oregon Scenic Waterways, Testimony, Statements, 1970 and 1972, Box 26, Ms. 2386, Oregon Environmental Council, OHS; Mrs. Virgil (Mary) Misener (submitted statements), April 21, 1972; John Murtha (p. 49); James Z. Snow (p. 50) in ibid, LaGrande, April 12, 1972; *Oregonian*, November 3, 1974; Duhnkrack, "Can Scenic Waterways Grow Up," pp. 8–9.

73. John Garren, "Protecting the Rogue: For Whom?" *Oregonian Northwest*, April 3, 1977, p. 31.

74. Verne Huser, "Floating Western Rivers," *High Country News*, June 11, 1971; Verne Huser, "Floating: Fastest Growing Outdoor Recreation," *American Forests* 79 (June 1973): 27–29; Verne Huser, "Floating Western Rivers" (part 2), *High Country News*, September 3, 1971; *High Country News*, May 21, 1976.

75. First quote: A.E.T. Rogers, "The Boat Trip up the Rogue River Rapids," *Better Homes and Gardens* 33 (June 1955): 163; second quote: W.T. Tankersley (submitted comments), May 17, 1972, Minutes of Public Hearings Held in Regard to the Rogue River ..., PAR 12, Scenic Waterways: Rogue, 1968–1971, File 2, DOTL. See Roger Keith Davis, "The Lower Rogue River Recreation Resources: Development and Potential," master's thesis, Oregon State University, 1966, p. 39; Lee Johnson to Robert W. Straub, December 26, 1974, 907.1, Rogue River, Administrative Correspondence, 1975–1978, Box 85, RG–90–A, RWSR; Don Holm, "The Rogue: A River for All Kinds of Boaters," *Oregonian Northwest*, April 17, 1977, p. 4; *Oregonian* October 6, 1974; Oregon State Game Commission, *Fish and Wildlife Plan, Rogue River: Oregon Scenic Waterways and National Wild and Scenic River Section* (Salem, 1971).

76. *Oregonian*, August 3, 1975.

77. Edward P. Morgan, *The 60s Experience: Hard Lessons about Modern America* (Philadelphia: Temple University Press, 1991), pp. 169, 171.

78. First quote: Wessel Smitter, "We Ride the Fighting Rogue," *Saturday Evening Post* 221 (October 23, 1948): 36–37, 105–06, 108, 110; second quote: Elton Welke, "Wild Rivers," *Better Homes and Gardens* 49 (March 1971): 119. See Ben Goldrath, "Sliding up a Damp Chute," *Field and Stream* 68 (July 1963): 46–47.

79. Wayne Thompson, "Rogue River Raft Trip Offers 22 Miles of Unpeopled Bliss," *Oregonian Northwest*, August 3, 1975, p. 11. See Cathy Howard, "Rogue River Trail Through Time," *Oregonian Northwest*, July 20, 1975, p. 28.

80. Mrs. J.R. Stauffer to *Oregonian*, October 23, 1974; Howard, "Rogue River Trail," p. 28.

81. First quote: Robert E. Pfister and Robert E. Frenkel, *Rogue River Study: Report 1—Field Investigations of River Use within the Wild River Area of the Rogue River, Oregon* (Corvallis: Oregon State University, 1976), pp. 1, 4–5; second quote: Davis, "Lower Rogue River Recreation Resources," p. 13. See *Earthwatch Oregon*, August–September, 1976; Davis, "Lower Rogue River Recreation Resources," p. 13.

82. Pfister and Frenkel, *Rogue River Study: Report 1*, p. 7; Pfister and Frenkel, *Rogue River Study: Report 2—The Concept of Carrying Capacity: Its Application for Management of Oregon's Scenic Waterway System* (Salem and Corvallis: Oregon State Marine Board and Oregon State University, 1975), p. 1; *Oregonian* October 6, 1974; John Shanklin, *Multiple Use of Land and Water Areas: Report to the Outdoor Recreation Resources Review Commission*, ORRRC Study Report No. 17 (Washington, DC, 1962), p. 15.

83. First quote: Albert and Irene Clay to Robert W. Straub, n.d., ca. January–February 1975, 907.1, Rogue River, Administrative Correspondence, 1975–1978, Box 85, RG–90–A, RWSR; second quote: Bill Pruitt, April 26, 1972, Minutes of Public Hearings Held in Regard to the Rogue River ... in PAR 12, 1968–1971, Scenic Waterways: Rogue, File 2, DOTL. See Mel Norrick in ibid, p. 3; Bill Norfleet in ibid, p. 3.

84. Clayton W. Dumont, Jr., "The Demise of Community and Ecology in the Pacific Northwest: Historical Roots of the Ancient Forest Conflict," *Sociological Perspectives* 39, 2 (1996): 282–84.

85. Cornelia Butler Flora and Jan L. Flora, "Creating Social Capital," *Rooted in the Land: Essays on Community and Place*, edited by William Vitek and West Jackson (New Haven, CT: Yale University Press, 1996), pp. 217–19; Dumont, "The Demise of Community and Ecology," p. 284.

86. Albert and Irene Clay to Straub, n.d., ca. January–February 1975, 907.1, Rogue River, Administrative Correspondence, 1975–1978, Box 85, RG–90–A, RWSR. See W.T. Tankersley (submitted comments), May 17, 1972, "Minutes of Public Hearings Held in Regard to the Rogue River ..." PAR 12, Scenic Waterways: Rogue, 1968–1971, File 2, DOTL.

87. U.S. Congress, *Wild Rivers System*, p. 335. See Bob Potter, memo to David G. Talbot, April 28, 1971, and Chas. A. Connaughton, memo, September 30, 1969, PAR 12, 1968–1971, Scenic Waterways: Rogue, File 1, DOTL; Harrison Loesch to Kessler R. Cannon, n.d, ca. February 6, 1970, Box 27, Ms. 2386, OECP; Larry Williams to Kessler R. Cannon (with clipping), March 24, 1970, Rivers: Rogue, Correspondence, 1970, Box 27, Ms. 2386, OECP; *Capital Journal*, January 8, 1970.

88. Kenny King to Robert W. Straub, November 8, 1976, 907.1, Rogue River, Administrative Correspondence, 1975–1978, Box 85, RG–90–A, RWSR. See Garren, "Protecting the Rogue," p. 31.

89. First quote: Garren, "Protecting the Rogue," p. 31; second quote: *Oregonian*, October 6, 1974.

90. Kenneth Mak in *Oregonian* August 31, 1974. See *Earthwatch Oregon*, August–September, 1976; *Oregonian*, August 25, 28, October 6, December 31, 1974; Thompson, "Rogue River Raft Trip," p. 11; Garren, "Protecting the Rogue," p. 31.

91. Pfister and Frenkel, *Rogue River Study: Report 1* pp. 40–41; *Oregonian*, December 31, 1974; Howard, "Rogue River Trail," p. 28; *Earthwatch Oregon*, August–September, 1976; *Oregonian*, August 2, 1975; Francis B. Russell (submitted comments), April 26, 1972, Minutes of Public Hearings Held in Regard to the Rogue River ... in PAR 12, 1968–1971, File 2, Scenic Waterways: Rogue, DOTL; John Garren to Robert W. Straub, January 20, 1977, 907.1, Rogue River, Administrative Correspondence, 1975–1978, box 85, RG 90-A, RWSR.

92. *Daily Courier* (Grants Pass), April 26, 1976; Robert W. Straub to Kenny King, November 24, 1976, 907.1, Rogue River, Administrative Correspondence, 1975–1978, box 85, RWSR. See *Oregonian*, August 25, October 6, 1974.

93. First quote: Maria Jackson to Robert W. Straub, November 23, 1976, 1130.1, Scenic Waterways, Administrative Correspondence, 1975–1978, Box 95, RG 90A–30, RWSR; second quote: W.T. Tankersley, Minutes of Public Hearings Held in Regard to the Rogue River ... April 26, 1972, p. 7, in PAR 12, 1968–1971, File 2, Scenic Waterways: Rogue, DOTL; third quote: Harold Ross, ibid, p. 5. See *Oregonian* August 25, 1974.

94. R. F. Rittenhouse to George Baldwin, January 2, 1973, PAR 12, 1968–1971, File 2, Scenic Waterways: Rogue, DOTL; Douglas P. Cushing (submitted statements), May 18, 1972, Minutes of Public Hearings Held in Regard to the Rogue River ... April 26, 1972, in PAR 12, 1968–1971, File 2, Scenic Waterways: Rogue, DOTL; Leo Grandmontagne (April 25, 1972), p. 5, Bill Cronenwitt (April 25), p. 5, W.T. Tankersley (submitted statements, May 17), in ibid. See Glen Wooldridge (submitted comments, May 18, 1972); Larry Williams to Supervisor, Siskiyou National Forest, August 19, 1970, Rivers: Rogue, Correspondence, 1970, Box 27, Ms. 2386, OECP.

95. George N. Baldwin to Robert F. Rittenhouse, June 1, 1972, PAR 12, 1968–1971, 2, Scenic Waterways: Rogue, DOTL; Thompson, "Rogue River Raft Trip," p. 11; Tommy L. Biggs to Robert W. Straub, December 28, 1975, and Straub to Biggs, January 7, 1976, 1130.1, Scenic Waterways, Administrative Correspondence, 1975–1978, Box 95, RG 90A–30, RWSR.

96. First quote: Ken Mak in *Oregonian*, November 2, 1974; second quote: *Oregonian*, December 20 1974. See ibid, August 31, October 6, February 22, 1975, October 15, 1977; *Daily Courier* (Grants Pass), April 26, 1976; *Earthwatch Oregon*, August–September, 1976; Pfister and Frenkel, *Rogue River Study: Report 1*, pp. 2, 22–23.

97. Pfister and Frenkel, *Rogue River Study: Report 1*, pp. 2, 22–23; Sid Pyle and Sons Guide Service to Robert W. Straub, February 10, 1975, 907.1, Rogue River, Administrative Correspondence, 1975–1978, Box 85, RG–90-A, RWSR; *Oregonian*, December 20, 1974; Gladwin Hill, "Oregon's Rogue: Lonely Example of Scenic Rivers Progress," *Oregonian Northwest*, September 4, 1975, p. 23; *Oregonian*, February 22, 1975.

98. Thompson, "Rogue River Raft Trip," p. 11. See Robert W. Straub to John Garren, January 28, 1977; Scott H. Pressman and Beverly K. Pressman to Straub, November 26, 1976, 907.1, Rogue River, Administrative Correspondence, 1975–1978, Box 85, RG–90-A, RWSR.

99. First quote: Robert K. Potter to B.A. Hanten, March 29, 1976, PAR 12, 1968–1971, File 4, Scenic Waterways: Rogue, DOTL; second quote: Irv Urie, Minutes of Public Hearings Held in Regard to the Rogue River ... April 26, 1972, pp. 5–6, in PAR 12, 1968–1971, File 2, Scenic Waterways: Rogue, in ibid; third quote: Thompson, "Rogue River Raft Trip," p. 11. See Paul Brown to Robert W. Straub, December 17, 1977; Loren Kramer to

Brown, December 14, 1977, 1130.1 Scenic Waterways, Administrative Correspondence, 1975–1978, Box 95, RG 90A–30, RWSR.

100. *Earthwatch Oregon*, November 1976.

101. Thompson, "Rogue River Raft Trip," p. 11.

102. Maine Legislature, *Legislative Record* (Augusta, 1966), pp. 327–28; *New York Times*, April 12, 1970; *Waterville Sentinel*, December 29, 1973; *Maine Sunday Telegram*, June 30, 1974. See David Vail, "Engines in the Wilderness: Governing Motorized Recreation in Maine's North Woods," manuscript copy, U.S. Society for Ecological Economics July 2001 conference paper, courtesy of the author; "Negotiations re: Allagash Plantation" (ca. 1968), Jerome Matus, Taxation, Allagash, MSA; Donaldson Koons in Aimé Gauvin, "Rumble Rattle Clank Roar: You're on the Allagash Nonwilderness Waterway," *Audubon* 74 (July 1972): 46, 48.

103. First quote: *Kennebec Journal*, May 19, 1977; second quote: Leigh Hoar in Francois Leydet, "Autumn Flames Along the Allagash," *National Georgraphic* 145 (February 1974): 183; third quote: Sam Jalbert in *Portland Press Herald*, February 10, 1975. See Director Herb Hartman in John W. Forssen, "More Protection for the Allagash," *Maine Fish and Wildlife* (Fall 1985); Gauvin, "Rumble Rattle Clank Clatter Roar," p. 51; Saltonstall, *Maine Pilgrimage*, p. 162.

104. Leonard Pelletier in Leydet, "Autumn Flames Along the Allagash," p 181. See *Bangor Daily News*, August 24–25, 1991.

105. Herb Hartman in John W. Forssen, "More Protection for the Allagash," *Maine Fish and Wildlife*, Fall 1985; Vail, "Engines in the Wilderness," p. 12; *Allagash Wilderness Waterway: Timber Harvesting and Sensitive Areas: Study of the Joint Standing Committee on Energy and Natural Resources* (Augusta, 1984): pp. i, ii; *Bangor Daily News*, July 1, 1983, January 20, 1984, November 2–3, 1996; *Portland Press Herald*, March 29, 1973; *Waterville Sentinel*, December 29, 1973; *Maine Times*, September 13, 1974, June 30, 1978.

106. Lew Dietz in *Maine Sunday Telegram*, April 28, 1974.

107. *Maine Times*, September 13, 1974, August 8, 1975; *Bangor Daily News*, May 24, 1977; *Maine Sunday Telegram*, July 7, 1974; *Portland Press Herald*, December 17, 1976.

108. First quote: *Maine Sunday Telegram*, December 3, 1978; second quote: *Portland Press Herald*, December 17, 1976; third quote: Elise Hawtin in *Maine Times*, August 4, 1978. See *Portland Press Herald*, December 17, 1976; *Waterville Sentinel*, November 21, 24, 1978; *Lewiston Sun*, September 10, 1974; *Maine Times*, September 13, 1974, August 8, 1975; Tom Cielinski in *Lewiston Sun*, September 10, 1974.

109. Burton Packwood in *Maine Times*, March 25, 1977. See *Maine Times*, June 30, 1978.

110. Burton Packwood in *Maine Times*, March 25, 1977. See Susan Vickery in *Maine Times*, October 3, 1975; *Maine Times*, June 30, 1978; *Maine Sunday Telegram*, December 6, 1973, August 3, 1975, November 26, 1989; *Bangor Daily News*, November 18–19, 1978, February 7–8, 1981, March 15, 1985; *Portland Press Herald*, June 18, 1981.

111. Joseph E. Brennan in *Kennebec Journal*, June 18, 1984; Richard Barringer in *Waterville Sentinel*, April 7, 1983. See *Waterville Sentinel*, January 17, 1982; *Kennebec Journal*, September 2, 1982, March 28, 1986; *Bangor Daily News*, February 3, March 31, December 31, 1982, May 10, June 18, 1983, July 12, 1984; *Maine Sunday Telegram*, April 25, December 26, 1982, May 29, 1983; *Portland Press Herald*, January 20, March 12, 1983, October 22, 1991.

112. Schwarz, Flink, and Searns, *Greenways*, pp. 3–5, 25, 123, 161–65 (quote at p. 165).

113. Thompson, "Rogue River Raft Trip," p. 11; Elton Welke, "Wild Rivers," *Better Homes and Gardens* 48 (March 1971): 119–24; Holm, "Rogue," p. 4.

CHAPTER 6: BRIDGE TO ECOTOPIA

1. Maine State Planning Office, *Maine Coastal Resources Renewal* (Augusta, 1971), preface, p. 141.

2. Press release, June 1, 1972, Willamette Valley Environmental Protection and Development Council, Governor Tom McCall's Records (hereafter TMR), Box 3, RG–G4, Accession 79A–14, Oregon State Archives (hereafter OSA); McCall, "Preface," Lawrence Halprin & Associates, *The Willamette Valley: Choices for the Future* (Salem, 1972); ibid, pp. 5–7, 8–9, 12–13, 21, 30–33, 38, 40, 41, 57, 62–69, 73, 78–79, 86, 92.

3. Richard Saltonstall, Jr., *Maine Pilgrimage: The Search for an American Way of Life* (Boston: Little, Brown, 1974), pp. 87–89, 93, 95, 148, 212, 247–51. See Saltonstall in "Maine Chance," *Time* 104 (August 26, 1974): 56; "Whither Maine?" *Down East* 25 (February 1979): 126–55; Commission on Maine's Future, *Final Report* (Augusta: Commission on Maine's Future, 1977), p. 43.

4. Ernest Callenbach, *Ecotopia: The Notebooks and Reports of William Weston* (Berkeley, California: Banyan Tree Books, 1975), pp. 18, 55.

5. Fred Bosselman and David Callies, *The Quiet Revolution in Land Use Control* (Washington, DC: Council on Environmental Quality, 1971), pp. 4–6, 54–56, 63, 67, 290.

6. *High Country News*, August 2, 1974. See John N. Cole in ibid, July 21, 1972; Bosselman and Callies, *Quiet Revolution*, pp. 315–18.

7. Russell Train, "Preface," Bosselman and Callies, *Quiet Revolution*, pp. i, ii, 2, 3. See John Dryzek, *The Politics of the Earth: Environmental Discourses* (New York: Oxford University Press, 1997), p. 63.

8. Bosselman and Callies, *Quiet Revolution*, pp. 320–24.

9. Gaylord Nelson in *High Country News*, August 31, 1973. See ibid, August 31, 1973; Noreen Lyday, *The Law of the Land: Debating National Land Use Legislation, 1970–1975* (Washington, DC: Urban Institute, 1976); Adam Rome, "Toward a Land Ethic: The Quiet Revolution in Land–Use Regulation," paper presented at the American Society for Environmental History conference, March 2000, Tacoma, Washington; U.S. Congress, 91st Cong., 2nd sess., *National Land Use Policy: Hearings on S. 3354 before the Subcommittee on Environment and Land Resources of the Committee on Interior and Insular Affairs*, Part 1 (Washington, DC: Government Printing Office, 1970), p. 393.

10. U.S. Congress, 92nd Cong., 1st sess., *Hearings before the Subcommittee on the Environment of the House Committee on Interior and Insular Affairs* (Washington, DC: Government Printing Office, 1971), p. 97.

11. Bosselman and Callies, *Quiet Revolution*, Appendix: "National Land Use Policy Act of 1971"; Lynton Caldwell, Lynton Hayes and Isabel MacWhirter, *Citizens and the Environment: Case Studies in Popular Action* (Bloomington: Indiana University Press, 1976), p. 10.

12. Jean Anne Pollard, *Polluted Paradise: The Story of the Maine Rape* (Lewiston, Maine: Twin City Printery, 1972). See Richard N.L. Andrews, *Managing the Environment, Managing Ourselves: A History of American Environmental Policy* (New Haven, CT: Yale University Press, 1999), p. 6.

13. Ursula K. LeGuin, *The Lathe of Heaven* (Cambridge, MA: Robert Bentley, 1971), pp. 26, 64–67, 126, 140, 171, 173–74.

14. *Maine Sunday Telegram*, November 26, 1967. See ibid, March 8, 1970; Charles E. Clark, *Maine: A Bicentennial History* (New York: W.W. Norton, 1977), p. 176.

15. Pollard, *Polluted Paradise*, pp. 110–11, 114; Donaldson Koons in *Maine Sunday Telegram*, January 17, 1971. See *Waterville Sentinel*, November 15, 1968; Richard Pardo, "Something Up, Down East," *American Forests* 75 (May 1969): 44.

16. Robert W. Patterson, "Introduction," *A Symposium: The Maine Coast, Prospects and Perspectives* (Brunswick, ME: Bowdoin College, 1967), n.p. See Center for Resource Studies, *The Maine Coast: Time of Decision* (Brunswick, ME: Bowdoin College, 1967), n.p.

17. Joseph Fisher, "Toward a Maine Coastal Park and Recreation System," *Symposium: The Maine Coast*, pp. 90–91.

18. Charles W. Eliot, "As Maine Goes ... Which Way?" *Symposium: The Maine Coast*, pp. 1, 6.

19. Eliot, "As Maine Goes," p. 6; Dennis O'Harrow, "Ecological Considerations: Wildlife," ibid, pp. 36–37.

20. Barnett Shur, "Practical Problems in Local and Regional Zoning," *Symposium: The Maine Coast*, pp. 83, 84.

21. Joseph Fisher, "Toward a Maine Coastal Park and Recreation System," *Symposium: The Maine Coast*, p. 91.

22. Orlando Delogu, "Wider and More Effective Use of the Zoning Tool," *Symposium: The Maine Coast*, p. 79, 81.

23. *Maine Sunday Telegram*, August 16, 1970; Holly Dominie, "Maine's Changing Landscape," in *Changing Maine*, edited by Richard Barringer (Portland: University of Southern Maine, 1990), p. 98.

24. Frank Graham, Jr., "They're Carving Up Old New England," *Audubon* 74 (September 1972): 73–93.

25. Maine Legislature, *Legislative Record* (Augusta, 1967), pp. 3373, 3275. See Paul B. Frederic, "Public Policy and Land Development: The Maine Land Use Regulation Commission," *Land Use Policy* (January 1991): 58.

26. Maine Legislature, *Legislative Record* (Augusta, 1967), p. 3277.

27. Maine Legislature, *Legislative Record* (Augusta, 1967), p. 3282. See ibid, p. 3372–78, 3282–83; *Maine Sunday Telegram*, November 24, 1968; Esther Lacognata, *A Legislative History and Analysis of the Land Use Regulation Law in Maine* (Augusta: Land Use Regulation Commission, 1974), pp. 1–7, 19, 22, 24, 67–68.

28. Bob Cummings in *Portland Press Herald*, June 2, 1971. See Maine Legislature, *Legislative Record* (Augusta, 1971), pp. 3777–79, 3785, 3795; Lacognata, *Legislative History*, p. 25.

29. *Waterville Sentinel*, November 15, 1968; Maine Legislature, *Legislative Record* (Augusta, 1970), p. 731. On the origins of the site location of development law, see Donaldson Koons to Kenneth Curtis, December 12, 1969, EIC correspondence file, Koons papers (hereafter DKP), in possession of Koons. In a personal interview (May 4, 2000), Koons affirmed that he and Curtis aides Peter Bradford and Andrews Nixon drafted the law.

30. Donaldson Koons to Kenneth Curtis, September 21, 1970; Koons to Curtis Hutchins, September 15, 1970, DKP.

31. Maine Legislature, *Legislative Record* (Augusta, 1971), p. 3246. See ibid, pp. 3244, 3250–51; *Maine Public Laws, 1971*, Ch. 535.

32. First quote: Maine Legislature, *Legislative Record* (Augusta, 1971), p. 3778; second quote: *Maine Sunday Telegram*, November 24, 1968; third quote: Clinton Townsend in *Maine Sunday Telegram*, May 30, 1971.

33. Maine Legislature, *Legislative Record* (Augusta, 1971), p. 3781. See ibid, p. 3784.

34. First quote: Maine Legislature, *Legislative Record* (Augusta, 1971), p. 3789; second quote: ibid, pp. 3790–91, 4054. See also ibid, p. 3780.

35. Maine Legislature, *Legislative Record* (Augusta, 1971), pp. 4059–60, 4063.

36. *The Allagash Group* (n.d., ca. 1971), Allagash Group, vertical files, Mantor Library, University of Maine Farmington; *Portland Press Herald*, April 13, 1971; *Portland Press Herald*, September 23, 1971.

37. Richard Barringer, *A Maine Manifest* (Portland: Tower Publishing 1972); Andrew Hamilton, "Maine: Finding the Promised Land (Without Losing the Wilderness)," *Science* 178 (November 10, 1972): 597; John N. Cole, "Letter from Maine: Oil and Water," *Harper's* 243 (November 1971): 48.

38. Philip Savage in *Bangor Daily News*, October 7, 1972.

39. Bosselman and Callies, *Quiet Revolution*, pp. 187–199. See *In re Maine Clean Fuels, Inc., Atlantic Reporter 2d.* (Maine Supreme Court, October 17, 1973), Vol. 310, pp. 736, 742.

40. *Portland Press Herald*, May 28, 1973. See Anne W. Simon, *The Thin Edge: Coast and Man in Crisis* (New York: Harper & Row, 1978), pp. 126, 130–135, and especially the chapter on Eastwell, a fictive Maine peninsular town.

41. Saltonstall, *Maine Pilgrimage*, p. 154; *Maine Times*, October 8, 1971; *Waterville Sentinel*, May 18, 1972; *Bangor Daily News*, July 15, 1972.

42. *Kennebec Journal*, March 6, 1972; *Maine Times*, July 12, 1974. See *Bangor Daily News*, July 20, 1972; *Waterville Sentinel*, July 20, 1972; *Kennebec Journal*, March 6, October 3, 1972.

43. *Bangor Daily News*, December 2, 1972; *Waterville Sentinel*, January 18, 1973; *Kennebec Journal*, January 23, 1973. See *Maine Sunday Telegram*, December 17, 1972; *Bangor Daily News*, November 15, 1972.

44. *Kennebec Journal*, March 6, 1973; *Bangor Daily News*, June 27, 1973; *Waterville Sentinel*, March 23, April 7, July 19, 1973.

45. *Kennebec Journal*, December 27, 1973; *Portland Press Herald*, June 2, 1973.

46. *Portland Press Herald*, April 20, 1974; *Kennebec Journal*, June 26, 1974. See *Kennebec Journal*, December 27, 1973.

47. *Maine Times*, July 12, 1974; *Portland Press Herald, Waterville Sentinel, Portland Press Herald*, July 2, 1974; *Waterville Sentinel*, July 4, November 28 1974, January 1, 1975.

48. *Waterville Sentinel*, March 7, 1975; *Kennebec Journal*, April 18, 1975; Richard W. Judd, Joel Eastman, Edwin Churchill, editors, *Maine: The Pine Tree State from Prehistory to the Present* (Orono: University of Maine Press, 1995), pp. 569–570.

49. *Portland Press Herald*, April 23, 1975.

50. *Portland Press Herald*, May 3, 1976. See *Maine Times*, May 30, 1975; *Waterville Sentinel*, October 2, 1975.

51. Richard Barringer in *Portland Press Herald*, May 3, 1976. See *Maine Times*, May 30, 1975.

52. Alec Giffen in *Maine Sunday Telegram*, June 1, 1975; Kenneth Cianchette in *Maine Times*, May 30, 1975. See *Maine Sunday Telegram*, November 5, 1978; *Bangor Daily News*, July 1, 1975; January 22, 1977; *Kennebec Journal*, May 14, September 17, 1976; November 30, 1977; *Waterville Sentinel*, August 7, 1976.

53. *Waterville Sentinel*, April 16, 1977.

54. Frederic, "Public Policy and Land Development," p. 58. See *Waterville Sentinel*, July 26, 1975.

55. Saltonstall, *Maine Pilgrimage*, p. 151.

56. John Martin in *Portland Press Herald*, September 16, 1982. See *Maine Times*, July 2, 1976.

57. David Vail, "Buying Back the North Woods," in "Property Rights and Sustainable Nature Tourism: Adaptation and Mal-Adaptation in Dalarna (Sweden) and Maine (USA)," edited by Vail and Lars Hultkrantz, *Ecological Economics* 35 (2000): 235–36.

58. "Maine Coast Heritage Trust: A Discussion," DKP; John N. Cole in *High Country News*, July 21, 1972.

59. Donaldson Koons to Arthur Johnson, December 1976; Koons, "Notes for an Oral Presentation on Draft Growth Strategy"; Koons to Richard Barringer, May 17, 1979, DKP.

60. Lois and Cliff Kenegy, August 14, 1995, Willamette Valley Oral History Project (hereafter WVOHP), transcript, Sound Recordings Collection 2641, pp. 29–32, 38–39, Oregon Historical Society (hereafter OHS), Portland.

61. Carl Abbott and Deborah Howe, "The Politics of Land-Use Law in Oregon: Senate Bill 100, Twenty Years After," *Oregon Historical Quarterly* 94 (Spring 1993): 4–35.

62. Hector Macpherson, February 25, 1992, WVOHP, 1121, pp. 39–40.

63. Land Use Subcommittee minutes, October 26, 1967, House Joint Resolution 53, Records of 1967–1969 Interim Committee on Agriculture (bound notebook), September 25, 1967, OSA.

64. Homer Conger in Land Use Subcommittee Minutes, October 26, 1967, House Joint Resolution 53, Records of 1967–1969 Interim Committee on Agriculture (bound notebook), November 27, 1967, OSA.

65. Land Use Subcommittee minutes, October 26, 1967, House Joint Resolution 53, Records of 1967–1969 Interim Committee on Agriculture (bound notebook), March 25, 1968, OSA.

66. Report of Findings and Recommendations, Interim Committee on Agriculture, December 1968, Interim Committee on Agriculture (bound notebook), OSA.

67. Brent Walth, *Fire at Eden's Gate: Tom McCall and the Oregon Story* (Portland: Oregon Historical Society Press, 1994), p. 246; Tom McCall, Arnold Cogan in Senate Agriculture Committee Exhibits on S.B. 10, February 24, 1969, OSA.

68. Keep Oregon Livable, *Editorial Digest*, Spring 1974; Governor's Office, Special Message on Land-Use Planning and Zoning, Senate Agricultural Committee Exhibits on S.B. 10, February 7, 1969, OSA.

69. Organizations, Oregon Shores Conservation Coalition, 1971–72, mimeo newsletter of OSCC (Save Our Shores), June–July 1971, Box 20, File 13, Oregon Environmental Council Papers (hereafter OECP), OHS; Halprin, *Willamette Valley*.

70. Senate Agricultural Committee to Tom McCall, March 12, 1969, Exhibits S.B. 10, on file with House Planning and Development Committee Records, 1969, OSA.

71. Walth, *Fire at Eden's Gate*, pp. 248, 249; Robert Logan to Edward Westerdal, October 28, 1969, Executive Department, Director's Correspondence, 1971–73, Local Government Relations Division, Item 2, Box 3, RG 75A–134; Activity Report, Interim Committee on Agriculture, Minutes, February 24, March 5, April 8, December 11, 1969, House Committee, Exhibits; Thomas Telford to House Planning and Development Committee, Minutes and Exhibits on S.B. 10, OSA.

72. Mrs. Lee (Agnes) James to Sen. Raymond, February 27, 1969, Senate exhibits on S.B.10. See Senate Minutes, February 26, 1969; Richard Tolleson (ZAMO), House Minutes

March 27, 1969; Jasper Grange, Resolution against Compulsory Zoning of Agricultural Lands in Oregon, Exhibits folder (separate), Senate Agricultural Committee, OSA.

73. Paul Ramsay, testimony, April 8, 1969, House Planning and Development Committee, Minutes and Exhibits on S.B. 10, OSA.

74. Mrs. Jim Banks, February 24, 26, 1969, League of Women Voters, Oregon Association of Realtors, Marion County Commissioners, March 27, 1969; Mrs. Alice Elshoff, March 25, 1969; Michael Shannon to Sen. Raymond, February 25, 1969, House Planning and Development Committee, 1969, Minutes and Exhibits on S.B. 10, OSA.

75. Walth, *Fire at Eden's Gate*, p. 249; Planning, Land Use, Ballot Measure 11 Correspondence, 1970, File 27, Box 21, OECP.

76. *Oregon Laws 1971*, Chapter 608 (S.B. 687); OSCC "Save Our Shores" (mimeo newsletter), June–July 1971, File 13, Organizations, Oregon Shores Conservation Coalition, 1971–72, Box 20, OECP; Planning, Land Use, OCCDC Correspondence, 1971–73, File 36, Box 21, OECP.

77. Richard Reynolds to Tom McCall (copy), January 15, 1972, File 36, Box 21, Planning, Land Use, OCCDC Correspondence, 1971–73, OECP.

78. Lois and Cliff Kenegy, August 14, 1995, WVOHP, 2641.

79. Kathleen J. Zachary, "Politics of Land Use: The Lengthy Saga of Senate Bill 100," master's thesis, Portland State University, 1978, pp. 3, 14; Abbott and Howe, "Politics of Land-Use Law," 4–35; Hector Macpherson, February 25, 1992, WVOHP, pp. 27–30, 1121.

80. Richard Cohan, "Key Points, Hector Macpherson's Land Use Planning Bill," September 1, 1972; Cohan, "Discussion Draft : August 9, 1972"; Cohan, "Macpherson Land Use Planning Bill, Rough Draft, LCC100, July 3, 1972"; Cohan, "Land Use Planning," August 10, 1972; Cohan "Land Use Planning," n.d., File 25, Planning, Land Use, Memos, Testimony, 1972–73, Box 21, OECP.

81. Zachary, "Politics of Land Use," p. 28.

82. Press Release, June 1, 1972, Willamette Valley Environmental Protection and Development Council, Box 3, RG–G4, Accession 79A–14, TMR; Halprin, *Willamette Valley*, pp. 57, 62–69, 73, 78–79, 86, 92.

83. Charles E. Little, *The New Oregon Trail* (Washington, DC: Conservation Foundation, 1974), p. 13. See ibid, pp. 11–14; Robert Cassidy, "The Last Oregon Story: Preaching the Good Life," *Planning* (October 1974), pp. 19–20.

84. "A Discussion Forum on Land-Use Planning," Fifth Governor's Conservation Congress, Portland, November 20–22, 1972, File 26, Planning, Land Use, Reports, 1972–74, Box 21, OECP.

85. *Oregonian*, February 4, March 7, 1973. See ibid, March 11, 1973.

86. *Oregonian*, March 13, 1973. See Diarmuid O'Scannlain in *Oregonian*, February 13, 1973; *Oregonian*, February 15, March 9, 1973; *Statesman*, February 16, 1973.

87. *Oregonian*, March 11, 1973.

88. *Oregonian*, February 24, 1973; L.B. Day in *Oregonian*, Februry 24, 1973. See ibid, February 13, 1973; *Statesman*, February 24, March 22, 1973.

89. *Statesman*, March 9, 1973; McCall in ibid, March 7, 1973; L.B. Day in *Oregonian*, February 24, 1973.

90. Carl Abbott, "The Oregon Planning Style," in *Planning the Oregon Way: A Twenty–Year Evaluation*, edited by Carl Abbott, Deborah Howe, and Sy Adler (Corvallis: Oregon State University Press, 1994), pp. 205–226 (quote on p. 209).

91. H. Jeffrey Leonard, *Managing Oregon's Growth: The Politics of Development Planning* (Washington, DC: Conservation Foundation, 1983), p. 5.

92. Oregon Land Conservation and Development Commission (hereafter LCDC), Meeting Notebooks for November 1973–March 15, 1974, Record of Commission Workshop, February 2, 1974, OSA.

93. Arnold Cogan to LCDC, "Report on Community Meetings," April 18, 1974, LCDC, Meeting Notebooks for April 26, 1974–July 26, 1974, Record of Commission Meeting held April 19, 1974, OSA, with clippings from *LaGrande Observer*, April 18, 1974; *Dalles Chronicle*, April 19, 1974; Abbott, "Oregon Planning Style," pp. 210–211.

94. First quote: LCDC Box 1, Notebook April 26, 1974–July 26, 1974; "Draft Agenda, LCDC Public Workshops: Round II," August 30, 1974, LCDC Box 1, Notebook, October 11–24, 1974; second quote: *World* (Coos Bay), October 1, 1974, J. Wesley Sullivan to L.B. Day, November 7, 1974, LCDC Box 2, Notebook, November 21, 1974–December 13, 1974. See Arnold Cogan to LCDC, memo October 11, 1974; Arnold Cogan to LCDC, memo October 10, 1974, "Summary of Round II Community Workshops."

95. Carl Abbott, Deborah Howe, Sy Adler, "Introduction," in *Planning the Oregon Way*, edited by Abbott, Howe, and Adler, p. x; Leonard, *Managing Oregon's Growth*, pp. 20–21.

96. Gerrit Knaap, "Land Use Politics in Oregon," in *Planning the Oregon Way*, edited by Abbott, Howe, and Adler, pp. xv, 7–12.

97. Abbott, Howe, Adler, "Introduction," pp. xv, xxi; Gerrit Knaap, "Land Use Politics in Oregon," in ibid, pp. 6, 7; Leonard, *Managing Oregon's Growth*, pp. 36–39.

CHAPTER 7: A VIEW ACROSS THE GOLF LINKS

1. John Gray, "Land Use: Profits and Environmental Integrity," typescript speech, National Association of Home Builders, Houston, Texas, January 8, 1973, File 26, Planning, Land Use, Reports 1972–74, Box 21, Oregon Environmental Council Papers, Oregon Historical Society, Portland. We would like to thank Brian Donahue for his thoughtful comments on an earlier draft of this chapter.

2. Edward Abbey in *High Country News*, October 22, 1976. See Hal Rothman, *Devil's Bargain: Tourism in the Twentieth-Century American West* (Lawrence: University Press of Kansas, 1998), p. 222.

3. Riley E. Dunlap and Angela G. Mertig, editors, *American Environmentalism: The U.S. Environmental Movement, 1970–1990* (New York: Taylor & Francis, 1992), p. 4; D.B. Luten, "Fading Away?" *Western Outdoors Annual* (Federation of Western Outdoor Clubs; Spring 1973), n.p.

4. Daniel D. Chiras, *Beyond the Fray: Reshaping America's Environmental Response* (Boulder, CO: Johnson Books, 1990), p. 129; Robert C. Mitchell, Angela Mertig, and Riley Dunlap, "Twenty Years of Environmental Organizations," in *American Environmentalism*, edited by Dunlap and Mertig, p. 23; Michael McCloskey, "Twenty Years of Change in the Environmental Movement: An Insider's View," in ibid, p. 81.

5. Richard N.L. Andrews, *Managing the Environment, Managing Ourselves: A History of American Environmental Policy* (New Haven, CT: Yale University Press, 1999), p. 209, 238, 252–55; Riley Dunlap, "Trends in Public Opinion toward Environmental Issues: 1965–1990," in *American Environmentalism*, edited by Dunlap and Mertig, p. 91, 238,

252; Mark Dowie, *Losing Ground: American Environmentalism at the Close of the Twentieth Century* (Cambridge, MA: MIT Press, 1995), p. 59.

6. Andrews, *Managing the Environment, Managing Ourselves*, pp. 237-38; Stephanie S. Pincetl, *Transforming California: A Political History of Land Use and Development* (Baltimore, MD: Johns Hopkins Press, 1999), pp. 186-87, 236; Hal K. Rothman, *The Greening of a Nation? Environmentalism in the United States Since 1945* (Fort Worth, TX: Harcourt Brace, 1998), p. 5; David Burner, *Making Peace with the 60s* (Princeton, NJ: Princeton University Press, 1996), pp. 217-19; Paul Lyons, *New Left, New Right, and the Legacy of the Sixties* (Philadelphia: Temple University Press, 1996), p. 101; William A. Shutkin, *The Land That Could Be: Environmentalism and Democracy in the Twenty-First Century* (Cambridge, MA: MIT Press, 2000), pp. 33-34.

7. Audrey Jackson in "The Oregon Eighties," *Earthwatch Oregon*, January/February 1980. For Maine, see Jay Davis, "So You Want to Move to Maine," *Down East* 24 (August 1977): 57; Commission on Maine's Future, *Final Report* (Augusta, 1977), pp. 3, 38-39; "North by East," *Down East* 21 (August 1974): 53; Lew Dietz, "The Way It Was, the Way It Is," *Down East* 26 (August 1979): 64-66, 108; Dietz, "Maine at the Crossroads," *Down East* 23 (May 1977): 67; *Maine Sunday Telegram*, February 5, 1978, January 4, 1981; Robert Deis, "Environmental Watch," *Down East* 25 (February 1979): 154-55; "Environmental Watch," *Down East* 26 (February 1980): 75, 124; Richard Condon, "The Tides of Change, 1967-1988," in *Maine: The Pine Tree State from Prehistory to the Present*, edited by Richard W. Judd, Edwin A. Churchill, and Joel W. Eastman (Orono: University of Maine Press, 1995), p. 568.

8. Holly Dominie, "Maine's Changing Landscape," in *Changing Maine*, edited by Richard Barringer (Portland: University of Southern Maine, 1990), pp. 99-100; Paul R. Ehrlich, "People Pollution," *Audubon* 72 (May 1970): 5.

9. B.J. Seymour in "Report on Annual Meeting," *Earthwatch Oregon*, January 1978.

10. "Report on 4th Annual Meeting," *Earthwatch Oregon*, January 1974; ibid, January 1975; Judy Roumpf, Dick Benner, Audrey Jackson, Ned Duhnkrack, David Hupp, Randal O'Toole, "The Oregon Eighties," in ibid, January/February 1980.

11. Larry Williams, "How Do I Say Goodbye?" *Earthwatch Oregon*, March 1978; *Earthwatch Oregon*, January, October (Bill Hutchinson), 1979; Judy Roumpf et al. in "The Oregon Eighties."

12. Roumpf et al., "The Oregon Eighties," pp. 8-12, 14, 15. See *Earthwatch Oregon*, January 1979. On free-market solutions, see Pincetl, *Transforming California*, p. 258.

13. David E. Shi, *The Simple Life: Plain Living and High Thinking in American Culture* (New York: Oxford University Press, 1985), p. 253; Charles A. Reich, *The Greening of America* (New York: Random House, 1970), pp. 6-10, 29, 35, 88-89, 90, 213, 273.

14. Reich, *Greening of America*, pp. 152, 162-74, 194, 204.

15. Reich, *Greening of America*, pp. 251, 225, 258.

16. Reich, *Greening of America*, pp. 237, 243.

17. Murray Bookchin, *The Limits of the City* (New York: Harper Colophon, 1974), p. 73; Evan Eisenberg, *The Ecology of Eden* (New York: Alfred A. Knopf, 1998), pp. 158, 426.

18. Burner, *Making Peace with the 60s*, pp. 222; Lyons, *New Left, New Right*, p. 133, 206-07.

19. Bookchin, *Limits of the City*, p. 95.

20. Richard M. Brown, "The New Regionalism in America, 1970-1981," in *Regionalism and the Pacific Northwest*, edited by William G. Robbins et al. (Corvallis: Oregon State University Press, 1983).

21. James T. Farrell, *The Spirit of the Sixties: Making Postwar Radicalism* (New York: Routledge, 1997), pp. 11, 209, 215; Shi, *Simple Life*, p. 249.

22. Joel Garreau, *The Nine Nations of North America* (Boston: Houghton Mifflin, 1981), pp. 272–73.

23. Rothman, *Devil's Bargains*, p. 14.

24. "L.L. Bean's Bonanza," *Down East* 1 (April 1955): 38–39 (from Tom Mahoney, *The Great Merchants*). See Leon A. Gorman, *L.L. Bean, Inc.: Outdoor Specialties by Maine from Maine* (New York: Newcomen Society, 1981), p. 7–8; Michael L. Johnson, "A Vertical Ride: Cowboy Chic and Other Fashions of the New West," in Johnson, *New Westers: The West in Contemporary American Culture* (Lawrence: University Press of Kansas, 1996), pp. 14–54.

25. M.R. Montgomery, *In Search of L.L. Bean* (Boston: Little, Brown, 1984), pp. 32, 38–39, 62.

26. "L.L. Bean's Bonanza," pp. 38–39. See Gorman, *L.L. Bean, Inc.*, p. 10; Richard Saltonstall, Jr., *Maine Pilgrimage: The Search for an American Way of Life* (Boston: Little, Brown, 1974), p. 262.

27. *Maine Times*, May 13, 1977; Gorman, *L.L. Bean, Inc.*, pp. 13–14, 17–8, 20–21.

28. First quote: Montgomery, *In Search of L.L. Bean*, pp. 172, 186–87; second quote: *Maine Times*, May 13, 1977. On western wear and regional identity, see Debra L. Donahue, *The Western Range Revisited: Removing Livestock from Public Lands to Conserve Native Biodiversity* (Norman: University of Oklahoma Press, 1999), p. 97, quoting Ann Fabian.

29. "The Vanishing Redwoods: An Album," *Audubon* 67 (November–December 1965): 359.

30. John M. Findlay, *Magic Lands: Western Cityscapes and American Culture after 1940* (Berkeley: University of California Press, 1992), pp. 73, 78, 97, 266–69.

31. Rose Scott, "The Center That Hahn Built," *Oregon Business*, January 1975, April 1981; *Earthwatch Oregon*, January, 1975. See Rutherford H. Platt, Rowan A. Rowntree, and Pamela C. Muick, editors, *The Ecological City: Preserving and Restoring Urban Biodiversity* (Amherst: University of Massachusetts Press, 1994), p. 30; *Oregonian*, August 11, 1974, May 13, 1975.

32. Saltonstall, *Maine Pilgrimage*, pp. 237, 239.

33. Win McCormack, editor, *Profiles of Oregon 1976–1986* (Portland: New Oregon Publishers, 1986); *Oregon Times*, September 1976.

34. James Keefe in Saltonstall, *Maine Pilgrimage*, p. 262; John Fraser Hart, *The Rural Landscape* (Baltimore, MD: Johns Hopkins Press, 1998), p. 357; *Oregon* 8 (October 1978).

35. Dayton Hyde, *Yamsi* (New York: Dial Press 1971), pp. 311, 312, 318.

36. David Vail, "Engines in the Wilderness: Governing Motorized Recreation in Maine's North Woods," manuscript copy, July 2001, courtesy of the author.

37. Lawrence Buell, *The Environmental Imagination: Thoreau, Nature Writing, and the Formation of American Culture* (Cambridge, MA: Belknap Press, 1995), pp. 15, 21; Robert Fishman, *Bourgeois Utopias: The Rise and Fall of Suburbia* (New York: Basic Books, 1987), pp. 147–48.

38. *High Country News*, August 2, October 25, 1974, March 26, 1976; Rothman, *Devil's Bargain*, pp. 231, 280–81.

39. David Cash, "Deschutes County: Where Sun and Mountains Meet," *Landmark: A Quarterly Journal of 1000 Friends of Oregon* 1 (Spring 1984): 20–22, in State Records, Oregon, Box 1, Samuel P. and Barbara Hays Environmental Archives, University of

Pittsburgh (hereafter SBHC); Raymond R. Hatton, *Bend in Central Oregon* (Portland: Binford & Mort, 1978).

40. Cash, "Deschutes County," p. 20; *Sunriver* magazine (http://members.aol.com/sunrivermg/sr.htm).

41. Harry W. Paige, "Leave if You Can," in *Rooted in the Land; Essays on Community and Place*, edited by William Vitek and West Jackson (New Haven, CT: Yale University Press, 1996), pp. 13–14; Hatton, *Bend*, pp. 122, 124, 134.

42. *Maine Times*, December 11, 1981; November 18, 1971, November 18, 1977; December 11, 1981.

43. *Maine Times*, December 1, July 17, 28, 1972, March 23, 1973. See ibid, October 8, 1971, October 31, 1980, September 24, 1982, January 21, 1983.

44. *Maine Times*, March 23, 1973, November 18, 1977. See ibid, July 17, 1972; January 21, 1983; Findlay, *Magic Lands*, especially p. 73.

45. John N. Cole, "Much of Maine Is at Stake" (editorial), *Maine Times*, October 8, 1971.

46. *High Country News*, January 31, 1975. See ibid, August 2, 1974, March 26, 1976.

47. "Oregon Environmental Council Newsletter," July 1970; John Gray, "Land Use: Profits and Environmental Integrity," *Sunriver* (http://members.aol.com/sunrivermg/sr.htm).

48. *Maine Times*, November 18, 1971, March 23, 1973.

49. *Maine Times*, November 18, December 17, 1971; Cole, "Much of Maine Is at Stake."

50. *Maine Times*, November 18, 1971.

51. Peter W. Cox, "Sometimes You're Lucky" (editorial), *Maine Times*, January 30, 1981.

52. *Oregon Times*, February 1972; Don Waggoner, "Guest Editorial," *Oregon Times*, June 1972.

53. *Oregon Times*, November 1972. See ibid, October 1973.

54. *Oregon Times*, March 1974; November 1974.

55. John Strawn, Tom Bates, "William Appleman Williams: In the Eye of the Revolution— An Interview," *Oregon Times*, June 1975; Bates, "Northwest Utopia," *Oregon Times*, September 1975.

56. *Oregon Times*, February–March 1976.

57. Bruce Hayse, "To Save Oregon's High Desert," *Earthwatch Oregon*, January 1978; Cash, "Deschutes County."

58. "High Country News" in *Maine Times*, July 29, 1983.

59. "High Country News" in *Maine Times*, October 25, 1974. See Margaret Lynn Brown, *The Wild East: A Biography of the Great Smoky Mountains* (Gainesville: University Press of Florida, 2000), p. 304.

60. Jack Hope, "$160 Boots, $66 Suites, and Wild Canyon: The Beautiful Ski Crowd Discovers Utah's Mountains," *Audubon* 76 (March 1974): 85–92.

61. Rothman, *Devil's Bargains*, p. 17.

62. "Whither Maine?" *Down East* 25 (February 1979): 140.

63. Michigan Society of Planning Officials, *Patterns on the Land: Our Choices, Our Future* (Rochester, 1995), pp. 21–22; Calvin L. Beale, "The Revival of Population Growth in Non-metropolitan America (Washington, DC: U.S. Department of Agriculture, 1975); Ray Rasker and Dennis Glick, "Footloose Entrepreneurs: Pioneers of the New West?" *Illahee: Journal for the Northwest Environment* 10 (Spring 1994): 34–39, in Regional List,

Box 1: the West, SBHC. See Alexander Wilson, *The Culture of Nature: North American Landscape from Disney to the Exxon Valdez* (Cambridge, MA: Blackwell, 1992), p. 17; Fishman, *Bourgeois Utopias*, p. 15; Bookchin, *Limits of the City*, p. 73.

64. Jack Lessinger, *Regions of Opportunity: A Bold New Strategy for Real-Estate Investment* (New York: Times Books, 1986), pp. 76, 88, 89, 93; Rasker and Glick, "Footloose Entrepreneurs."

65. Evan Richert, "Maine's Changing Population and Values," in *Changing Maine*, edited by Richard Barringer (Portland: University of Southern Maine, 1990): p. 30–36; *Maine Times*, December 13, 1974; Rasker and Glick, "Footloose Entrepreneurs"; Commission on Maine's Future, *Final Report*; Bookchin, *Limits of the City*, p. 72–73 (the term "splatter" comes from a marginal notation in Fogler Library copy, University of Maine, p. 73). See Dietz, "Maine at the Crossroads," p. 66; Leonard Lutwack in Johnson, *New Westers*, p. 308.

66. Richert, "Maine's Changing Population," p. 37–38.

67. William G. Robbins, *Colony and Empire: The Capitalist Transformation of the American West* (Lawrence: University Press of Kansas, 1994), pp 190–92, 194.

68. Murray Morgan, *The Northwest Corner: The Pacific Northwest: Its Past and Present* (New York: Viking, 1962), p. 28.

69. Terence O'Donnell, *Cannon Beach: A Place by the Sea* (Portland: Oregon Historical Association, 1996), pp. 100–101, 104, 110 113.

70. O'Donnell, *Cannon Beach*, pp. 122–23.

71. *Maine Times*, June 16, 1972. See ibid, December 17, 1971.

72. *Maine Times*, February 18, 1972, June 4, November 26, 1982.

73. *Maine Times*, August 27, 1976. See ibid, May 17, July 12, December 13, 1974, August 27, 1976; *Portland Press Herald*, September 8, 1971; *Maine Sunday Telegram*, August 16, 1970; Richard Pardo, "Something Up, Down East," *American Forests* 75 (May 1969): 45.

74. *Maine Times*, February 18, 1972. See ibid, November 12, December 10, 1982; Louis Ploch in ibid, August 27, 1976; *Portland Press Herald*, September 8, 1971; Charles E. Clark, *Maine: A Bicentennial History* (New York: W.W. Norton, 1977), p. 176.

75. *Maine Times*, February 18, 1972, and June 4, July 2, November 12, 1982.

76. *Maine Times*, February 18, 1972. See ibid, March 14, 1975, August 5, 1983; Kenneth P. Hayes, *As Envisioned by Maine People: A Report to the Commission on Maine's Future* (Orono, ME: Social Science Research Institute, 1976), pp. 85–86, 90; Dietz, "Maine at the Crossroads," pp. 43, 66–67; Commission on Maine's Future, *Final Report*, p. 51.

77. *Maine Times*, November 12, 1982, and June 17, July 8, 1983.

78. Gordon A. Reims, "Where's the Ocean? A Photographer's Guide to Maine," *Down East* 23 (July 1977): 76–82; Mary Bolte, "Alna: Paradise on the Sheepscot," ibid 24 (September 1977): 68, 75. See *Maine Times*, August 4, 1972.

79. "Cottage Slums: Our Northwoods Blight," *Michigan Riparian*, Winter 1968, in State Records: Michigan, Box 7, SBHC; Evan Richert, "Maine's Changing Population," p. 30–35. See Shirley Elder and Richard Lyons, "How It Got This Way: The Shores of Winnipesaukee, Then and Now," *New Hampshire Times*, July 27, 1985, in State Files, Group II, Box 3, Indiana–Maine, SBHC. *Maine Times*, December 13, 1974; Rasker and Glick, "Footloose Entrepreneurs"; *High Country News*, July 7, 1972; Commission on Maine's Future, *Final Report*, pp. 19–20; Cornelia Butler Flora and Jan L. Flora, "Creating Social Capital," in *Rooted in the Land*, edited by Vitek and Jackson, pp. 217–18.

80. James L. Machor, *Pastoral Cities: Urban Ideals and the Symbolic Landscape of America* (Madison: University of Wisconsin Press, 1987), p. 169; Keith Montgomery Carr, "Changing Environmental Perceptions, Attitudes, and Values in Oregon's Willamette Valley, 1800-1978," master's thesis, University of Oregon, 1978, pp. 116-17. See P.C. Kennedy to *Oregonian*, August 1, October 15, 1965; Charles DeDeurwaerder, "The Fight against Ticky-Tacky," *Oregonian Northwest*, November 16, 1969, p. 12; Oral Bullard, "The View of the Artist," ibid, pp. 8, 10; Louis Harris and Associates for Pacific Northwest Bell Telephone Co., "The Public's View of Environmental Problems in the State of Oregon," study no. 1990, n.d., January 1970, typescript, Knight Library Oregon Collections, University of Oregon; Central Lane [County] Planning Council, *Crisis: Water: An Examination of Water Quality Conditions, Programs and Needs in Oregon's Willamette Valley* (Eugene, OR, 1968), p. 7; Findlay, *Magic Lands*, p. 2.

81. Shutkin, *The Land That Could Be*, chapter 4.

82. Richard Britz, "The Edible City," *Seriatim: Journal of Ecotopia* (Winter 1978): 43-49, in Periodicals/Serials Consolidated List 2, Box 20, SBHC. See Bookchin, *Limits of the City*, pp. 128, 130-31, 137-38.

83. Jeffrey Jacob, *New Pioneers: The Back-to-the-Land Movement and the Search for a Sustainable Future* (University Park: Pennsylvania State University Press, 1997), pp. 7, 20-21; Jay Davis, "It Works Real Nicely Here in Town," *Down East* 24 (October 1977): 44. See Mildred Loomis, "Introduction," in John Seymour and Sally Seymour, *Farming for Self-Sufficiency: Independence on a 5-Acre Farm* (New York: Schocken Books, 1973), p. 10.

84. Wilkie Wilkinson, "Sunburst Farms," *Seriatim: Journal of Ecotopia* (Autumn 1978), in Periodicals/Serials: Consolidated List 2, Box 20, SBHC; Arthur Stein, *Seeds of the Seventies: Values, Work, and Commitment in Post-Vietnam America* (Hanover, NH: University Press of New England, 1985), p. 62; Bob Kohl, "Cerro Gordo," *Seriatim: Journal of Ecotopia* (Autumn 1978): 16, Kohl, "Cerro Gordo: Alive and Building," ibid (Winter 1978): 82-84, both in Periodicals/Serials Consolidated List 2, Box 20, SBHC.

85. First quote: Seymour and Seymour, *Farming for Self-Sufficiency*, p. 9; second quote: Paul Goodman, "Introduction," in Helen Nearing and Scott Nearing, *Living the Good Life: How to Live Sanely and Simply in a Troubled World* (New York: Schocken Books, 1954, 1970), p. ix; Nearing and Nearing in ibid, pp. xvii, 3, 6, 35, 49, 146. See Saltonstall, *Maine Pilgrimage*, p. 211.

86. Davis, "So You Want to Move to Maine," pp. 41, 55. See Saltonstall, *Maine Pilgrimage*, pp. 56-57; Dietz, "Maine at the Crossroads," pp. 43, 66-67; Davis, "'It Works Real Nicely Here in Town,'" p. 41.

87. Jacob, *New Pioneers*, p. 89, and especially chapter 4; Stein, *Seeds of the Seventies*, p. 31.

88. *High Country News*, July 4, 1975. See ibid, October 10, 1975; Nearing and Nearing, *Living the Good Life*, pp. 11, 26; *The Oregon Advocate*, April 1980, Ephemera, State Records, Oregon, Box 1, SBHC; Seymour and Seymour, *Farming for Self-Sufficiency*, p. 13; Jacob, *New Pioneers*, chapters 6-7.

89. Donella H. Meadows, *The Limits to Growth: A Report for the Club of Rome's Project on the Predicament of Mankind* (New York: Universe Books, 1974).

90. *High Country News*, April 11, 1975; "Whither Maine?"; Pincetl, *Transforming California*, pp. 190, 222, 236.

91. Shi, *Simple Life*, pp. 262-68; Farrell, *Spirit of the Sixties*, pp. 243-44; *Seriatim: Journal of Ecotopia*, in Periodicals/Serials Consolidated List 2, Box 20, SBHC.

92. See *Maine Environment*, March 1975, August 1977, February 1978.

93. Many of the insights on these late-1970s developments come from conversations and correspondence with Brian Donahue.

94. See *Maine Environment* (Natural Resources Council of Maine), April, June 1977; *Wild Oregon: The Voice of the Oregon Wilderness Coalition*, July 1978, May/June 1979, January/Feburary 1980; *Earthwatch Oregon* (Oregon Environmental Council), 1974.

CHAPTER 8: THE ENVIRONMENTAL IMAGINATION AND THE FUTURE OF THE ENVIRONMENTAL MOVEMENT

1. Riley E. Dunlap and Angela G. Mertig, "The Evolution of the U.S. Environmental Movement from 1970 to 1990: An Overview," in *American Environmentalism: The U.S. Environmental Movement, 1970–1990*, edited by Dunlap and Mertig (New York: Taylor & Francis, 1992), p. 1; Anthony Downs in Richard N.L. Andrews, *Managing the Environment, Managing Ourselves: A History of American Environmental Policy* (New Haven, CT: Yale University Press, 1999), p. 237, and Andrews, p. 238; Denis Hayes, "Earth Day 1990: Threshold of the Green Decade," *Natural History* (April 1990): 56.

2. Michael McCloskey, "Twenty Years of Change in the Environmental Movement: An Insider's View," in *American Environmentalism: The U.S. Environmental Movement, 1970–1990*, p. 81.

3. Andrews, *Managing the Environment*, p. 367.

4. Stephanie S. Pincetl, *Transforming California: A Political History of Land Use and Development* (Baltimore, MD: Johns Hopkins Press, 1999), p. 292.

5. Andrews, *Managing the Environment*, p. 370.

6. Andrews, *Managing the Environment*, p. 371; Daniel D. Chiras, *Beyond the Fray: Reshaping America's Environmental Response* (Boulder, CO: Johnson Books, 1990), chapter 7.

INDEX

ABOUT THE AUTHORS

Richard W. Judd is a professor of history at the University of Maine. He is author of *Common Lands, Common People: The Origins of Conservation in Northern New England; Socialist Cities: Municipal Politics and the Grass Roots of American Socialism;* and *Aroostook: A Century of Logging in Northern Maine, 1831–1931.*

Christopher S. Beach is an associate professor of history and the humanities at Unity College in Maine. Trained in both history and law, his research and teaching interests include conservation law and history, environmental law, the interpretation of natural and cultural heritage, and environmental stewardship.

PHOTOGRAPHY CREDITS
Identified by page number